Advanced Composite Techniques

Lightweight Moldless Techniques for the Aircraft Homebuilder

by Zeke Smith

Aeronaut Press, Napa, California

Advanced Composite Techniques

Lightweight Moldless Techniques for the Aircraft Homebuilder

by Zeke Smith

Published by: Aeronaut Press, 1076 Stonebridge Dr., Napa, CA 94558 USA

First printing 2005. Printed in the United States of America

Order from:

Pathway Book Service, 4 White Brook Road, Gilsum, NH 03448 USA
Tel. 800/345-6665, Fax 603/357-2073

Publisher's Cataloging in Publication

Smith, Zeke,

Advanced Composite Techniques: Lightweight Moldless Techniques for the Aircraft Homebuilder/ by Zeke Smith

p. cm

Includes index

ISBN: 0-9642828-4-4

1. Airplanes - Materials 2. Composite Materials I. Title

629.13334

Table of Contents

Advanced Composite Techniques

Advanced Composite Techniques

Introduction

Read This First!

This is a teaching how-to-do-it book for the amateur home builder who wishes to learn how to use light weight, high performance, composite materials, without a large investment in tooling and facilities. The book will emphasize aircraft applications, but race-car builders and high-performance boat-builders will use the same techniques.

No previous composite experience is required. This book will give the builder the confidence necessary to undertake high performance projects.

What's Different

It is important to make quite clear that the techniques described here differ considerably from those we are familiar with in the large commercial world of "fiberglass" boats, bathtubs, and advertising signs which comprise the overwhelming proportion of the tonnage of manufactured "composite" products. Also, we will not attempt to match the performance achieved in the fields of advanced aerospace and very high end sporting equipment. Such highly-engineered products as spacecraft, Olympic skis, and the F-35 Joint Strike Fighter require very expensive materials, tooling, and process control, which are not available to the home builder.

What We Will Do

We will show how the home builder can achieve surprisingly good performance from a combination of relatively cheap fibers (almost always glass), matrix resins (generally epoxies sold for the aircraft trade), and cheap core materials (generally light weight foams). The structures to be demonstrated make extensive use of relatively thick ***sandwich panels,*** consisting of thin glass skins tightly bonded to a very light-weight core.

The resulting panel is considerably lighter and stiffer than a comparable thinner panel of conventional single-surface "fiberglass."

We will demonstrate a family of composite layup techniques:

1) "Conventional" open layups, in which glass fibers (usually cloth) are stacked on a release surface to a desired thickness, wetted out with resin, and allowed to cure in place

2) "Sandwich" layups, in which the builder forms stiff panels consisting of glass skins on each side of a light-weight foam core

3) Vacuum-forming techniques, in which vacuum is used to produce high-quality, light-weight sandwich panels.

We will also demonstrate a ***wide range of techniques for forming and bonding large composite panels*** into useful structures.

Please note that this book only discusses a family of specific structural techniques. ***A wide range of special skills is necessary to design and build a successful airplane, racecar, or boat and the reader is directed to other references for this other necessary knowledge.***

Particularly in the case of aircraft, the author strongly recommends that the builder ***acquire the plans and instructions applicable to successful and well-tested designs.*** *Such plans are offered by designers who advertise in "enthusiast" magazines appropriate to the reader's interest. This book (especially if the reader goes through the detailed Exercises) will give the reader the experience and insight necessary to undertake modifications of other designs. If the reader is interested in one of the many composite* ***kit aircraft*** *that are currently available, this book will be a valuable supplement to the instructions that are typically included.*

Whatever the source of technical or design information, it will be necessary to ***fully test any structure intended for flight.***

The book consists of four main parts:

1) A brief overview of composite technology as applied by the home builder.

2) Detailed descriptions of specific techniques and requirements for construction and tooling.

3) A series of teaching exercises to demonstrate specific techniques, from the most basic open flat layup through the construction of full-scale tooling for a hollow "cored skin" wing. The intention is that the reader will actually construct these exercises in his home shop to acquire practical skills that can be directly applied to an aircraft project.

4) A collection of appendices containing background material that is "good to know," but is not directly applicable to the development of the "how to do it" information.

SPECIAL NOTE ON REFERENCES TO THE WITTMAN TAILWIND:

This book will make many references to the ***Tailwind*** aircraft, which will be used as a reference design to illustrate alternative ways of accomplishing a design goal through use of composites. ***The Tailwind is a very successful, popular, and well-established small homebuilt aircraft, which uses classical construction techniques - not composites.*** This book does not suggest that a complete and fully-tested redesign of the *Tailwind* in composite technology is presented. The examples and analysis given are for a hypothetical composite aircraft which will be called the ***Crosswind***. Excellent detailed plans for the original *Tailwind* are available from Aircraft Spruce and Specialty Company.

Acknowledgements

Steve Rahm, the designer of the composite *Vision* aircraft (the plane on the cover), has contributed comments, useful criticism, and his great expeience to this project almost from the beginning. Steve's concept of the *Vision* was to design a simple, low cost, modern aircraft of good performance that was suitable for the beginning builder in composites and to provide plans that fully disclosed every detail of construction. In principle, the new builder could construct the entire aircraft with only the information given with the plans; it would not be necessary for the builder to buy parts or subassemblies that were only available from a kit manufacturer's proprietary tooling. American Affordable Aircraft (see www.visionaircraft.com) sells the plans for the *Vision.*

The community of *Vision* builders has an active Internet list through which members exchange a wide range of important information. The author has received generous input from too many to mention all of them. *Vision* builders Glen Shearer, George Lendich, Tom Kendall and Joel Walworth (of Applied Vehicle Technology) deserve special mention.

Frank Roselli, of Roselli Engineering Group, was very helpful in the area of adhesives and related items.

My wife, Jean, was patient and supporting while a "two-year" project turned into a five-year project. Many good ideas weren't so good and she shared the disappointment. When something worked, she shared the pleasure.

The Experimental Aircraft Association provides the environment that makes Experimental Aviation work. The author has had the honor to be president of local chapters three times over twenty years.

Zeke Smith

AeronautPress.com

Napa, California

November, 2004

Chapter 1

Basic Principles of Composites

Composite Materials

A composite material is a mixture of at least two components. One component must be "strong" and stiff and the second component (the "matrix") must be somewhat less stiff, and surrounds the "strong" particles with an intimate bond. In general, the best performing composite materials contain the largest proportion of the "strong" component and the least amount of the "softer" matrix, preferably just enough to wet the fibers and fill the voids. The entire large mass then ***approaches the strength and properties of the strong material alone.***

For our purposes, the ***"strong" component is nearly always in the form of fibers,*** which are very long compared with their diameter. Some practical stiff fibers are glass, cotton (cellulose), and graphite. The ***matrix can be any material which can completely wet and surround the fibers,*** bonding them securely. In this book, we will be almost exclusively concerned with the epoxy family of plastics, that is applied as a thin fluid which hardens in place to bond the structure. Practical structures are then built up from as many layers of composite material as needed to carry the design load.

Composite Skins

For aircraft applications, we will make much use of composite materials which are fabricated into thin sheets, roughly similar to familiar sheets of thin aluminum. These sheets may be quite strong in tension (like a rope), but, like any thin sheet (or a rope), they bend easily under load and usually require additional support in a practical structure. ***A thin sheet of any practical material cannot support significant compression*** or bending loads.

Advanced Composite Techniques

In this book, we will use thin composite sheets as load-carrying surfaces, or ***skins***, where the thin sheets are bonded to a thick ***core*** of lightweight material to make a ***sandwich. The core acts to constrain the thin skin from buckling under compression. If a thin strong skin is bonded to each side of a thick core, the resulting light-weight panel will be found to be highly resistant to bending and twisting, while the tensile strength remains the same.***

A Brief Digression on "Stiffness"

*Practical materials typically used in successful aircraft structures cover quite a wide range of "stiffness." The reader will repeatedly find discussions in the aviation literature where authors have observed that aluminum has approximately twice the stiffness (modulus) of glass composite and so reach the generalized conclusion that a composite structure must always be more flexible than a corresponding aluminum structure. The fact is that properly designed structures do **not** correspond and therefore **different designs must be used for different materials.** An argument based solely on the differences in modulus does not account for the wide success of low-modulus wooden aircraft. The WW II German V-2 rocket successfully used a skin of very thin high modulus steel welded to a dense mesh of supporting stiffeners. We do not see steel aircraft today.*

Composite Structures

Large hollow ***structures***, like wings, cowl covers, and fuselages, can be assembled from a few sandwich panels which may be accurately formed into large subassemblies. Usually, adhesive bonding of the subassemblies is used, so ***loads are distributed*** over large bonding areas, not concentrated as when rivets or bolts are used.

When a load force is applied to any practical object, the object must elongate or deflect, however slightly. The movement due to the deforming force is called STRAIN. If a fly walks over a steel rail, the steel rail must deflect just enough (strain) to develop a counterforce to carry the weight of the fly. With very sensitive instruments, we can weigh the fly by measuring the strain.

If a load force is applied to an area, the ***force per unit area is called STRESS.*** If a tensile force of 2,000 pounds is applied to a shaft with an area of 2 square inches, the ***stress*** is 1,000 pounds per square inch.

We will refer to stress in structures a very great deal in this book, but we should keep in mind that we can ***only measure strain*** when we apply a test force to an object.

The ***ratio of unit stress to unit strain*** is a characteristic of a material and is called the ***modulus of elasticity***, often called ***Young's Modulus***, the ***"stiffness"*** of the material, or just the "modulus." Diamond is a material with a very high "modulus" indeed. Rubber has a very low modulus. Spring steel is stiffer than hardened aluminum, which is stiffer than cured fiberglass. The characteristic is very widely published in engineering handbooks.

Assume for discussion that the “strong” fiber chosen for a composite is an ideal material which has ***infinite stiffness and therefore does not deform under load,*** so that ***all*** of the deflection (strain) must be passed to the matrix that grips the fiber. Assume further that the structure is made up mostly of “ideal” fibers, packed closely. The less-stiff matrix, then, will form a thin film which completely fills the small spaces between fibers. As the surrounding matrix deforms all along the ideal stiff fibers, it passes shear forces to neighboring ideal fibers, which, in turn, pass this strain throughout the structure. ***All of the displacement (strain) in the loaded object then occurs in the matrix,*** but is distributed throughout the huge area of contact between the fiber and the matrix.

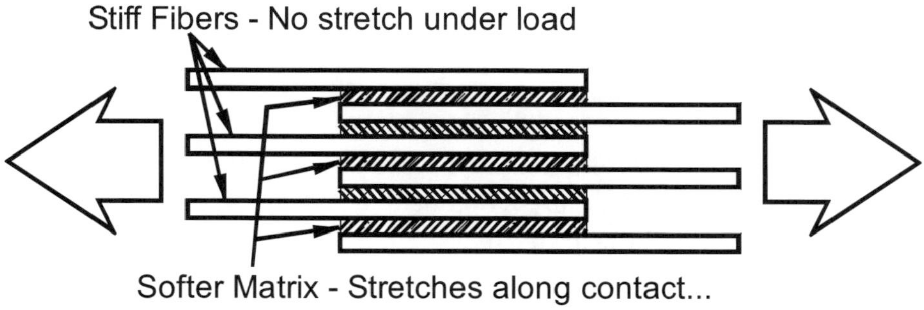

How Fiber Composites Work

In a practical composite, the fibers cannot be ideal, and must deform under load to some extent; but the principle works if the fibers are stiffer than the matrix. In practical composites, the strong component is typically at least two to four times stiffer than the matrix.

Examples of Composites

Composites are very common. Concrete is a composite. The very best concrete consists of a densely packed array (no voids) of nearly round, flawless, granite stones and pebbles, graded down to sand, surrounded by a matrix of Portland cement. Ideally, there will be a minimum thickness of Portland cement required to completely cover every granite particle, so that the well-mixed concrete will have the most granite and the least cement. The resulting structure then approaches the properties of solid granite, except that it has the considerable advantage of being able to be poured into place!

Tough metal alloys and abrasive grinding wheels are composites. In both cases, crystals of extremely hard material are bound in a less-stiff matrix. Asbestos fibers have very high tensile strength and, when mixed in a matrix of Portland cement, make a very good composite that was used for many years in furnace flues until the toxic effects of asbestos became known.

For aircraft applications, we will be mainly concerned with composites made of finely drawn glass fibers (fiberglass) embedded in a matrix of epoxy resin. Other fibers and matrix materials are also important.

Wood as a Composite

Wood is both a composite and a structure. Under a microscope, wood is seen to be a complex structure of tiny hollow cells. The cell walls are mostly made of strong cellulose fibers embedded (with air and water) in a resinous matrix called lignin. Wood has "grain"; we know that wood is much stronger "with" the grain than "across" it. Bamboo is an excellent example of nature's efficiency at making very long cellulose columns with a minimum of material. Bamboo is a complex structure of highly oriented fibers, embedded in lignin, and arranged into a hollow tube, with stiffening "panels" at regular intervals.

Panels

Thin Panels

Light-weight aircraft structures require much use of thin panels that cover large areas. In general, some kind of supporting structure is required that collects and concentrates the loads. This is evident in classic fabric-covered aircraft, where most of the skin area of the aircraft consists of strong cloth stretched tightly between rigid supports. Flight loads impressed on the fabric are concentrated and transferred to a system of formers, ribs, bulkheads, and spars. ***A sheet of stretched fabric, like a rope, cannot support a compression load, only tension,*** so necessary compression loads must be carried elswhere by the supporting structure.

All forces at any point on a free body, including an entire aircraft, must balance to zero. This means that any ***tensile forces in the body must be balanced somewhere in the structure by resulting compression forces.*** In the case of aircraft with wing struts, like the smaller *Cessnas* or the homebuilt *Tailwind*, flight loads will tend to pull the wings up, placing a lifting load on the strut attachment, so the ***lifting force produces tension in the strut.*** The strut is connected to the attachment point at an angle, so there must then be a resulting compression load in the wing (spar), pushing in toward the cabin.

Similarly, where there is a pure bending load, such as outboard of the strut attachment, this bending must be opposed by compression in the top surface of the wing (or spar) and by tension in the bottom surface. The designer must provide sufficient material in the right places and with the right properties to carry all these loads. It happens that ***wood is quite weight-efficient in carrying compression,*** which accounts in part for the success of the highly-evolved wood-frame aircraft designs of World War I.

The Buckling Problem

We know that if we push on the ends of a long, thin, column (like a bamboo stick or a long piece of steel piano wire) it will bend out of the way with a very small load, long before the inherent compression strength of the material is reached. Similarly, if we place a piece of cardboard on a flat surface and push the edges together, the cardboard sheet will wrinkle or bend off of the table easily. This is ***buckling.***

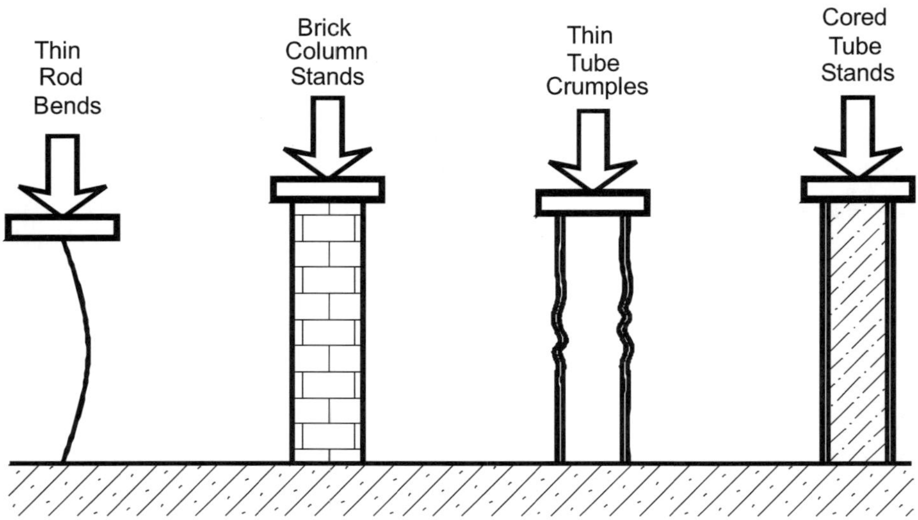

Essentially the same argument can be made for modern light aircraft, which use thin aluminum skins over an aluminum supporting structure: The thin metal skin cannot carry significant compression without buckling, hence the designer must use stiffeners (like "hat sections") to reduce the effective panel size or must carry the compression load elsewhere than in the skin, as with fabric-covered aircraft.

One obvious way to deal with compression buckling is to make the skin thicker. Unfortunately, this adds weight very quickly if the same material is used throughout the thickness of the skin.

We can make very useful "thin" skins from composite materials, and we will give quite a bit of attention to that very goal in this book. These composite thin skins approximately correspond to thin aluminum and, like aluminum, are ***subject to buckling*** when compressed in thin, long sections. Practical structures using thin composite skins can be - and are - designed, but only for applications without compression.

Torsion

We "wring out" (twist) a wet towel because we wish to generate a high ***compressive*** force on the fabric and drive out the unwanted water. We could squeeze the wet fabric directly between our fingers, but we soon learn that ***twisting generates more compression*** throughout the volume.

Twisting, or torsion, must be dealt with frequently in aircraft applications and is often the limiting factor in a practical design. In fabric-covered aircraft wings, the designer may choose to carry torsion with a series of internal steel bracing wires which run diagonally between the top and bottom of adjacent ribs. In aluminum aircraft, the designer may use stiffening members attached to the thin skin, so that local compression forces will not cause it to buckle.

> ***Torsional rigidity of a structure is sharply dependent on the cross-sectional area of the structure. Therefore, it is beneficial to make as much use as possible of the material at the periphery of the structure — the skin — and that the skin must be capable of carrying a compression load.***

Sandwich Structures

A ***sandwich structure is a beam*** or panel which uses a thin, strong, skin on ***both*** sides of a core, which is typically of a much lighter material. Modern high-performance skis and snowboards are examples of sandwich construction. Most present-day skis use non-metallic composite skins over a variety of light cores. The first successful "composite" ski, by Head, used aluminum skins bonded to a lighter wooden core, a valid sandwich. (It is interesting that Mr. Head, a non-skier, was an aircraft engineer.)

For aircraft design, the use of sandwich construction gets around the buckling problem by making the effective panel thickness very large without significantly increasing the weight. Basically, the "thin" panel is made "thick" in a very weight-efficient way.

The essential point of the sandwich is that a very thin (strong) skin is supported by the light-weight core over a very large span so that the skin does not buckle under compression. This allows the designer to use sandwich panels in many places in an aircraft where compression loads must be carried, usually simplifying the structure. The effectiveness of this design strategy is generally surprising to the newcomer, who may be familiar only with conventional thin materials. (See Appendix 2 for a discussion of elementary beam theory and an explanation of the theory of the sandwich.)

In this book, most of the attention will be given to the many applications of the sandwich principle and how very well adapted it is to the use of composite materials.

Attachments - Distributed Loads

The use of composite construction allows the builder to think about the problem of attaching parts in a different way. In a practical aircraft, there are many areas where large subassemblies must be attached to each other: like wings to wingtips or doors to hinges. The ***classical builder thinks in terms of highly stressed parts*** passing concentrated loads, like high-grade bolts in bushings, lines of rivets, and small-area welds. The ***composite builder can often distribute a load over a wide area, greatly reducing local stresses*** and usually avoiding complexity.

Composite designs lend themselves to the use of very large one-piece subassemblies, such as half of an entire fuselage or a control surface which contains no intermediate fasteners at all. These subassemblies may be joined by adhesive bonding over very large areas which permit low stresses.

Environmental Factors - Postcuring

Composite materials have the great added advantage of being essentially immune to corrosion. They do not rust. They do not absorb water to a significant degree (hence their great success in boats). They are dimensionally stable. They do, however, have lower temperature limits than metals and may be subject to deterioration due to ultraviolet light, unless simple precautions are taken.

The composite materials to be described in this book for use by the home builder require an awareness of temperature limitations. Modern composite materials, ***post-cured*** as described below, retain safe strength in very hot environments, as long as the aircraft is not stored at a temperature which approaches the softening temperature of the matrix. In practice, this would allow storage up to about 140 degrees F. Most builders will agree that this is not an important limitation.

Importance of Postcuring

The home builder will not have access to the expensive tooling and curing ovens used by the mainline aircraft industry. Layups must be made and cured at room temperature. Fortunately, the amateur can get almost as good properties in a cured composite as can the well-tooled professionals by heating the finished part in a temporary home-made oven for several hours, even using solar heating. Postcuring can be done at any time after the initial room-temperature cure, so builders usually find it convenient to postcure several previously cured parts at one time.

Briefly: the process of curing a two-part resin never quite completes over time. Toward the end of the curing process, as the matrix becomes quite solid, any remaining unreacted molecules of the two parts find it much harder to move to combine with a partner. The initially-cured matrix, like most non-crystalline plastics, will have a temperature at which it just begins to flow. Materials scientists call this the ***glass transition temperature***, or tg, comparing the effect to the softening of glass. For the epoxies typically used in aircraft, tg will be about 40 degrees F higher than the temperature at which the sample was cured. ***Postcuring consists of raising the temperature of the workpiece to just below the tg point and holding it for several hours.*** The additional curing that takes place at the higher temperature raises the tg to a new, higher level. (During postcure, the part is softened, so it is subject to deformation and must be held in position by appropriate means.)

Basic Materials

In this book, all of the Exercises and most of the discussion will involve the following materials, all currently widely used for the light-weight structures typical of composite aircraft home building:

Reinforcing Fibers

The most widely-used fiber is common bottle glass, known as "E-glass," which is drawn into very fine fibers, then specially coated for adhesion with the matrix, grouped into yarns, and delivered as woven cloth in various weights and types of weave. (See Appendix 3 for a detailed discussion of the fibers and weaves used, including special high-performance fibers, such as S-glass and carbon.)

Matrix Materials

We will use one of the several two-part epoxy products sold by the aircraft supply houses for the homebuilder trade. (See Appendix 3 for a detailed discussion of matrix materials, including a comparison with the polyester matrix materials widely used in boatbuilding, but not used here.)

Core Materials

Foams: We will use very light-weight rigid foams as the primary core materials in a structural sandwich. These foams will be used in two ways: (1) to serve as forms that define a shape for a skin which is added after the foam is shaped, and (2) included as a core when forming a sandwich skin, using a mold or "tool".

Other Mixes

Syntactic foam is a mixture of the "regular" matrix material used and tiny hollow glass "microballoons," usually called "glass bubbles" in catalogs. The builder first mixes up a batch of epoxy matrix and stirs it well, then adds several volumes of the extremely light microballons, depending on the application. Builders refer to both the balloons themselves and the mixture as "micro," for short. Essentially, the microballoons act to dilute and extend the liquid matrix while displacing the much heavier epoxy. The epoxy hardens to create the equivalent of a foam. ***Slurry*** is a thin mixture of epoxy and micro, about 1:1 by volume, which is used to fill and bond the interface between the core and the skin. Slurry can be applied with a paint brush. ***Medium or wet micro,*** 1:2 to 4 by volume, is a mixture with about the consistency of heavy cream which is used to bond pieces of foam as needed. ***"Dry" micro*** or "***paste***," 1:5 to 6 by volume, is a thick mixture which will stand up in the mixing cup like peanut butter and is used to repair foam and to fill any voids in a core. It also may be used as a surface filler. (See ***Fillers***, below.) Dry micro may be applied with a trowel or squeegee.

"Flox" is a special case of a core material which is made by mixing cotton linters (essentially individual cellulose fibers) into premixed epoxy to the consistency of peanut butter. The result is a true composite which may be troweled into position and which hardens to form a very strong and tough structure. (Again, builders refer to both the raw, unmixed fibers, and the material when mixed with epoxy, as "flox.") Flox is much too heavy for general use as a core, but is used in special applications, such as pass-throughs for bolts, or for high stress corners. In some applications, weight can be traded for strength by mixing an equal volume of micro with the cotton flox.

Adhesives

Structural adhesive is a specially formulated two-part epoxy which has the property of extremely good adhesion to cured epoxy. It is also somewhat less stiff than the glass structure it bonds, so that the shearing stress on the joint can be passed into the bonding area. (Chapter 7 explains why this is important.)

Fillers

The cured surface of an open layup always leaves the rough pattern of the cloth weave and has high and low areas that need to be filled and adjusted to level with a light-weight, easily sanded filling material. The automotive industry offers a wide range of surface filling systems, some of which are applicable to aircraft finishing. Aircraft supply houses offer filling systems specifically designed for composite aircraft.

We will make extensive use of ***automotive polyester body filler***, sold everywhere under the "Bondo" brand, but we ***will not use this material to carry structural loads.*** We will use it ***only*** for temporary jigging and clamping of parts. Variations, which are easier to sand out, are sold as "Non-Clog Lightweight Filler" or "Finishing Paste." The unmixed material is a paste about the consistency of peanut butter, to be mixed with a small amount of catalyst in very small batches, as needed. Setting time is about three minutes after mixing. ***All these materials are styrene-based and will dissolve polystyrene foam core material if they come in direct contact.*** They are entirely satisfactory on poly***urethane*** foam. Polystyrene foam must first be coated with epoxy, to avoid attack by the styrene.

Very dry micro, briefly described above, is an excellent filler, and is light and cheap, but it is quite difficult to handle as a surface filler and most builders try to use some other filling method for surfaces.

Mold-making Plasters

We need a quick, cheap way to form complex shapes like wingtips and fairings. This shape can then be used to support a layup directly or to make an accurate mold to shape the part. Home supply stores carry a wide range of water-catalyzed products like plaster of paris and its derivatives, patching plaster and drywall compound. Generally, the mold builder will want to use a fast-setting, non-sandable, stable material like plaster of paris or Fixall which has a good proportion of sand in it as a first rough approximation to the desired surface. This surface will be followed by a layer of ***drywall compound, which is readily sandable.*** The expensive auto body finishing materials can then be used to form the final surface.

Gelcoats: Why Everybody Else Uses Them

In the main-line world of high-production commercial molded composites of such products as boats, shower stalls, car bodies, trash bins, and pretty colors, what the buyer sees as "fiberglass" is the surface ***gelcoat***.

In commercial production, the gelcoat is a ***non-reinforced*** layer, almost always polyester, which is sprayed into the mold as a liquid and forms what will be the finished outside surface. After an initial setup, whatever fiber-reinforced structure required by the design is added. The gelcoat is usually a special formulation and contains color pigments and ultra-violet inhibitors as well as fillers to reduce cost and increase "stickiness."

The advantages of the gelcoat are so great that its use is almost universal in commercial production. Any dust or dead bugs that may be around can be safely incorporated in the gelcoat. The fragile mold release film is protected from any shop damage which might be caused by adding the fiber layers. In fact, a large part of commercial fiberglass production is made by "chopper gun," in which chopped glass fiber and catalyzed resin is sprayed into the mold, much like making a gunite swimming pool. Craftsmanship may not be the first priority and weight saving is not thought of. The molding process produces a smooth surface with very low labor cost. Shop manuals for the general trade describe gelcoat tools and techniques at great length.

Gelcoats are used in some kit aircraft production, but with greater attention paid to thickness control and a higher level of craftsmanship.

We choose not to use gelcoats for high-quality homebuilt aircraft for several equally important reasons: weight being one. A small aircraft has a lot of surface area and a 12-mil gelcoat will add about fifty pounds of essentially non-structural weight. Another reason is that we use the epoxy family of matrix materials - rarely polyester - and most available gelcoats are not compatible.

The "price" paid for not using gelcoat is basically more work. Mold release is not as easy and we don't get a "free" finished surface. The reward is lighter, stronger, accurate structures.

The following chapters will describe the special techniques useful in high-performance gelcoat-free homebuilt structures.

Chapter 2

Overview of Techniques

Introduction

This chapter is an overview of the elementary skills that are required to make basic composite structures, using the facilities of a home shop. The builder will need to become quite familiar with the basic skills of preparing a working surface, cutting and positioning glass cloth, and wetting out the cloth with mixed resin. In fact, if the builder completes the Exercises described in Part 3, he will find that these basic operations have become so familiar that they are almost instinctive and go quite quickly.

Handling and Cutting Glass Cloth

Cloth Weaves

Each fiber of a composite can carry a significant load only along its length. Ideally, we would place all of the fibers in a composite structure exactly where they can best carry the design loads. This is, in fact, done for such special applications as rocket casings and pressure bottles, where the fibers are accurately wound in place with special machinery - somewhat like a silk cocoon.

The handling and ideal placement of individual fibers is hardly practical for the home builder, so the principle raw material for the builder is in the form of woven cloth, in which the fiber orientation is an approximation to the ideal. (See Appendix 3 for a description of the many variations of fibers and weaves used by the aircraft builder.)

The most widely-used form of cloth has the same number of fibers running lengthwise as crosswise and is referred to as ***bidirectional***, shortened to ***BID***, and capitalized by convention. Another very important form of glass fiber is not "woven" at all, but is delivered as a sheet of continuous fibers, running lengthwise, which is held together with a bare minimum of crosswise supporting fibers. This form is ***unidirectional***, shortened to ***UNI***, and is used frequently when the designer wishes to place all available fibers along a load path, as with spar caps.

It is necessary, therefore, to support the entire working area of cloth while cutting and to very gently roll the cut piece into a cylinder to avoid pulling it out of shape. The cut and rolled piece should then be put into a clean, labeled container, such as a paper grocery bag, until it is needed for the intended layup.

Cutting Table

It follows that the builder needs cutting tablea with large area, preferably in a clean place. Unfortunately, the typical composite shop tends to be dusty and crowded with other things, so the "large clean area" may be hard to find. The best compromise solution must be left to the builder.

Marking Cloth and Plastic Films

Felt-tip marking pens with permanent ink (never wax!) are convenient markers for glass cloth, plastic film sheets, and for many other applications which require a bold line.

Cutting Marked Glass Cloth - Dry Cloth

The most general-purpose cutter is ordinary dressmaker's scissors with an offset. Glass wears scissors fairly rapidly, so the builder may choose to buy cheap scissors and treat them as expendables.

A faster and more convenient large-volume cutter for ***dry cloth*** is the carbide-edged dressmaker's wheel, or "pizza cutter," sold in fabric stores. To use, the builder inserts a strip of somewhat resiliant backing plastic under the work area, then rolls the wheel against the backing plastic along the marked line. The result is a fast cut and a very clean edge.

The Problem of the True 45-degree Cut

The beginning builder will quickly find that it takes a bit of experience to unroll cloth from the tube onto the cutting table and preserve the orientation of the weave. Most cuts will be at a 45-degree angle to the roll and small areas can be devilish in sliding into other orientations. If the builder ignores this "ooching" effect, and marks the cloth for cutting just the way it falls, the orientation error may be significant. ***The weave pattern of the cloth is a good guide to a true cut line.*** (Some trials on scrap glass will make this clear.) It may be better to cut the glass following the weave line and "ooch" the glass as needed after the cut. This method is an argument for the use of scissors.

Cutting Wet Cloth

Scissors may be used for general cutting of either dry cloth or cloth that has been previously wetted out with resin. ***(We will make considerable use of the technique where uncut cloth is wetted inside a sandwich of marked plastic film and the entire sandwich is cut along the marked lines.)*** The advantage of the offset scissors is that the cloth need not be lifted far from the table surface to make the cut.

Measuring and Mixing Resin

The epoxy resins used almost exclusively in aircraft construction are ***two-part***. The two separate parts contain chemical components that must react in exactly the intended ratio if they are to form a strong final solid. Any excess of one component cannot find a partner and must appear as a dilutant in the final solid. A familiar example is water (H_20) where exactly two hydrogen atoms must find one oxygen atom to form stable water. Any excess of one or the other will remain unreacted. (See Appendix 3 for a description of resin systems.)

The hardening reaction begins immediately after mixing the two components. The allowed working time after mixing is called the ***pot life***. Pot life varies with the particular formulation and is typically from five minutes to two hours, depending heavily on the temperature. The chemical reaction is ***exothermic***, which means that ***the reaction produces heat.*** If the mixed batch cannot dissipate this heat quickly enough, the temperature will rise rapidly, creating a risk of fire and spreading toxic fumes. This is not a problem if mixed batches are limited to about 6 to 8 ounces at a time and/or are applied to the layup very shortly after mixing, before the exotherm has had a chance to grow.

Resin manufacturers usually formulate their product so that the components are to be combined by ***volume*** in a whole numerical ratio, such as 3:1 or 4:1.

The resin components can be measured by volume or by weighing, but whatever method is used must be quick and easy, and not involve calculation by the busy builder. The two most popular methods are the ratio pump (volume) and the ratio balance (weight), described in Chapter 3 and Appendix 1.

Wetting Out

The purpose of the wetting out operation is to wet the reinforcing material (usually glass cloth) with just enough liquid resin to completely fill the voids between fibers and no more. Excess resin adds weight and cost and does nothing to add strength. Basically, the goal is to drive out air and replace it with the liquid resin. Fortunately, glass and resin are transparent and it is easy to see areas where there is inadequate wetting, especially if the mix is tinted (Chapter 3). Dry air spots appear opaque white, like the air in ice which makes snow white.

Mixed resin is usually applied by pouring it on the work area from the mixing cup and distributing it with a squeegee or paint brush. Often, the builder will want to improve wetting by locally heating the work area with a heat gun or hair dryer to reduce the viscosity of the resin. Air is driven out by using the edge of the squeegee, or rolling with a grooved roller or bristle roller.

By far the largest areas in an aircraft use thin skins of glass, usually three layers or less. For example, one possible combination in wing design is to place one layer of medium-weight BID at 45 degrees, covered by a second spanwise layer of UNI, followed by a surface layer of lightweight BID at 45 degrees.

Some layups require many layers of glass. The number of layers that can be wetted out in one pass is determined by the thickness of the glass layers, but for most cases, three layers at a time is a practical limit.

Other glass layers are then added as needed and wetted in turn until the number of layers required is completed.

In any case, ***it is highly desirable to complete any layup in a single working session, so that there will be the best possible bonding between successive layers.*** In the case of laying up very large parts like wing skins, it helps a great deal to have at least one assistant and make it a social occasion and/or team effort.

In this book, we will describe several wetting out techniques, but they all come down to one thing: ***it is necessary to move the liquid resin relative to the fibers to drive the resin into the mass of fibers and drive the air out.*** In general, this means that some amount of excess resin must be used, and, since we are intending aircraft-quality work here, that excess should be removed, as much as is practical.

Peel Ply

Peel ply is smooth woven dacron polyester or nylon cloth which may be specially treated so that it does NOT bond to the epoxy matrix material. It is a temporary surface layer which may be laid up and wetted out in the same operation as the structural layers, but, after curing, the layer can be "peeled" from the layup. Alternatively, dry peel ply may be applied to a wet surface, where it will absorb some excess epoxy. Peel ply protects the underlying surface against contamination and shop damage. Removal of peel ply leaves a completely fresh and clean surface with the slightly "toothed" pattern of the cloth weave, which is very suitable for bonding. Because it does not bond, ***peel ply must ALWAYS be removed from a layup.*** For safety, it is important to mark peel ply on the ***OUTSIDE*** surface with a ***wax*** crayon, so that it will not be left in place by error. (See also Appendix 3.)

Handling Core Materials

We will make extensive use of sandwich panel structures which use very light-weight core materials, usually closed-cell foams. (See Appendix 3 for a more detailed description of foams and their characteristics.)

It is very important to understand that the core is not just a "spacer" that separates two load-bearing skins; the core carries shear forces

between the skins which are essential for the sandwich panel to sustain a bending load.

If the core of a panel is inadequate, the panel will fail in bending no matter how strong the skins are. (See Appendix 2 for a much more detailed discussion of the working of elementary beams.)

Other things being equal, a properly designed thick beam or panel will be very much stronger in bending and compression than a thin beam of the same weight and materials (as is shown in Appendix 2), and bending and compression are usually the limiting loads in practical structures. ***The thick panel allows the use of thinner skins and lighter core material for a net saving in weight.***

Plywood as a Core

Wood, usually plywood, is widely used as a "core," even though many builders do not recognize that this is what they are doing when they "glass" plywood. A panel of plywood with a single layer of glass cloth on each side is much stiffer than the plywood alone and is only slightly thicker and heavier. If such a "glassed" panel is loaded in bending until one of the glass skins buckles or tears, failure of the plywood core may follow as a consequence, but ***it is the stiff glass skin - not the wood core - that determines the limit load.***. The conclusion is that the the plywood, in this case, is an inefficient core because it is much heavier, at about 22 lbs/ft^3, than it needs to be to carry the required shear loads between the skins. Other materials, exist (principally foams) which serve exactly the same function with much less weight, and we will have a great deal to say about that subject.

One very important reason why we see so many examples of "glassed" plywood, mainly in boatbuilding, is that plywood makes what appears to be a sufficiently strong structure all by itself. The builder may see that the glassed structure becomes stiffer with one glass skin and a ***lot*** stiffer with a second inside skin to form a sandwich, but much of the faith in the final structure comes from the use of the strong plywood core.

Foams as Cores

In all the examples in this book, we will make extensive use of two very light (less than 2 lb/ft^3) closed-cell foam core materials: polystyrene and polyurethane. (See Appendix 3 for a much more detailed discussion of the characteristics of these materials.) Both foams appear to the newcomer to be quite fragile and brittle, especially polyurethane. It is true that these materials are ***entirely unsuitable for use as structural materials by themselves***, and must ***always*** be used with a skin. In fact, we will ***never*** leave a surface of foam uncovered in a completed aircraft structure.

If a newcomer to aircraft composite technology picks up a piece of 1/2-inch thick, "2-pound" polyurethane, she will be impressed by the obvious light weight and stiffness. If she attempts to bend the specimen, she will also quickly discover that she can't bend it very far before it snaps; the material has a very limited "***strain to failure,***" i.e., the allowable stretch under load is small. Polystyrene does a little better, but not much.

Fortunately, we will use these materials ***only*** inside a sandwich, where ***the only strain they will see in service is the strain passed from the skins*** bonded to them.

Foams as Forms for Skins

Bare foams may not appear to be very strong and may be somewhat limited in bending, but they can be shaped into complex forms, like the entire bottom of a fuselage, and held in place with very simple supports. This temporary form is entirely adequate to hold its shape while it is covered with a glass skin. After the layup is cured, the now-stiff structure is removed from the support and the second skin is added to the other side of the sandwich. (Following chapters will discuss several techniques for doing this, in detail.)

Forming Gently Curved Surfaces

Gently curved shapes such as fuselage bottoms can be made by first building a simple form, or "jig" of wood strips bent over plywood bulkheads. Sheet foam strips are then "tacked" to the wooden form with string or dabs of hot melt glue. Cracks, flaws, and damaged areas are then filled with scrap foam and/or micro and the foam surface is next shaped to close to the desired final surface. The foam is then covered with a glass skin, ***forming the outer surface***, with the layers of glass required by the design. After curing, the now much stiffer structure can be removed from the wooden form and the ***inner skin is then added in a second operation***. This method has the advantage of allowing the beginning builder the ***easiest possible access to the working surface*** and the best opportunities for recovering from mistakes.

An ***alternative*** process is to do the above skinning operations in the reverse order: First, make a female support jig from wood strips and "tack" the core material on the ***inside***. Next, prepare and ***skin the shaped core from the inside***. Remove the skinned core from the form. Shape and repair the stiffened core as needed and skin the outside. This method has the advantage of placing the finished outer surface more closely to the exact design, because it differs by only the thickness of the skin from the surface defined by the form.

The foam core can also be used with ***no external support whatever,*** as in the extraordinarily successful "moldless" designs by Burt Rutan. In this case, a ***single block of solid foam is shaped*** to very near the final form, usually by hot wire cutting or carving, and is then covered with an appropriate skin. Flight surfaces of the Rutan family of designs are made by this method.

> *All of the above methods require that inevitable irregularities in the cured surface produced by an open layup must be* ***filled with an appropriate light-weight filler and sanded out to produce the final surface.*** *This filling and finishing operation may create a lot of messy work and other methods should be seriously considered.*

Foam Added To Previously Formed Skins

In the following chapters, we will make considerable use of techniques in which one skin is shaped by a smooth form, mold, or "tool," and the foam core is bonded to the formed skin. The second skin is added in a following operation, but the ***accuracy and smoothness of the final surface is determined by the tool.***

Cutting Foam

Foams come from the aircraft supply houses as sheets or blocks. One of the important rewards in working with this family of materials is that foams are extremely easy and quick to cut. Sheet foam can be quickly cut to measure with an ordinary utility knife. For straight cuts, it is convenient to use the knife to cut somewhat past the middle of the sheet, then bend the cut sheet over the edge of a table to crack the part free. Curves can be cut with a jigsaw, spiral saw, or even a bread knife. Rough trimming can be done with a wood rasp or "cheese grater" forming tool.

In the particular case of polystyrene foam, it is quite convenient to cut this material using a hot wire. (See Chapter 4 for details of hot wire cutting techniques.)

Important warning: Polyurethane foam MUST NOT be cut with a hot wire. Extremely toxic gas is produced from heating polyurethane to cutting temperatures.

While polyurethane foam cannot be hot-wired, it is very easy to cut and to carve into complex shapes.

Bending Foam Sheets

Aircraft surfaces often make use of gentle curves formed by straight line elements. In these cases, foam sheets can simply be formed in place by slight bending. Many current aircraft designs make extensive use of cores of 1/4-inch thickness and the designers call for forming to typical fuselage curves by simple bending.

Obviously, significant work is saved if the builder can use flat sheet core stock by simple bending and designers typically assume this.

For curves of shorter radius, foams can be bent if heated to the softening temperature. Unfortunately, ***foams are excellent heat insulators*** (which explains their wide use as refrigerator insulation), so it is very difficult in practice to make heat flow to the interior of a mass of foam without melting at the point of application.

Thin sheets, 1/8- to 1/4-inch thick, can often be bent by local heating with a heat gun while the builder pushes the foam into position, *keeping the foam under compression.* It may take a few tries to develop reasonable skill at this operation without cracking the foam, but, with patience, it can be done.

Bending Thick Foam Sheets

Larger and thicker sheets cannot be bent to a short radius without heating the foam to the softening point, which is extremely difficult to control.

Thick foams can also be bent by ***slitting the sheet part-way through***, as discussed in Chapter 4, locally heating the remaining thin web with a heat gun, and bending the heated area as required. Complex curves can be formed in this way. It then remains to fill the open slots with replacement foam made from dry micro. The resulting core weighs a little more than the same core made from unfilled foam, but it is a lot less work.

Heat-soaking in a low temperature oven at somewhere near 200 degrees F will allow forming sheet stock up to 1/2 inch thickness. Appendix 1 describes a "steamer" made from a standard steel garbage can which can be used for heating several sheets at a time. The author has never met a builder who was happy about the nasty business of heat-bending thick foam sheets and most prefer to solve the problem another way.

In the following chapters, we will make extensive use of panels of relatively small area, so that there will not be too much bending applied to each separate panel.

Use of Vacuum Bagging

In general, in order to assure that the fibers are wet through, the builder must provide an excess of resin in the layup. With experience and good technique, an experienced builder can achieve an acceptable ratio between glass and resin, but the less-experienced builder will usually have a significant excess of resin, which does not add strength, but just adds weight and costs money.

Applying a vacuum works similarly to the act of squeezing the water out of a wet sponge. A force is applied to the sponge, which tends to reduce its volume and drive the water out through the holes in the sponge. In our case, atmospheric pressure is used to provide the squeezing force, which can be quite large when applied over a big area. A simple mechanical pump, using the techniques to be described, typically develops a vacuum of over 20 inches of mercury, or a useful force of around 1,500 pounds per square foot, which is used to drive out excess resin and trapped air.

Vacuum bagging is the process by which a "bag" of a soft compliant material, such as nylon film, is sealed around an object and the enclosed air removed with a vacuum pump. The result is that atmospheric pressure is applied over the entire surface of the object.

Professional composites fabricators make wide use of vacuum bagging techniques, but the professionals use special high temperature materials and expensive tools that are out of reach for the amateur.

Unlike the professional, the amateur works and cures at room temperature, thus has no need of special high-temperature equipment or materials. This greatly simplifies the equipment requirements and allows the amateur to use common hardware-store materials to do vacuum bagging.

Basic Vacuum Bagging Components

Basic Setup

The sketch below shows the basic vacuum bag setup as might be used to make a simple flat glass layup. Important variations of this basic method are described in detail in Chapter 9.

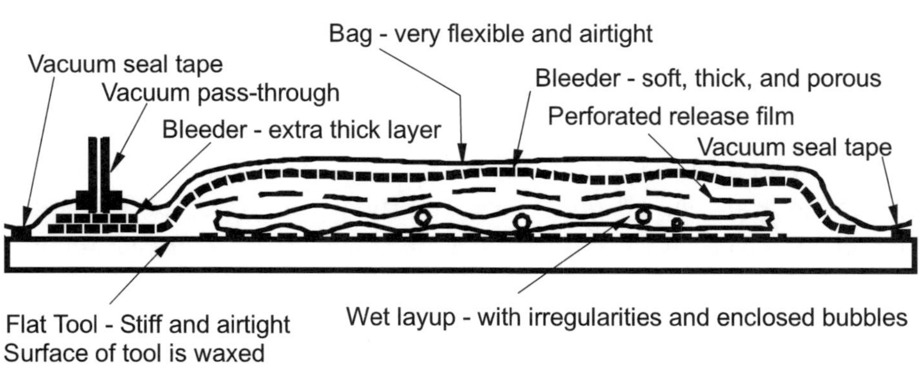

Basic Vacuum Setup - Before Vacuum

The ***mold***, or "***tool***" is a rigid surface that does not leak air, such as a piece of smooth plastic, glass, or metal.

As shown, the "***release***" is a thin film of wax or one of a number of commercial release agents sold for the trade, which will allow the layup to be removed from the tool after curing.

The "***work***" or "***layup stack***" is one or more wetted out layers of fiber, that is presumed to have some amount of excess resin present which we wish to remove.

The "***perf layer***" is a layer of non-stick plastic which has many holes punched in it so that liquid resin can flow through the holes when the volume of the chamber is reduced by the application of vacuum. A very important and widely-used special case is ***peel ply***, that is a layer of cloth which is wetted out with the structural layers, but which ***does not stick*** and can be removed ("peeled"). Peel ply is permeable, so it can be used either with or instead of perforated plastic release film.

The ***"bleeder"*** is a layer of disposable permeable material, such as cotton quilting or rags, which will allow some horizontal flow of air or resin when the material is compressed. The basic requirement for a bleeder is that it must not compact completely under vacuum, but will leave some porosity.

The ***"bag"*** is a layer of thin flexible plastic film which must be airtight over its entire area.

The ***"sealant"*** is any means used for making an air-tight connection between the bag and the tool.

Effect of Vacuum

The bag, bleeder, and perforated release layer act like a piston with many small holes in it - rather like a garlic press or espresso machine. Air and excess resin is squeezed out of the work, through the holes in the release layer, and into the voids in the bleeder. The compaction process continues until the materials in the chamber balance the vacuum load, which will happen when the glass fibers are pressed into contact, closing the voids to a minimum, and/or when the viscosity of the resin limits further flow. The soft bag crumples and folds as it conforms to the work, the tool, the sealing surface, and itself. Excess resin can be seen through the transparent bag as it soaks into the bleeder material.

The sketch below shows the work after vacuum has been applied and the entire system has been brought to minimum volume.

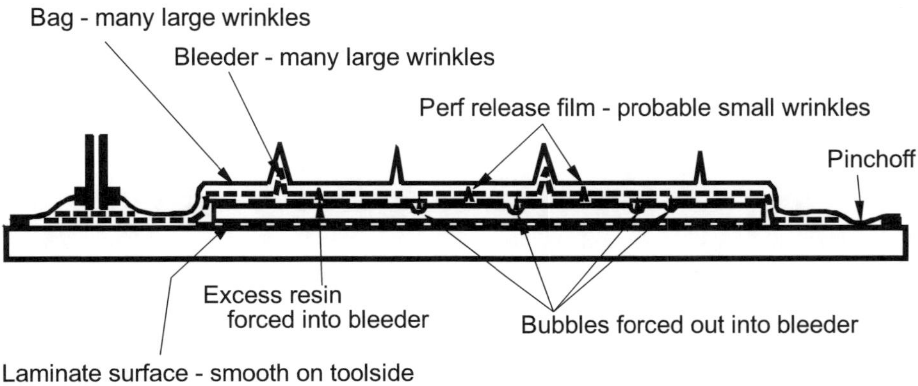

Basic Vacuum Setup - After Vacuum

Chapter 3

Basic Materials Handling

Introduction

This chapter will discuss the "nitty-gritties" of working with composite materials in a home shop: handling, storing and cutting fibers; handling and measuring matrix materials; and general workroom conditions.

Shop Safety - Important!

In our work with composites, we will need to handle potentially dangerous liquid chemicals and dusts. With elementary precautions, nearly every builder can work with these materials for decades without problems. But, as with powerful cutting power tools like table saws, it only takes one serious mistake to ruin your day. You will quickly become familiar with the elementary materials and processes and it is then that the temptation to cut corners on safety will appear. Pay attention!

Dusts

Certain operations, particularly surface finishing, produce large quantities of fine dusts of potentially nasty materials. The need to wear a mask and to vacuum off dust is so obvious in finishing operations, that any builder with sufficient resolve to complete an airplane will surely protect himself with a mask and long-sleeved garments.

The temptation to cheat on mask-wearing comes during short operations which produce only small quantities of dust. Sawing operations (such as those using the bandsaw, spiral saw, or jigsaw) all make dust and put some small amount into the air. Even the shop-cleanup vacuum stirs up dust with its exhaust and the bag cannot catch all the tiny particles. If you make dust or move dust, wear the mask. Don't trust dust-catchers.

Epoxy Safety

The two-part epoxies we use today (see below and Appendix 3) are far safer than those used in the early Rutan designs, but they still must be treated with respect. Early epoxies caused quite severe, even disabling, allergic reaction in a significant number of individuals. The combination of individual susceptability and the casual practices common at the time caused too many builders to abandon their projects. Even today, we still see cases of acquired extreme sensitivity.

The basic rules of epoxy safety are: ***Avoid direct contact and avoid breathing the vapor.***

Protecting Skin: Most builders use latex rubber gloves bought in packets of 100 and changed frequently as needed. The aircraft supply houses sell protective creams that allow the builder to work without gloves. ***The opinion of the author is that it is best to be extra safe and always use gloves.*** If a glove breaks, you know it immediately and can deal with the problem at once.

Vapor: Both components of a two-part epoxy must be considered to be hazardous, but the hardener is the more dangerous. Fortunately, the odor of hardener is distinctive. When pouring hardener indoors, wear a standard OSHA safety mask, like those sold in paint stores. If the mask is working properly, you ***should not smell the hardener***. After the hardener is poured and mixing with the resin has begun, the effective exposure to vapor is greatly reduced. If the workroom is well ventilated, the mask can be removed after mixing has started. The test for safety is whether or not you smell the hardener. If you do, ***wear the mask.***

Storing Epoxy

Storage Bottles

The standard screw-top cans used to ship epoxies are very inconvenient for handling and accurate pouring. The handle provided is almost useless and the cutoff stream is always trapped by the lip of the can, making an unacceptable mess. The solution is to use the widely available ***plastic bottles used for liquid laundry soap and bleach.*** These bottles have a very effective sharp-edged pouring spout and provision for leading dripped material back into the bottle, without escaping the threaded seal. They also have excellent, properly-located, handles that are easy to wipe off when necessary. They are also essentially free. It is important that new bottles be ***rinsed very well*** before use for epoxy storage. After two or three thorough hot water rinses, the new bottles should be filled once more with hot water and placed in the sun to soak. The bottles will be clean enough when the soap odor has disappeared. The author uses ***100 ounce bottles for resin and 44 ounce bottles for hardener.***

Short Term Storage

If the epoxy is to be used for a single working project, typically for a few days at a time, the builder will find that it is convenient to have the materials ***available at a quite warm temperature,*** usually above the temperature of the workroom. The viscosity of warm resin is much less than when cold so it is ***easy to measure and mix warm resin.*** It is worth the builder's trouble to provide a small storage chamber for resin bottles that can be kept at a chosen temperature. (Appendix 1 describes a simple storage box that is heated with a thermostatically controlled light bulb.)

Long-term Storage

Long-term storage is not critical. Some manufacturers recommend that the materials be used within six months after shipment, though this is conservative. They also recommend that freezing be avoided.

Cleanup

Much of the literature of the composite field recommends the use of acetone as a cleanup solvent to remove mixed epoxy from tools, brushes, squeegees, etc. A frequent suggestion is to use two 2-pound coffee cans where one is the “clean can” and the other is the “dirty can.” The builder makes a rough rinse in the “dirty can,” then moves to the “clean can,” shakes off excess acetone as best he can, then puts the tool somewhere to dry. Plastic coffee can lids reduce evaporation loss and odor. After several cleaning cycles, the old “clean can” becomes the new “dirty can,” and a new “clean can” begins with fresh acetone.

The author went along with the acetone program for a while, but found acetone to be much too volatile, odorous, and potentially dangerous for general use in the shop. The author’s cleanup process is as follows:

1) First cleaning: Rinse tools in the cheapest possible grocery store ***white vinegar.*** This step will remove 90% of the resin. If the builder is in the process of doing an extensive run of layups, start a "dirty vinegar" container, otherwise discard the used vinegar. Don't bother to water rinse. This process produces a curd-like deposit of removed epoxy.

2) Second cleaning: Use "***dirty alcohol***" in a coffee can until the wet tool appears to be quite clean.

3) Third cleaning: Quick rinse in "***clean alcohol," using the least amount practical. Place used alcohol in the "dirty" container.***

4) Water rinse. Dry.

Vinegar is kept in the plastic bottles it comes in. Alcohol is kept in very well rinsed plastic liquid laundry soap bottles (see above.)

Measuring Epoxy

Ratio Pumps

Any composite shop, whether a home shop or a large-scale commercial operation, finds it necessary to quickly and reliably measure and mix many small batches of material in an easy and efficient way.

All large shops and many home builders measure resin components with a ***ratio pump.*** A ratio pump delivers ***very accurately measured volumes*** of the two components into a common mixing cup. A large commercial shop may have the resin components piped directly from storage drums to the ratio pump mechanism. The pumps sold by the aircraft supply houses to home builders have storage chambers attached to the pump. Ratio pumps are extremely convenient; the user merely strokes a pump handle and delivers exactly the quantity required, which significantly reduces waste. Ratio pumps are convenient when a very small amount of mixed epoxy is required.

Resin manufacturers usually formulate their product so that the components are to be combined by ***volume*** in an even numerical ratio, like 3:1 or 4:1.

The only real ***disadvantages*** of ratio pumps are that they take up precious shop space, are quite difficult to clean, are a bother to fill, and are relatively expensive. ***Calibration should be periodically checked***, especially if the pump is idle for many days. Crystallization can occur in the chambers and sticky valves can cause error.

Care must be taken not to knock over the rather top-heavy pump and storage chambers and cause a nasty cleanup problem. Still, the ratio pump is the choice of most active builders. (See Appendix 1 for a more detailed description of the tools used in composite construction, including resin measuring methods.)

Ratio Balances

The resin components can be measured by weighing, but, to be practical and competitive with the ratio pump, the method must be quick, easy, and not involve any calculation by the busy builder.

Many experienced builders use a ***ratio balance.*** The ratio balance requires that the user keep ***several sets of carefully calibrated weights***, where one weight represents a quantity of resin and a second corresponding weight represents the corresponding quantity of hardener. To use, the builder (1) balances the empty cup against a balance beam, (2) places the "resin weight" somewhere on the balance beam and pours resin to balance, (3) adds the "hardener weight" and pours hardener to balance again and complete the measurement.

To use a ratio balance accurately, it is necessary to have very good control of pouring. The ***plastic bottles used for liquid laundry soap or bleach fluid are excellent for handling epoxy*** components. With these bottles, it is possible to control a very fine stream of poured material. They also provide ***airtight storage*** of the materials, which reduces contamination and exposure to moisture in the air.

Usually, the builder is in a hurry and tends to overpour the resin component a bit. The scale can be quickly rebalanced and then the "hardener" weight is added. Hardener is less viscous than the resin component, and it is easy to pour the hardener very accurately.

Ratio balances are not quite as quick to use as ratio pumps, but they have the advantages that they ***never have to be cleaned***; are small, light, and cheap; and they take up very little space in the shop.

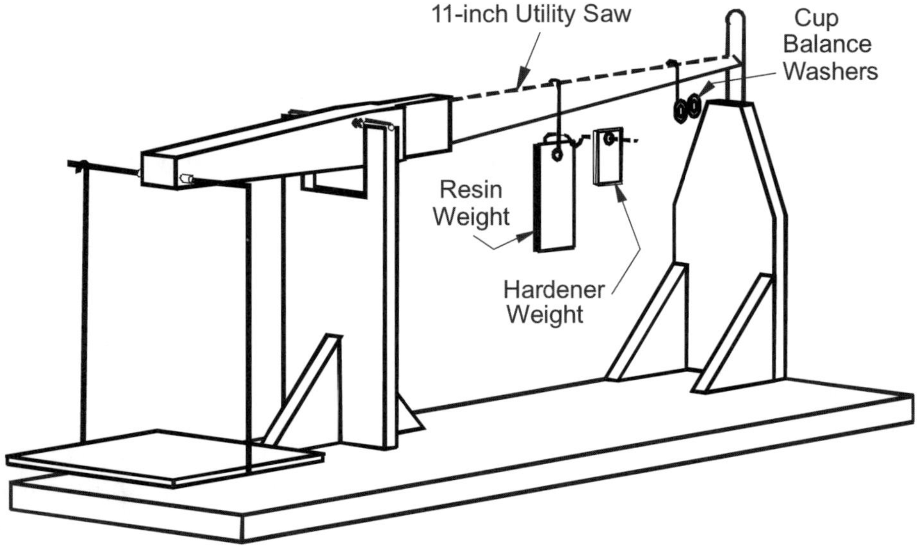

The Author's Ratio Balance

Appendix 1 shows a drawing of a simple ratio balance built around a common utility saw blade. ***The author prefers the ratio balance***, after measuring by both methods for several years.

Other Measuring Methods

Some builders use other methods for measuring epoxy and Appendix 1 describes them in greater detail. These methods include medical hypodermic bodies (without needles) and weighing with electronic postal scales.

Glass Cloth

Storage

Glass cloth is delivered from the aircraft supply houses as rolls, wound on cardboard tubes, usually from about 38 inches to 60 inches in width. Glass rolls should be stored where they can be protected from dust and contamination. Plastic garbage bags are convenient as dust covers. One form of space-saving storage is to hang a dedicated shelf from the ceiling of the workroom and use this shelf for storing rolled fabrics to be used in the work. The ends of all the rolls should be clearly labeled so that time is not wasted shuffling rolls.

Cutting Surface

If space is available (which is not often likely), a separate clean cutting area can be provided, in which there can be a smooth-topped table of generous area where the glass fabric can be unrolled over the surface.

If space is somewhat tight, the builder must use some ingenuity for his own cutting-table situation.

Most cuts are made at 45 degrees to the weave, so it is useful to mark the cutting table with a series of guide-lines at 45 degrees to the sides.

Marking Cloth and Plastic Films

The most convenient markers for glass cloth and plastic films are the felt-tipped pens sold as "fine-point permanent markers." The kind with "super permanent ink" is preferred, but the difference is not important. The ink is dissolved by liquid epoxy and becomes a part of the layup, staining the work slightly, but there is no practical disadvantage. ***Do not mark the glass cloth with wax crayons. Wax is a release agent!***

> ***Red*** *marker ink tends to creep through the layup, even through primers, and into final paint.*

Cutting Marked Glass Cloth

Offset ***dressmaker's scissors*** may be used for general cutting of either dry cloth or cloth that has been previously wetted out with resin. (We will make considerable use of the technique in which uncut cloth is wetted inside a sandwich of marked plastic film and the entire sandwich is cut along the marked lines.) The advantage of the offset scissors is that the cloth need not be lifted far from the table surface to make the cut. The author prefers to buy packs of cheap offset scissors and treat them as expendable tools. Other builders may prefer to buy professional quality scissors, clean them thoroughly after a working session, and sharpen them as needed.

A fast and convenient cutter for volume cutting of ***dry cloth*** is the carbide-edged ***dressmaker's wheel***, or ***"pizza cutter,"*** sold in fabric stores. To use, the builder inserts a strip of resiliant backing such as window plastic or plastic-faced Masonite, under the cutting line, then rolls the wheel against the backing plastic along the marked line. The result is a fast cut and a very clean edge. The only significant ***disadvantage*** appears when the brittle edge of the cutting wheel gets chipped in service and ***fails to sever all the fibers in a single pass.*** Rather than attempt a second pass, which never exactly matches the first pass, it is best to clean up the cut, as needed, with a razor blade. When the tool gets dull or too chipped, replace it. Be prepared to replace glass-cutting tools fairly often: Dull tools waste time.

Handling Bias-cut Cloth

After cutting, roll the piece of glass cloth into a tube shape, being careful not to distort the tube by elongating it. Bend the rolled tube double and place in a labeled container, such as a brown paper grocery bag.

Woven cloth is nearly always applied to a structure so that the fibers in the weave run at 45 degrees to the length of the structure. (Appendix 3 explains why this is so.)

The beginner will quickly find that a strip of dry ***bias-cut*** glass cloth, cut at 45 degrees to the weave, can be ***very frustrating*** to handle. If the cut strip is pulled from the ends, without confining the sides, it elongates very easily because the sides pull in to compensate and the horizontal and vertical weaves of the cloth are no longer at right angles to each other. Some builders call this ***"ooching."*** The beginner can best appreciate this effect by attempting to bias-cut a small scrap of dry glass and testing the effect, which will be instantly obvious. It will also be obvious that it is ***essentially impossible to handle narrow strips of bias-cut dry glass cloth without making a terrible mess of it.***

In practice, the builder will ***cut wide strips*** of dry glass from the roll at 45 degrees, then handle the wide strips with care until he can ***move the glass to "cutting plastic."*** The uncut dry glass is placed between two sheets of 2- or 4-mil polyethelyne plastic, the plastic is marked for cutting with a felt-tipped marking pen, and the entire dry sandwich is cut with scissors in a single operation. The PE plastic is extremely cheap, certainly cheaper than wasted glass, and handling is greatly simplified.

Tinting Resin

When clear glass is wetted with clear resin, white areas (bubbles of air) may be hard to see, especially for the beginner. Some workers use a variety of tinting agents, including some sold for the purpose. The author has found that ***ordinary carpenter's chalk*** (red or blue - not yellow), works very well to make bubbles contrast when mixed with the resin in very small quantities. ***Bubbles show up plainly*** and can be attacked and eliminated quickly with the squeegee, the grooved roller or the roller brush. It is much better to be sure about a properly wet layup than to repair or reject dry spots in a cured layup. ***The author very strongly recommends the general use of tinting and the use of tinting in mixed resin is assumed throughout this book.***

Thixotropic Fumed Silica

Fumed silica, sold in bulk under the trade name Cab-O-Sil, is used commercially as a neutral additive in paint to allow wet paint to ***spread freely when brushed, but to stick to a vertical wall without running down, when brushing stops.*** This property is called ***thixotropy*** and is also found in printer's ink and most greases. When disturbed by wiping or shearing, a thixotropic material flows and wets easily, but forms a non-moving gel when the shearing force ends.

Wet epoxy may be viscous and sticky, but it is ***not thixotropic***. Fumed silica is used in the composite industry to produce thixotropy. It is

frequently necessary to cure wetted fabric in a vertical position and the wet epoxy will slowly ooze downward, out of the wetted cloth, leaving too-dry areas at the top and too wet at the bottom, and appearing on the shop floor. In following chapters and Exercises, we will make a great deal of use of the technique of ***taping***, where wet bias-cut cloth is cut into strips and placed into a wide variety of vertical corners. In this case, it is very important that the wet epoxy stay where placed.

The particles of fumed silica are extremely small compared to glass fibers, but they are, in fact, ***discrete particles which bond to the epoxy in the same manner as the fiberglass to form a composite.*** We may look at fumed silica in our mixed epoxy as serving a similar function to sand in mixed concrete, which must also contain rocks and large pebbles in the aggregate. ***Practical fiberglass weaves must leave gaps between adjacent yarns which fill with unreinforced epoxy and are essentially wasted.*** Fumed silica, in addition to making the epoxy thixotropic, also serves the very useful function of adding somewhat to the effective fiber-to-matrix ratio, so there is actually a gain in useful strength. In use, fumed silica may be added to mixed epoxy in the ratio of one volume of fumed silica to two volumes of epoxy, or even equal volumes. The test is when the mixture just starts to remain on the stirring stick without dripping off. It is also cheaper than epoxy, so we will use it quite extensively in the work to follow. ***The material is a very light and very nasty DUST. Use a mask when handling and avoid the risk of air currents.***

A minor disadvantage is that the material tends to form lumps in the mixing cup and the builder must be a bit patient to remove them. An ordinary ***kitchen fork makes a good tool for breaking up the lumps***.

Working Temperatures

Suppliers of epoxies typically used in aircraft construction recommend that the workroom temperature be maintained at a minimum of 65 degrees F, with the strong recommendation that ***70 to 80 degrees, or higher, is best.*** A room temperature of 100 degrees is harder on the builder than it is on the work, but the builder must be aware that pot life will be sharply reduced. Boatbuilders, or aircraft builders who must work in a hangar, may find that it is necessary to have a workroom temperature below 65 degrees. Hardening time for typical aircraft epoxies at 65 degrees is approximately twice that at 75 degrees.

The viscosity of the mixed liquid epoxy is sharply dependent on the temperature and it is ***highly desirable to have the lowest practical viscosity during wetting-out.*** This low viscosity can be produced temporarily by ***locally heating the working area*** with an electric heat gun, such as a common ***hair dryer***, until wetting out is complete. When the builder moves to wet another area, the region just wetted out will cool, but the fabric will have been safely wetted.

After wetting out is complete, the entire layup must be brought to an appropriate curing temperature. If the workroom can be heated sufficiently, that problem goes away. If not, a temporary housing or tent made of plastic film to cover the work area will be required. An electric heater can be used to heat the temporary tent. (Canadian builders have reported the ability to work year round using this scheme.) ***Small work can be warmed under a large cardboard carton and heated with a light bulb*** (see Appendix 1).

A common queen-size ***electric blanket*** is convenient for curing large flat layups, especially when vacuum is used. After the vacuum setup is going, the builder spreads the blanket, covers the work with a comforter, then turns the control up to "max."

If it is not possible to heat the work room adequately, the builder can often "cheat" by getting a complete wetout with local heating, then letting the layup harden, accepting a long setup time. After the work has hardened to the point where it can be handled safely, it can be moved to a warm place for a normal complete cure.

In cooler climates, it is convenient to ***store unmixed epoxy in a small insulated box which can be heated*** to a good working temperature. Warm epoxy greatly facilitates mixing. (See Appendix 1.) This is particularly convenient if the builder is using a ratio balance, because ***only the bottles themselves require warm storage.*** The box must be significantly larger if a ratio pump must also be housed. The warmed bottle box can be "borrowed" as a curing chamber for small work.

Exotherm Risk - A Special Case

The chemical reaction of mixed resin is ***exothermic***, which means that the reaction produces heat. If the mixed batch cannot dissipate this heat quickly enough, the temperature will rise rapidly, creating a risk of fire and the spreading of toxic fumes. This is not a problem if mixed batches are limited to less than about 8 ounces at a time and are applied to the layup very shortly after mixing. When spread out in a thin layer, the reacting resin can pass heat to the environment easily and avoid a runaway "exotherm." If the builder has help, perhaps at an EAA "layup party" for a large structure like a wing skin, where a large quantity of resin can be mixed and spread quickly, mixing as much as half a gallon at a time is safe.

In general, we are working with thin skins with large surface area which will dissipate the heat of curing adequately, so the builder can work at the highest workroom or curing temperature that is convenient to produce. In special cases, such as thick spar caps or landing gear beams, where the layup must use many layers and/or is ***surrounded by heat-insulating foam***, curing must be slowed to ***avoid runaway heating*** from exothermic heat. In these cases, the workroom can be kept cooler and/or ***longer-setting epoxies can be used.***

What to Do If an Exotherm Occurs

In most cases, the exotherm will happen in the mixing cup. The warning condition is if the ***cup starts to feel warm to the touch.*** If so, get the cup outdoors quickly. If possible, place the warming cup in a metal container. Remember that the ***heating process does not depend on the presence of air,*** though the components are themselves combustible in air. If convenient, use water to prevent or extinguish a fire and resolve to mix smaller batches next time, to work faster, or both.

If the exotherm occurs while curing a large layup, such as a deep wing spar with many layers, there is little that can be done to save the structure and the builder's concern becomes ***personal safety.*** Treat the situation as a fire emergency and take necessary steps to limit the damage to the structure. Ventilate the workroom as much as possible and use water on the exotherm area.

Chapter 4
Handling and Cutting Foam
Introduction

Aircraft home builders use several core materials in composite sandwich construction, including plywood, end-grain balsa wood, and others, but we will be concerned here almost exclusively with very lightweight closed cell foam. The aircraft supply houses offer a variety of foams as called for in current homebuilt designs, and each foam type has its own advantages and disadvantages, as discussed in greater detail in Appendix 3.

Aircraft home builders are generally unfamiliar with the structural foams used in composite construction and even the basic techniques are rather off-putting to a builder brought up in aluminum, steel, and wood. These foams are “obviously” weak, “dusty” (urethane), and frustratingly brittle. This chapter will give detailed examples of the basic techniques for handling and forming the most widely used light, closed-cell, 2 lb/ft^3, structural foams:

1) ***Small-cell polystyrene/styrofoam:*** Medium blue. Fine-grained sponge-like appearance. Fairly brittle, but can be tapered to a sharp edge. Easily cut with a hot wire (see below). Dissolves in petroleum-based and styrene based solvents, but otherwise inert to most solvents. Can be damaged by long-term ultraviolet radiation from strong sunlight, so must be stored in a sheltered place. Usually available as sheets or in blocks up to 4 inches thick. (***Do NOT confuse with the soft non-structural*** blue or pink foam widely sold for heat insulation.)

2) ***Large-cell polystyrene/styrofoam:*** Usually medium blue. Coarse sponge-like appearance which always requires filling of the open cells with slurry to form a bonding surface. Fairly brittle, but can be tapered to a sharp edge. Easily cut with a hot wire. Dissolves in petroleum-based and styrene-based solvents, but otherwise inert to most solvents. Can be damaged by long-term ultraviolet radiation from strong sunlight.

Usually available in blocks up to 10 inches thick. Cheap. Used in huge quantities for boat dock floats and refrigeration insulation.

3) ***Polyurethane:*** Pink, tan, or green. Fine-grain sponge appearance. Dusty to the touch. Quite brittle. A sharp edge is very fragile. ***MUST NOT BE CUT WITH A HOT WIRE.*** Extremely easy to carve, sand, and cut, but dust is hazardous. Extremely inert and unaffected by fuel, styrene-based resins, and common solvents. Particularly sensitive to embrittlement by ultraviolet light and must be protected, even in a temporary shop storage environment.

Cutting Sheets

Marking

The best way to mark foam sheets for cutting is with the marking pen, though a soft graphite pencil can be used. Do not mark with any wax, which will contaminate the layup. Where appropriate, a carpenter's string chalk line can be "snapped" to define a long straight line with chalk.

Knife Cutting

A standard utility knife can be used to cut about half-way through a foam sheet, supported on the worktable. It is often convenient, but not necessary, to clamp a steel or wooden guide next to the cut. After the scoring cut, the work is moved so that the cut line is over the edge of the table. A gentle pressure on the free edge of the foam causes a clean crack all along the line.

Saw Cutting

Most sawing operations work very well with foams. In fact, the materials cut so easily that the builder needs to be particularly careful. An ordinary circular ***table saw*** is appropriate for cutting straight, clean-edged, lines with a fence guide. A ***band saw*** will cut almost any curve. Where access is a problem, a ***spiral saw*** will go where a band saw cannot. Sharp corners are handled by ***utility saws.*** The small Japanese pull-to-cut ***Ryoba hand saw*** is excellent. Fine cutting into a sharp corner can be done with a ***hacksaw blade*** clamped into a handle sold for the purpose. (Again - don't breathe foam dust.)

Carving Foam

We will define "carving" as any forming operation, usually with a block of foam, requiring curves or odd angles, whether freehand or done with a guide.

The most useful tools for carving either blue polyethylene or urethane foam are the steel "***cheese graters,***"such as the ***Stanley "Surform,"*** sold in several sizes and shapes for wood-working. These tools are very fast-cutting and will usually complete an operation in seconds. Rough carving, like cutting off a square corner, can be done with a common ***kitchen bread knife.*** A few minutes ' practice on scraps of foam will very quickly give the builder a good feel for the tools and materials. Urethane foam is particularly easy to carve, so easy that the beginner might have doubts about the suitability of this strange material as a structural core.

Bevels

In this book, we will make much use of core panels with sharp beveled edges. Bevels can be cut free-hand, but, if there is much work to be done, the builder will be very well advised to make a ***beveling tool*** as shown below and described in detail in Appendix 1. See that the tool has steel edges, as shown.

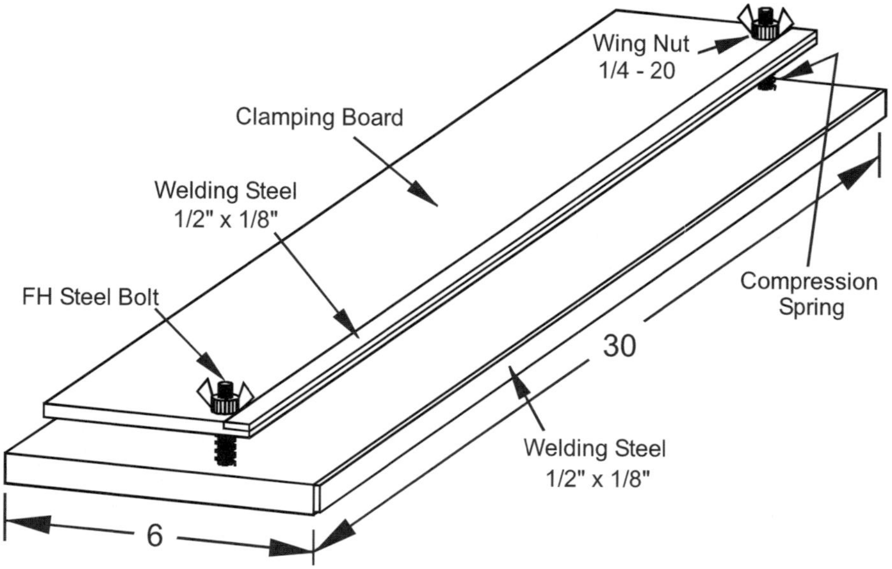

Foam Beveling Tool

The bevel is required to approximate a smooth transition of compression forces in the skin when a panel is bent, compressed, or twisted. The principles are discussed in Appendix 1.

To use, the builder (1) places a sheet of foam into the beveling tool so that the desired edge of the bevel coincides with the edge of the tool, (2) tightens the wing nuts to clamp the foam, (3) ***cuts the bevel with a "cheese grater," by working against the steel edges as guides.*** The author can cut a perfectly straight bevel edge in urethane in about 15 seconds, using this tool. (This is a good job to do outdoors, particularly with urethane, because it makes copious dust.) ***If the foam is polystyrene, the bevel can be cut easily with a hot wire.*** It may take a few minutes to set up the hotwire rig, but most jobs require several panels to be cut in one work session, so the hotwire is quick and efficient.

Perforating Foam Sheets

When a layup is to be made against the surface of a mold or "tool," and a foam panel is to be added to a wet or cured skin, it is necessary to be sure that all air has been displaced from between the skin and the foam core. This is difficult to do if the core panel is of large area and all trapped air must be passed out to the side of the panel. This problem is particularly important in a vacuum bagging operation (see below), but can also be a factor in a plain layup without vacuum.

The solution is to ***perforate the foam*** with a special tool, as shown on the opposite page. (Construction details are in Appendix 1.) Slurry must be applied to the face of the core panel so that all voids are filled, plus just a little extra to be sure that coverage is complete. When the panel is put into place and held with either weights or vacuum, ***air and slurry will be forced up through the holes without having to travel all the way to the edges.*** This technique also serves to prevent contamination of the laminate with micro, by allowing excess micro to exit through the holes and become part of the core, rather than oozing out onto the bonding surface.

To use the perf tool, the builder presses the tool against the material to be perforated, while using a ***soft backup, such as a piece of carpeting,*** to prevent damage to the nails. After each operation, the builder squeezes the handles together and the ***stripper plate forces the tool away from the work for another operation.***

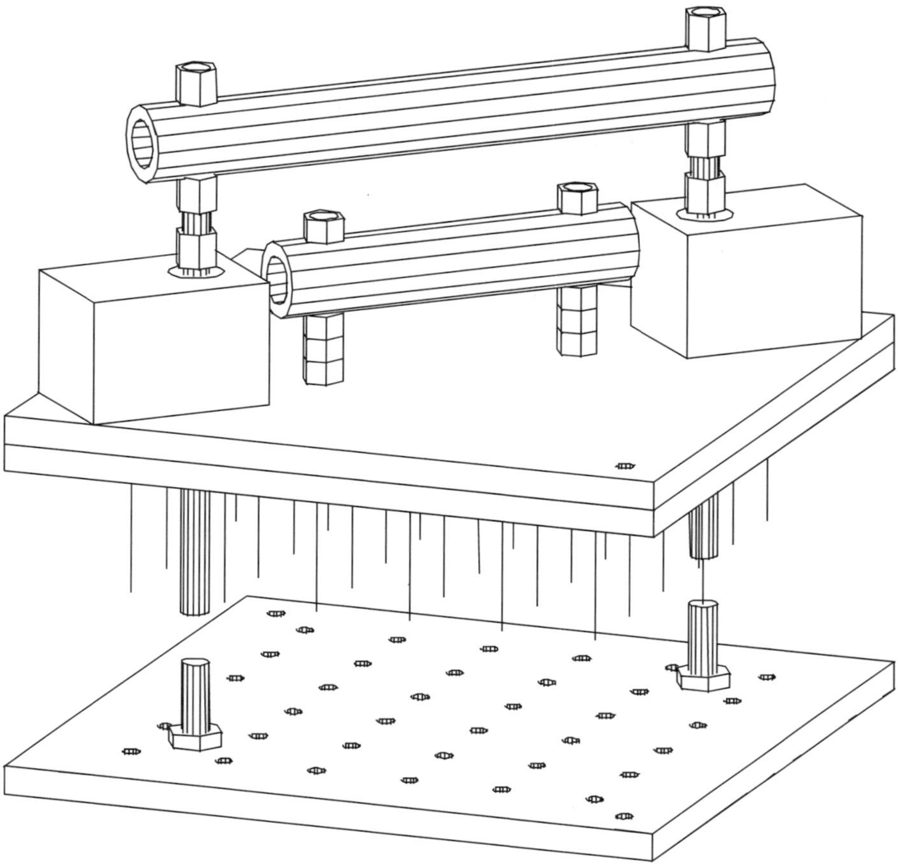

Perforation Tool

The edge of the tool is the same distance from the edge of the nails as the spacing between the rows of nails, so the edge of the tool can be used to "sight" each pass. The tool will make 36 holes in 1/2 inch foam in ***less than ten seconds per pass.***

The builder will find that the perforating tool will be used frequently in the shop and that the time spent building one will be well rewarded. The dimensions shown in the appendix are approximate and will serve for perforating plastic films and foams ***up to 1/2 inch thickness.***

If the builder anticipates the need for several pieces of perforated foam, the easiest way to handle it is to perforate an entire sheet at one time, before marking or cutting. ***Perforation should be done before beveling because the unsupported bevel edge is very fragile.***

Attaching Foam to Temporary Forms

Temporary structures are used to generate the shape of fuselage halves, cowlings, turtlebacks, and the like. Usually, designers specify a system in which plywood or chipboard bulkheads are rigidly attached to a floor or strongback frame. Light wood or metal stringers are run across the bulkheads to define the desired shape, often a complex curve. The ***problem is to reliably and accurately attach foam to the stringers*** so that the work can stand the bumps and bruises of the shop, foam finishing operations, application of a skin, and - finally - removal of the cured shape from the form.

It is important to assure that the foam does not become firmly bonded to the stringers by stray epoxy that has worked its way past the foam. Brown plastic mailing tape or duct tape, applied to the stringers, will reliably prevent sticking.

Lacing

Several current designs call for the use of flat aircraft fabric ***rib lacing twine or heavy button thread*** at about one-inch intervals to secure the foam. A large darning needle or a curved aircraft lacing needle can be used to pull the thread through the foam. The edge of a wooden tongue depressor can be used to crush the thread down slightly below the foam to assure a smooth surface. A short piece of piano wire can be ground to a sharp point and used as a needle to make holes. After the core is skinned and cured, the threads are cut with fingernail nippers or a razor blade to remove the work from the stringers. The string left inside the core is ignored. Flaws, dents, and needle holes are filled with micro. This can be quite a bit of work, but so is classic rib-stitching.

Hot Melt Glue

One very good alternative is the use of small blobs of ***hot melt glue***, applied with a glue gun, to secure the foam. The hot melt glue will stick to the mailing tape on the stringers sufficiently to hold the foam, but will break away when necessary to remove the work. The ***low-temperature version is preferred***. When the cured layup is removed from the form, each little glue blob will tear out a bit of foam and leave an easily-filled flaw. This is a good method when access to the inside of the form to tie the lacing twine is not convenient. A disadvantage is that working time before the adhesive cools is short, only a few seconds, so handling large sections will be difficult.

Paneling Adhesive

If more working time is needed, another good alternative is to use small blobs of ***paneling adhesive***, sold in home supply stores for use with a caulking gun, for "quick grab" attachment of interior trim and paneling. Paneling adhesive ***bonds to brown mailing tape*** sufficiently well to hold the foam, but lets go of the foam when needed. To use, the builder places small dabs on the tape-protected stringers, presses the work into place, removes the work for one to three minutes, then presses the work to adhere.

Hot Wire Cutting Polystyrene Foam

Hot wire cutting is a ***very powerful work-saving technique*** for carving foam and is almost a good enough reason by itself to choose polystyrene foams. Basically, stainless steel resistance wire - usually ordinary standard aircraft .032 inch safety wire - is heated by a low voltage electrical source to just below red heat and used to melt its way through the foam. The cutting speed for polystyrene foam is about one-half inch per second to one inch per second. This process makes use of the property of excellent heat insulation because the effect of the hot wire is limited to collapsing the foam very near the wire itself, so it has almost no effect on the foam even a small distance away. The method leaves a clean, smooth edge.

CAUTION: PolyURETHANE foam MUST NOT be cut by the hot wire method. The high temperature used causes a chemical change which releases a very poisonous and dangerous gas.

Hot Wire Equipment

The most convenient way to produce an adjustable low voltage is to use a standard electronic ***household lighting fixture dimmer in the primary of an old television filament transformer***, as shown in Appendix 1. The satisfactory transformer may not be conveniently available, but the aircraft supply houses sell these items as a kit, ready to assemble. The output of the low voltage controller can be adjusted over a wide range by turning the dial of the dimmer.

The builder should make up three sizes of the wire frames, as shown in Appendix 1, which will serve a very wide range of applications. The sketch below shows one suggested medium-sized design.

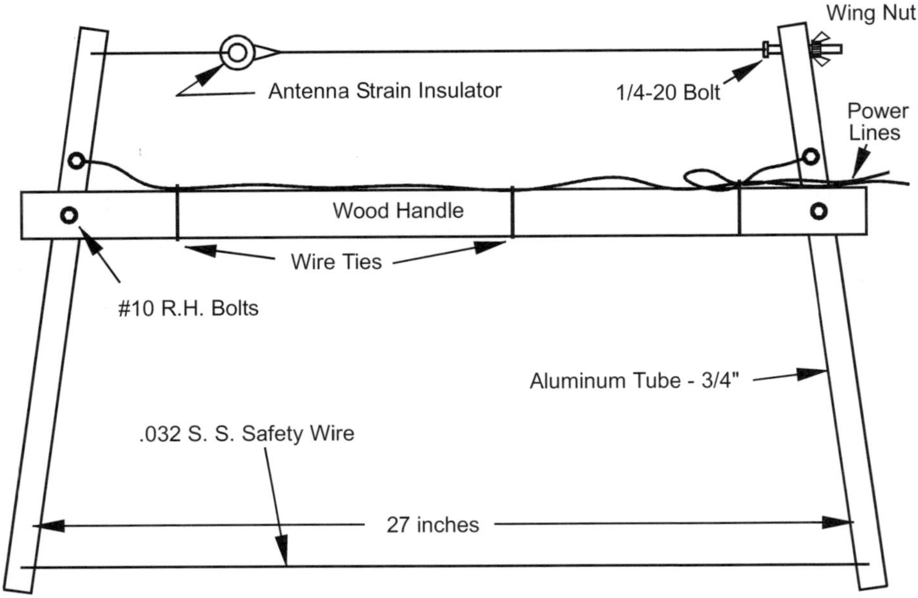

Typical Hot Wire Foam Cutter

Cutting Wire

Standard stainless steel aircraft safety wire (0.032 inch) makes excellent cutting wire. The disadvantage is that it ***tends to stretch appreciably at the low red heat cutting temperature***, causing inaccuracies due to "sag" if cutting curves. (Safety wire is perfectly satisfactory for making straight cuts, as in splitting blocks of foam.) The beginner will likely break safety wire a few times until he develops a feel for temperature adjustments and the forces that the hot wire will take. A little practice with scraps of foam will quickly teach the technique. ***"Sag" is not a problem when making straight cuts or if the wire is short.***

Where sag must be limited, as with short radius curved cuts (as in the Rutan designs), it is well worth it to buy ***special non-stretching nickle-chromium ("Ni-chrome") hot wire*** (.032 inch), sold for the purpose by the aircraft supply houses. Ni-chrome wire is expensive, but almost never breaks and greatly reduces the fuss and bother of making cuts without delay.

Splitting Polystyrene Foam Blocks

The most economical way to buy polystyrene foam is in the form of large blocks. Hot wire cutting may be used to split foam sheets from the block as needed very quickly and easily.

The sketch below (and in Appendix 1) shows a simple splitting board. Steel ***drywall corner edging*** with a 1/2 inch flange is screwed to the sides of a sheet of chipboard to form a pair of smooth rails exactly 1/2 inch high. To cut a sheet of foam, the ***block is placed on the splitting board as shown and is weighted in place with scrap steel and/or sandbags.*** The hot wire frame is then pressed against the rails and passed through the block at about one inch per second. The result is a very smooth straight cut, at exactly the height of the rails. This takes a bit of practice, but not much. After about three trials and adjustments of the wire temperature, the beginner can get nearly perfect cuts. A slight sawing action while cutting is helpful.

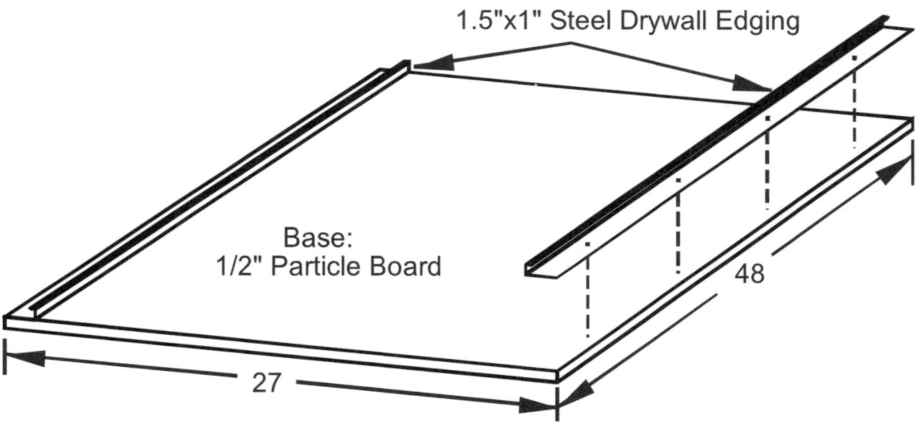

Foam-splitting Board

The ideal temperature adjustment is reached when the cutting wire produces fine "hair" at the end of the cut. Again, this becomes obvious after a few tries.

It is very easy to produce sheets of other thickness, or even a taper, by placing any kind of shim or spacer on the splitting table as needed. For example, if a sheet of 1/4 inch plywood is placed on the 1/2 inch splitting table, the resulting cut sheet will be 1/4 inch. Very good sheets down to 1/8 inch can be made this way. The tendency of thin plywood spacers to warp is removed by the use of ***sufficient sandbags*** to hold the work down.

Right-angle Slicing

The fixture shown below and in Appendix 1, the "baloney slicer," is useful for trimming small pieces and for establishing an accurate 90 degree angle, as needed. It can be helpful in free-hand trimming by providing a secure guide for the cutting wire while the work is held against the base with one hand.

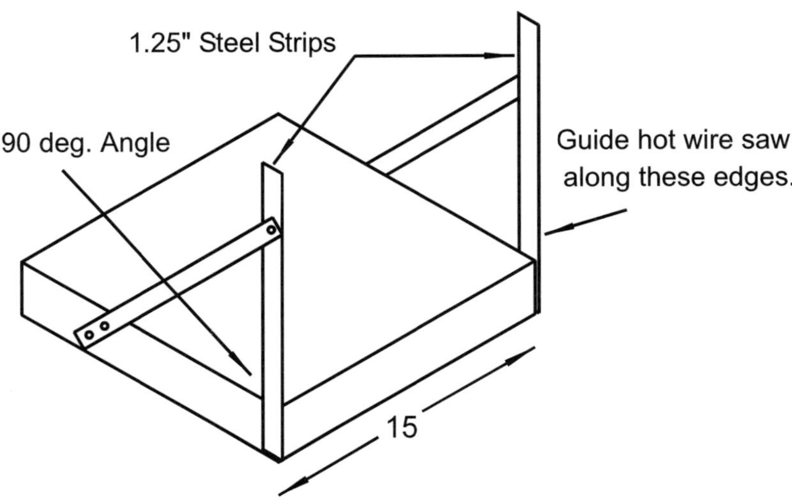

"Baloney Slicer" Foam Trimming Table

Free-hand Cutting

Polystyrene foam is excellent in many applications for making odd-shaped forms and cores. The builder makes up a very rough shape from blocks and scraps of blue foam and makes the first rough cut to shape by free-hand cutting with a small hot wire cutting frame.

Curved Shapes - Wings

In the method pioneered by Burt Rutan, a solid block of foam is cut to a complex shape, like an airfoil, by using a hot wire guided between two templates. This can be done quite accurately, if wire "sag" is allowed for or if ***inconel wire*** is used. The resulting solid core is then skinned with glass as required to carry the load. Spar caps and glass shear webs can be imbedded as the designer requires.

Cutting Foam Blocks Between Templates

The hot wire is guided by being ***stretched between a pair of shaped templates,*** as shown below for the case of an airfoil. If convenient, the templates should be made of a material, such as sheet aluminum, which can stand indefinite contact with the hot wire. Masonite or Formica will work as templates, but are subject to some damage from the heat if the hot wire is allowed to remain in contact at one spot.

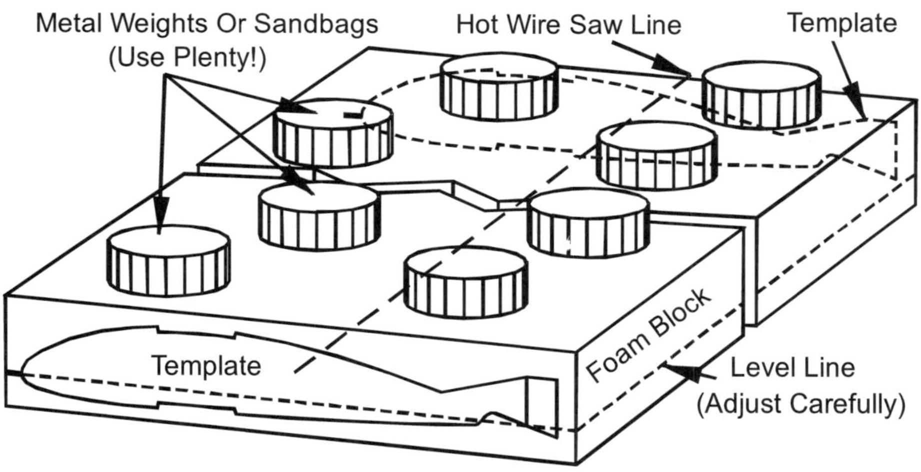

Hotwire Cutting An Airfoil Core

Both templates are marked with a corresponding array of "calling numbers." The templates are attached to each side of a block of foam with ordinary nails. The foam block is firmly weighted in place with scrap iron and/or sandbag weights.

To make the cut, the builder and a helper bring the cutting wire to the block and cut in until they BOTH contact the steel tip of the template. At command of the builder, they then work together to move along the tip at about ***one inch in two seconds*** to the first "calling number." In unison, they then advance from number to number, called out in turn, until they have gone all the way around the templates and returned to the tip. Wherever there is a sharp corner, the two workers stop for a called pause of about two seconds, to let the "sag" in the hot wire straighten, before advancing to the next calling number. They then cut their way out of the block and remove the wire. As an alternative, it is possible to make the cut as two operations, one for the top half, the second for the bottom.

> *Note that the **cutting action must not be allowed to stop,** because the cut area will continue to grow around the hot wire and appear as a groove in the desired shape. If the cut is delayed for any reason, the wire must be moved away from the desired cut line. If the damage is not too great, the groove can be filled with micro.*

The result is a quite good copy of the form defined by the templates. Note that the method will make any shape defined by a system of straight lines, so it is ***just as easy to make a tapered wing*** and/or add washout as it is to make a straight one. If the cut is not too fast where there is significant curvature, as near the leading edge, the error due to wire sag is usually acceptable. This method is used in hundreds of successful aircraft of Rutan design and derivative designs which follow Rutan's methods.

Attaching Foam to Foam

Splicing Sheets and Blocks

It will be necessary from time to time to splice two similar pieces of foam together to make a continuous piece for some following operation. If the joint between two sheets is a straight line, the two pieces to be joined are first taped tightly together with brown mailing tape, as shown on the next page. The joint is then opened, like a book, and filled with enough medium micro to assure complete coverage. The "book" is then closed, forcing excess micro out to be picked up with a squeegee. After cure, the mailing tape can be removed.

To join two thick blocks, the builder (1) applies slurry to both sides of the joint, then (2) applies wet micro in a pellet in the middle of the joint. The pieces are then pressed and wiped together so that ***the wet micro displaces the air in the joint.*** Excess micro is picked up with a squeegee. If the joint gap is more than about 1/10 inch, the joint faces should be squared up. Slivers of foam cut with the "baloney slicer" can be used to fill out the joint. ***In no case should an air void be left.***

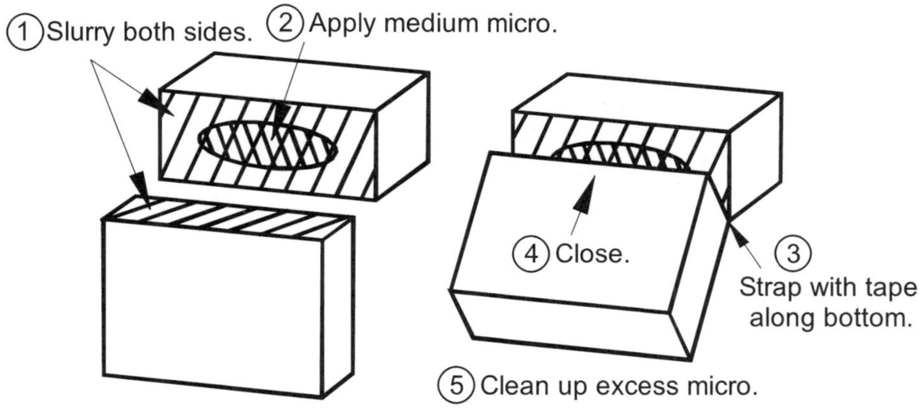

Splicing Foam Blocks

The Problem of Sanding Hard Micro

Cured micro is significantly harder than the foams we use and any attempt to sand a joint with both micro and foam together ends with over-sanding the foam and leaving nasty ridges of hard micro standing up like lava veins in sandstone. It is therefore necessary to ***grind down any micro lines to BELOW the line of the foam, BEFORE attempting to shape the foam.*** This can be done quickly with a small cylindrical cutter in a Dremel tool or with a spiral saw with the shaft held at a 45 degree angle. The operation is not critical and any small amount of extra foam removed will be replaced with new micro when the foam surface is filled.

Hot-wire Cutting Block Joints

A hardened micro joint ***cannot*** be accurately cut with a hot wire. The hard micro stops the wire and the foam around the hot wire melts. The builder must then pull the wire out of the foam block before the situation becomes any worse. The part may be salvaged by making the cut from the ***other*** side of the joint to meet the wall of micro. This leaves a thin "fin" of foam and micro which can be removed by free-hand cutting with a spiral saw or Dremel tool. The damaged area can then be sanded out and patched with dry micro.

When large blocks are required, as when building a Rutan wing, better practice is to ***temporarily attach the blocks with 3-inch finishing nails***, thoroughly weight everything down, and make the cut. After the cut, the nails are removed and the matched ends are attached with micro in the regular way. Care must be taken in placing the nails so that the cutting wire will not contact them.

Temperature Limitations

Whatever structure is to be made with a foam core, eventually the skin must be post-cured at higher temperatures. The foam and the epoxies used can stand post-curing temperatures. Most common glues that might be cut cleanly with a hot wire cannot, so they will not be acceptable if the part will carry a significant load.

There are applications where post-curing is not required, where the completed part will not be heavily loaded and/or where the convenience of carving complex shapes is more important than achieving maximum strength. In making ***plugs*** (Chapter 13), the foam core will be not be flown, will not be postcured, and will be ***used only as a form***, so a free-cutting joint is desirable. Several alternatives are described below.

Attachment With Polyurethane Glue

Foam blocks can be also be attached with ***polyurethane glue***, sold as a general purpose wood glue. This material forms a clean glue line that CAN be cut with a hot wire and its use should be considered as a compromise which may be useful in making mold patterns or other complex shapes. Polyurethane glue is only moderately effective as a filler, so the attachment faces should fit fairly closely before applying the glue.

The material is water-catalyzed by moisture in the air, so does not actually "dry," as by evaporation of a solvent. Open working time is quite short. When bonding foam blocks, a good procedure is to apply the glue quickly from the tube, wipe the bonding faces together to spread the glue evenly (or use an expendable paint brush), separate the blocks, exhale moist breath over the wet glue surfaces to "charge" the glue, and reattach the blocks.

Once the container is opened, this glue has a short shelf life.

Polyurethane glue ***produces a gaseous product while hardening.*** This makes a line of tough, leathery bubbles all around the edge of the joint, as the gas leaves the joint. The ***expansion force can be large,*** so the parts to be bonded must be ***very firmly clamped*** until the reaction is complete. This method should not be used for very thick blocks because the builder cannot be sure that gas is not trapped in the joint.

The hot wire cut line over the joint is not perfect, but clean-up is not difficult- much easier than trying to deal with a micro joint.

Attachment With Cabinet Bonding Cement

The building trades use a calking-gun material sold as ***heavy-duty construction and remodeling adhesive***, recommended for permanent assembly of cabinets and countertops. This material works very well as a quick-acting adhesive for both small-cell and large-cell foam. The joint cuts perfectly with a hot wire and requires no further attention.

Attachment With Spray Adhesive

When the attached parts are not to be flown as a working core, but will be used only to make a mold or plug, ***spray adhesive,*** such as 3M Super 77, as sold in office-supply stores, can be used to attach foam blocks. To use, the builder covers each face of the joint with a generous coat of cement, waits 30 seconds for the solvent to evaporate, then mates the parts. Note that ***you get one chance - the joint cannot be "wiped"*** into position.

The joint cuts cleanly with a hot wire and requires no further attention.

Bending Unskinned Foam Sheets

In most situations requiring a gentle single curve, such as in forming a fuselage side, the flat sheet foam is easily bent and held to the desired form while a skin is laid up. Obviously, the thinner the sheet, the greater the angle it can be bent to before the ***limit strain is reached at the outer convex surface*** and the piece cracks.

If thicker sections of bent foam are required, special operations will be needed. One example is the urethane core wing strakes which serve as fuel tanks for the *Cozy* aircraft. The bent cores, before layup, look like formed taco shells and have much the same sense of brittleness.

Some designs call for bending a sheet of foam into a desired form using a simple jig and temporarily attaching the foam to the jig to hold it until a skin can be applied. This is rarely a problem for the gentle curves typically required. Shorter radius curves, such as around a seat and seat back, may be near the limit for bending. The builder will quickly find that she must be particularly careful to ***avoid bending the foam so that the convex side is in significant tension***. Heat can be applied to the ***outside*** of the curve to induce expansion. Urethane foam, particularly, will crack on the tension side, so the builder must be careful to push from the side to assure adequate compression and avoid tension. Many beginning builders have found that ***bending 1/4-inch foam, as required by some current designs, is difficult, but it can be done***, with care.

Bending Thick Flat Panels

In this book, most of the examples and Exercises make use of ***1/2-inch thick foam cores***. Sandwich panels made with such thick cores are quite stiff, which is, after all, the point. If a sandwich panel with a 1/2-inch core is made up on a flat tool, the flat panel will be very difficult to bend to a sufficient angle as typically required for a small aircraft. ***Flat sandwich panels made with 1/4-inch foam cores are probably a practical limit*** where even small bending of the finished panel is required for assembly.

Bending Slotted Sheets

Where the core thickness is too large for the required bend, it will be necessary to cut slots in the foam ***so that the remaining web is made thin***, of the order of 1/16- to 1/8-inch. The builder must then fill the slots with soft micro before placing the sheet in the final bending form. Excess micro will ooze out of the slots and can be removed with a squeegee. (See the sketch on the following page.) The builder can proceed with laying up the final skin before the micro has hardened. (See also Exercise 10, the Mailbox.)

The sketch shows an example, drawn to scale, of the effect of using saw cuts or V-cuts to make slots in 1/2-inch foam. Both are very quick and easy to make. The builder simply marks the slot lines with a marker pen and the ***slots can be cut freehand*** along the lines. The process is not critical. See that ***rectangular slots tend to close when bent***, but this is not a problem if the slots are filled before bending. The ***V-slot is easier to fill*** to the bottom. The builder will find that it is possible to be quite aggressive in forming short curves with slotted foam.

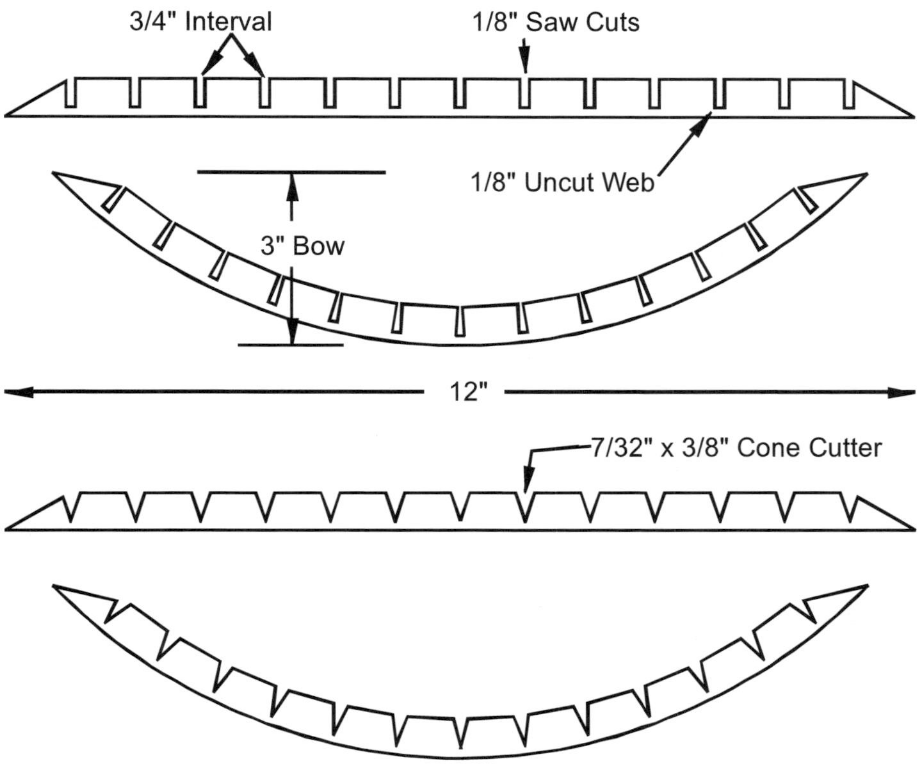

Bending 1/2-inch Slotted Foam

Single Curves

A ***table saw*** may be convenient for the slotting operation. Large sheets can be slotted easily with a small ***cordless circular saw, spiral saw, or Dremel tool***. Note that it is necessary to ***control the depth of cut accurately*** to avoid cutting through. Some experimentation may be needed to choose the depth of cut correctly. The voids left by the saw cuts must be filled with an appropriate filler, like "soft" micro, to close voids and restore the integrity of the core before laying up the skin.

Many single-curved forms, such as seats and seat backs, are defined by straight lines. One easy way to place foam on such a form is to saw-cut the foam in the high bend areas on about 3/-inch to one-inch centers with the saw cuts on the side *away* from the first skin. After curing, the work can be turned over, the saw cuts filled with micro, and the second skin applied, usually without waiting for the filler to cure.

To give the reader a feel for the bending problem, the Dremel tool can be used with a Type 125 ***cone-shaped cutter***, which makes a "V"-shaped slot 7/32-inch wide and 3/8-inch deep. The depth of cut can be set to slightly greater than 3/8-inch, leaving a web in 1/2-inch foam of a little less than 1/8-inch. ***If a series of slots is cut freehand in 1/2-inch polystyrene on about 3/4-inch centers, the resulting sample can be easily bent to a concave radius of 2 1/2-inches, without adding heat***. The slot tends to close upon bending, but enough opening is left to allow void-free filling by forcing soft micro in with a squeegee.

Similarly, for ***convex curves***, a small ***straight cutter*** can be used in either a Dremel tool or a spiral saw to make narrow slots which *expand* upon bending. A small battery-operated circular saw (3 3/8-inch), with a thin steel blade, is very fast and convenient. (The standard carbide blade is typically too wide.)

Double Curves - Pre-fitting Slotted Foam

In most circumstances, the design will require only a limited area where a small-radius curve will be required for a thick core. In these cases, it may be convenient to make the required foam core somewhat oversize and bend it to approximately the required concave shape over an external form, such as a coffee can. If the web left from the slotting operation is about 1/8-inch, gentle application of a heat gun will safely allow a radius of about 2 1/2 inches. The formed core can then be trial-fitted to the skin and trimmed as necessary until the fit is very close. The fit need not be perfect, because the web will tolerate the required small amount of bending when the core is put into position.

Bonding Slotted Cores

To apply the slotted foam, the builder applies slurry to the "web" side in the standard way and presses the foam gently into position against the curved skin. Securing the foam while it cures may require some ingenuity by the builder. Because these tightly-curved areas are usually small, application of sandbags may be sufficient. In other cases, a local vacuum bag may be preferred to assure a reliable bond (see Chapter 9). After the core bond has cured, the open slots can be filled with soft micro and smoothed into the desired final shape.

If the core has a quite short radius, like 2 1/2 inches, it may be desirable to fill the slots with soft micro BEFORE placing the slurried core in its final position. This will assure that the micro has completely filled the slots and allowed no air voids.

Filling with micro BEFORE bending will assure that the fill goes all the way to the bottom of each slot. Excess micro oozes up and out to the side where it can be picked up and removed. (See Exercise 10, the Mailbox, for a specific example of a "closing" foam bend.)

Double-Slotting Foam - The "Tile Technique"

Small ceramic bathroom floor tiles are usually shipped in arrays which are bonded to a temporary backing of light cotton threads. The tile-setter then places the entire array as though it were a single unit. The loose binding allows the array of tiles to conform to any irregularities that may be present. Much the same technique can be used by the aircraft builder by making ***both horizontal and vertical slots*** in the core.

Where the required curve is a complex double curve as in wing fillets or transitions in fuselages, the scoring can be done in ***two*** directions, as with a "cube steak." This technique will generally require some piecing and ingenuity to complete a light core without excessive voids, but the work goes quickly and it is easy to form a smooth surface with dry micro fill between the "cubes" of foam. After the micro has cured, it will be necessary to clean up the core. ***Cured micro resists sanding much more strongly than bare foam, so care will be necessary*** to avoid cutting down the foam too much. A small belt or disk sander, patiently applied, is useful here. (Careful of foam dust!)

Bending With Heat

If a foam sheet is to be bent to an angle beyond the cold-bending limit of strain, the bend area must be heated to the point of plastic flow but not to the point where the foam cells collapse. This is not a convenient operation and ***is almost impossible if the piece is large compared to the area that can be locally heated and maintained at bending temperature.*** If the piece is small, a hand-held heat gun can be used. Larger pieces can be bent over a hot pipe, such as cabinet makers use for bending wood. A serious practical problem is that ***foams are extremely good heat insulators,*** so it is easy to have the surface melting and collapsing while the interior is still too cold to bend (as with hot-wire cutting). The practical limit for local heat bending is less than 1/4-inch.

The most controllable way to heat foam for bending without overheating is to use a steam box and heat-soak the foam before bending. The determined builder can throw together an "asparagus cooker" made from a steel garbage can, but most builders will prefer to solve the curve problem another way. (*Cozy* strake cores are relatively thin and are bent with the same steam gun used to clean wine barrels.)

Laminating Cores

It is possible to make short-radius curves in fairly thick cores by side-stepping the problem and ***laminating*** the core to the desired thickness. This is a practical technique for builders who are using polystyrene foam because it is so ***quick and easy to make thin slices*** of almost any size by slicing a block of raw foam with a hot wire.

Assuming, for example, that the builder wished to maintain a core thickness of 1/2 inch over a tight radius, he might first perforate a 1/2-inch slab of foam, then make three slices of 0.16 inch. These slices can then be coated with slurry, as described in Chapter 3, and bonded into the desired form under vacuum or sandbags. The ***perforation holes of each layer will not align,*** but there is enough cross-wise flow of the slurry between layers to remove air and get a void-free bond.

Chapter 5

Basic Layup Techniques

Introduction

Adding mixed liquid resin to dry fiberglass is probably the most important skill that the composite builder will need. In following chapters, and in the Exercises following, we will describe rather complicated combinations of processes and conditions which will take this basic skill for granted. Some practice is required before the builder will develop the judgement necessary to estimate quantities and choose the most effective tools. ***If the reader is seriously considering building an airplane, he should do the Exercises described in Part 3*** to gain that necessary experience. In this chapter, we will describe specific techniques for placing dry glass cloth on a surface and wetting it out quickly and efficiently. These skills will apply almost universally.

There are two basic cases where wetting-out glass cloth is required:

1) Making a glass skin directly on a mold, or "tool," where the tool determines the final shape and surface finish of the cured skin. The finished structure will be removed from the mold after curing.

2) Adding and bonding glass to a supporting surface, usually a foam sandwich core, where the bonded core will become part of the final structure and a final surface finishing operation may be required.

Basic Wetting-out Techniques

Basic Operations

The purpose of the wetting-out operation is to add just enough wet epoxy to drive out the air and fill the voids between fibers in the fabric. There is absolutely no substitute for experience and practice to develop the judgement and individual preferences in tools to do this quickly and efficiently. Different builders do it different ways but, whatever the choice, the problem is to displace the air with a practical minimum of excess epoxy.

Exercise 1, in Part 3, takes the beginning builder through a detailed description of a simple multi-layer layup.

The description that follows assumes that the problem is to wet out a stack of several layers over a flat surface, or tool, similar to the example of Exercise 1. The use of tinted epoxy is assumed.

1) Set up the work area as shown. The release film can be cheap PE painter's plastic or industrial quality release film. Stretch out smoothly.

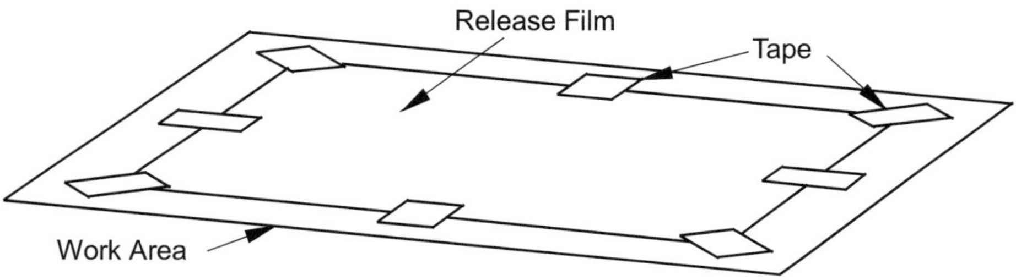

Release Film on Work Area (Tool)

2) Place the first glass layer on the dry working surface and position it as desired. Note well that dry glass is easy to slide into position and that this ***opportunity to reposition easily is lost once wetting has begun.*** As an alternative, one or more spots of resin can be poured on the supporting surface and the glass can still be slid into position without disturbing the weave too much.

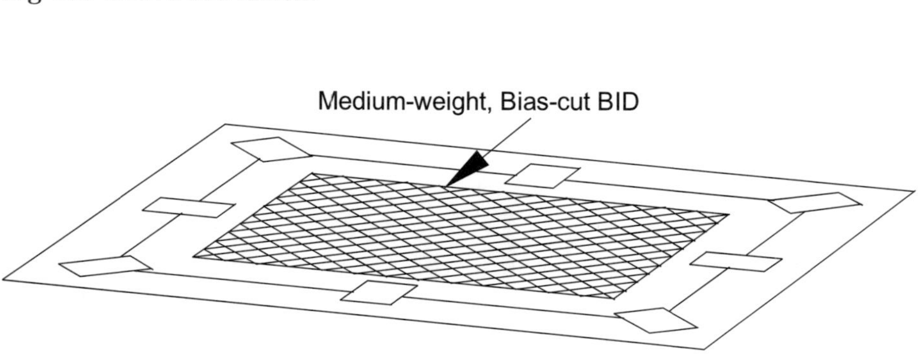

First Layer in Place for Wetting

3) ***Position the second and third layers*** of glass. Add a little epoxy if necessary to secure the position.

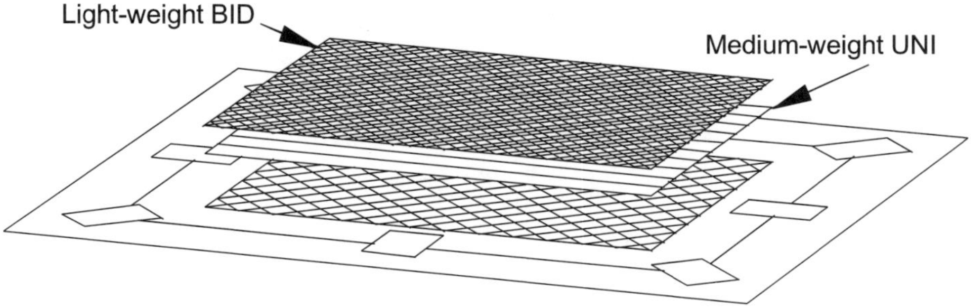

Place Following Layers for Wetting

*4) **Add epoxy to the layers*** by slowly pouring a stream of epoxy over the surface of the layup.

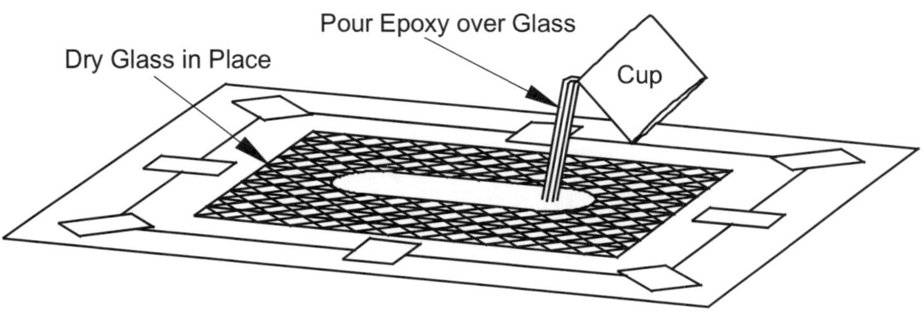

Applying Mixed Epoxy

*5) **Spread the tinted resin*** over the work (see the sketch). A very good "quick and dirty" way to make the initial spreading is by placing a piece of 4-mil PE plastic over the wet area and ***working through the plastic with a dry squeegee.*** This goes very quickly. It will be obvious which areas are too wet or too dry. If more resin is needed, the edge of the wetting plastic can be lifted and more resin added and "chased" to the dry areas with the squeegee. With a little practice, the builder can move large air bubbles out from under the plastic without driving out too much resin. If the work area is a little cool, a ***heat gun*** can be used to warm the resin through the plastic.

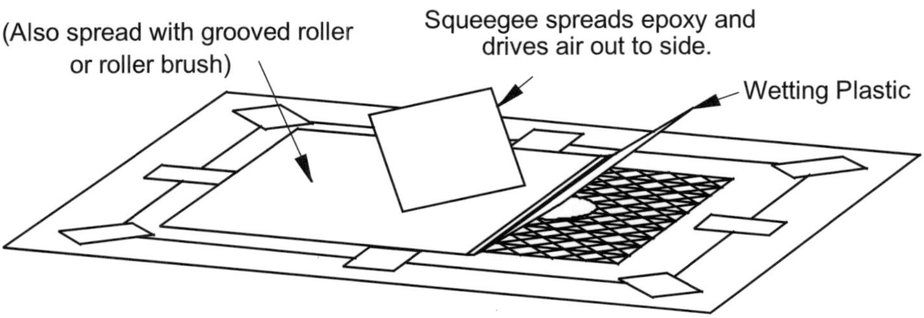

Spreading Wet Epoxy

6) Remove the "wetting plastic" used in the step above. Small ***air bubbles will appear as white*** areas and can be removed by "stippling," or repeatedly pressing down with a cheap expendable paint brush to push the wet layup against the tool. (Some builders shorten the bristles of the stippling brush by trimming them to about one inch with a bandsaw.) Use of a ***grooved roller or a roller brush***, as sold for the trade by the aircraft supply houses, is a faster way to consolidate the layup (see Appendix 1). The beginning builder should ***not attempt to wet out more than three layers*** at a time because it will be more difficult to find and drive out bubbles, even with the used of tinted epoxy. Add more epoxy as needed to dry areas and inspect with a ***shop light held at the side at a low angle.***

7) (This step is optional.) ***Add a layer of dry peel ply to the wet stack***, as shown in the following sketch. Cut the peel ply so that it hangs over the work with an adequate margin to assure covering. The builder will find that the smooth, tightly woven, peel ply is easy to smooth out with a squeegee and roller. The first layers should be a little too wet and the excess resin brought up in Step 6, above, may be sufficient to wet the final layer of peel ply. The goal - again, an operation that improves with practice - is to distribute the poured epoxy without disturbing the position of the cloth. Add epoxy as needed to be quite sure that no part of the peel ply layer is left too dry. ***The peel ply layer serves as an added inspection.*** After cure and removal, it also serves to provide an excellent fresh bonding surface for following operations and to remove excess epoxy from the underlying layup.

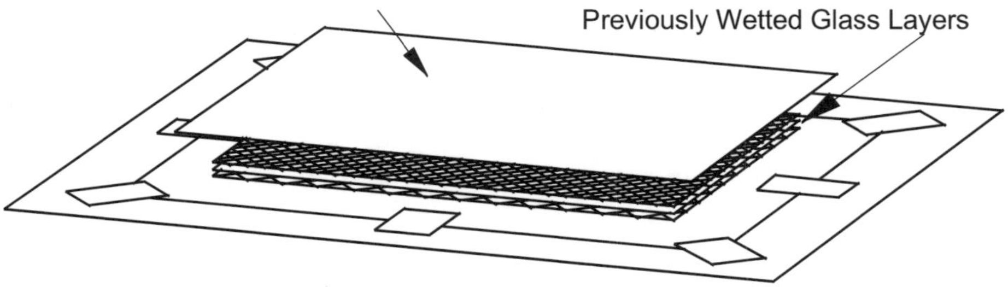

Application of Peel Ply

The "Prepreg" Sandwich

It is frequently necessary to place wetted glass on a surface that may be difficult to reach or may be vertical or even inverted. In these cases ***it may be practically impossible to place the glass accurately, keep the weave straight, and wet it out.***

The solution is to wet one or more sheets of glass ***inside a sandwich of release film***, such as cheap 3- or 4-mil PE plastic. This is very easy and quick to do on the bench. (The term "prepreg" refers to the general industrial practice where uncured resin is added to the woven fabric in a precision factory operation. Most composite kit manufacturers use prepregs routinely.) The wet glass can be handled and cut along marked lines quite easily when protected by a sandwich of plastic. The outside surface of the plastic will be dry.

To apply the wetted glass to the work, the builder removes one layer of plastic, then places the exposed wet glass as needed. The wet glass will stick quite well to the supporting plastic and can be accurately positioned. The remaining layer of plastic is then peeled off, leaving the glass in place.

Wet Glass Tape

It is also frequently necessary to prepare ***narrow strips of bias-cut wetted glass*** (taping) that are to be placed accurately on the layup. One typical application is placing a binding layer over a joint between two cured panels. The builder will find that this operation is ***maddeningly near*** impossible to do with dry glass or wet unsupported glass while preserving the orientation of the weave, smooth alignment, and the temper of the builder.

Wetting a Prepreg

Usually, the builder will require several wetted "prepreg" strips at one working session. The best way to do this is to cut a sheet of PE plastic large enough to fold over the sheet of dry glass that is to be wetted. Mark the OUTSIDE of the plastic sandwich with a marking pen. (Remember that epoxy dissolves the ink.)

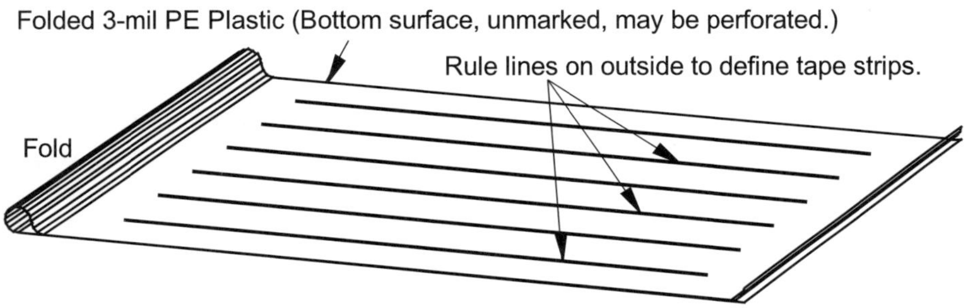

Marking Plastic for Wetting and Cutting Tape Strips

Turn the folded plastic over so that the ***marks are on the bottom***. (The marks will be visible through the wet glass.) Place the glass and wet out as above. It will be necessary to lift the plastic from time to time to get the air out from under, but a good job can be done by squeegeeing through the plastic. This method may waste some epoxy out to the side of the layup, but time, too, is money. The layup can be lifted from the bench, turned over, and worked from the other side. (We will assume that the resin mix has been tinted.)

> ***Important Option:*** Better control of wetting can be made if the ***unmarked surface of the wetting plastic is perforated*** with a perforating tool. Excess epoxy is removed through the perforations when the layup is rolled and squeegeed out. Sometimes the excess epoxy can be "recycled" to fill dry areas.

The builder then ***cuts the wet strips as needed with scissors.*** One layer of plastic is removed with the gloved hand, exposing the wetted cloth. The remaining layer of plastic will support the strip of glass and prevent the weave from losing orientation. The builder places the strip as needed and works it into position as she removes the final layer of plastic. This operation goes almost as quickly as the time it takes to describe. Get your mistakes out of the way on the Exercises in Part 3 before using these techniques on an airplane.

Bonding Fiber Skins to Foam

It is essential that the skin of a composite sandwich be bonded to the core of the sandwich securely, without voids. (See Appendix 2 for details on why this is true.) The ideal bond would evenly distribute shear forces from the load-carrying fibers of the composite to the edges of the small cells that form the foam without stress concentrations or excessive weight.

The practical solution is to fill the open cells at the surface of the foam with a compatible medium. If the cells are very small, pure epoxy can be used, the same epoxy as used in the rest of the layup. The builder paints the foam surface with epoxy, squeegees it in to drive out air, and waits for 10 or 15 minutes while the epoxy drives out air to wet the open-cell surface. ***A heat gun helps wetting considerably*** at this stage. When wetting is complete, there will be very little excess epoxy showing on the surface of the foam, just a little gloss when the foam is viewed with a light at a low angle. After wetting, the layup proceeds as described above.

"Slurrying" Foam

The process of laying up on foam starts with the slurrying operation.

Both "2-pound" foams we use in this book have sufficiently large cells to benefit from the use of a "syntactic foam" of mixed epoxy with added micro balloons, as mentioned in Chapter 1. This foam is much lighter (and cheaper!) than pure epoxy, but flows well enough to reliably wet the rough foam surface while still filling the voids without excessive weight.

Before applying slurry, the foam core should be cleaned using a shop vacuum with a soft brush head. This is particularly true with polyurethane foams, which tend to dust off easily.

Slurry is a mixture of epoxy and micro in a non-critical ratio of about 1:1 by volume. This mixture flows freely from the mixing stick and handles much the same as pure mixed epoxy. To apply, the builder paints the slurry on with a paint brush. (***The use of tinted epoxy is particularly effective with slurry. Use a little blue chalk for pink foam, red chalk for blue foam.)*** It takes a little practice, but the goal is to cover the work with just a little more than enough to fully wet it. It is necessary to drive out air to the extent possible and the quickest way to do this is with a ***plastic auto body squeegee.*** The squeegee causes very high local turbulence at the sharp edge, even with the viscous medium, to displace air. Local heating with a heat gun is highly desirable to give quick wetting. Large-cell foam may produce a faint crackling sound when the slurry penetrates the cells.

Wait about fifteen minutes, then proceed to apply the glass cloth and continue with the wetout.

Applying Dry Glass to Foam

The following detailed description assumes that the job is to lay up the skin of a large, nearly level surface area, such as a wing skin.

1) Assure that the foam, cut glass, workroom temperature, and all ***needed tools are ready. Inspect*** the setup and ***verify alignment,*** if necessary. See that each piece of cut glass, peel ply, or release film needed for the job is placed in a labeled grocery bag.

2) Mix, then tint a cup of epoxy. (The use of carpenter's chalk as a tinting agent will force discipline in adequately mixing the epoxy. It takes what feels like "unnecessary" stirring to break up and distribute tiny clumps of powdered chalk.)

3) Repair nicks and dings with very dry micro. Use a little of the mixed epoxy in a second cup and add more micro to make the required dry micro filler.

4) Add micro to the mixed and tinted epoxy in the ratio if about 1:1 or 1.5:1 to ***make slurry.*** The mixture should run freely from the stirring stick, be brushable, but be sufficiently viscous to be removable with a plastic squeegee.

5) Spread the slurry over the work and squeegee it in. Heat locally with a heat gun to improve wetting. Wait about fifteen minutes. (Use some of the waiting time to mix and tint more epoxy for the next stage.)

6) Lay on the cloth, carefully, and position as needed. If the piece is large, a second pair of hands or a transfer sheet (see below) will be helpful. Pour on the mixed epoxy and distribute with a squeegee. Distribution of epoxy is often helped by working through a piece of PE drop cloth plastic, about one foot square. Hold the plastic in one hand and work the squeegee with the other. A small amount of experience will quickly teach how to move the wet epoxy.

The Transfer Sheet

It is very difficult to keep the weave alignment correct while handling and applying large areas of dry glass, because it is not practical to adjust the position of the glass after it has made contact with the wet slurry. The solution is to use a ***smooth-surfaced sheet of flexible material*** such as linoleum, 1/8-inch Masonite, or Formica, to ***support the dry glass*** while it is brought into position. The builder then ***slides the glass off*** of the transfer sheet while slowly removing the transfer sheet. If one or two helpers are available, this operation goes very quickly (see below).

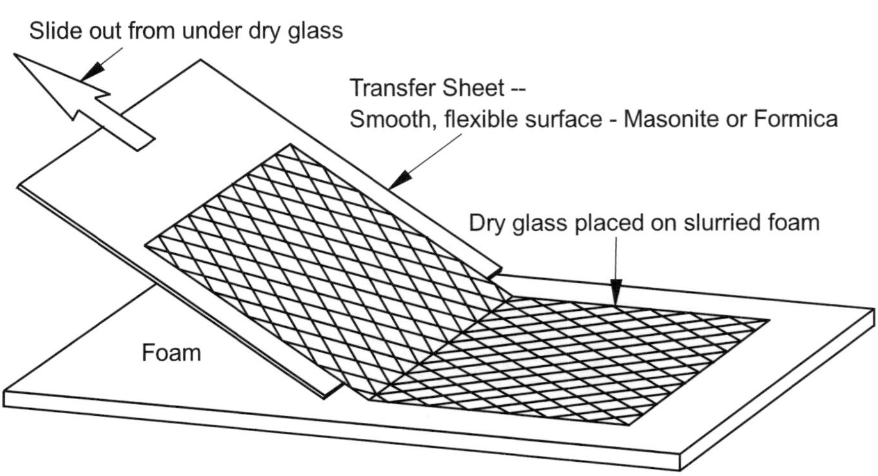

Use of Transfer Sheet to Apply Dry Glass

7) Inspect the work, preferably with a shop light held at a low angle. ***Drive out air bubbles*** as needed with the edge of the squeegee, a grooved roller, or a roller brush. Heat locally, as needed. The tinting should make bubbles obvious. Move any excess epoxy to dry spots. Add more epoxy as needed. Rough trim the edge of the cloth with scissors.

8) If there is another layer of dry cloth to be added, it will be very difficult to apply over an already-wet layer. Place the new layer on a "transfer" sheet of marked 4-mil PE plastic and wet sufficiently to stick to the plastic sheet. Trim as needed and put the new layer in place. If the new piece is large, a second pair of hands will be useful. Work it down with the squeegee to pick up excess epoxy from the previous layer. Add and distribute tinted epoxy as needed. Repeat for more layers, if required.

9) At the last layer, ***inspect*** the work carefully. Be sure that wetting is complete, with no dry spots or bubbles. Squeegee off excess from wet spots. Wet spots are identified by being smooth and glossy, not showing the texture of the cloth. Trim off excess glass with scissors, leaving no more than about a 1/2-inch margin. ***CAUTION: If a large margin of excess wet glass is hanging over the edge of the layup and left to cure, the weight of the hanging glass will cause the sheet of glass to pivot over the edge and lift off the surface, making a bubble. Trim off the excess, leaving only a small margin.***

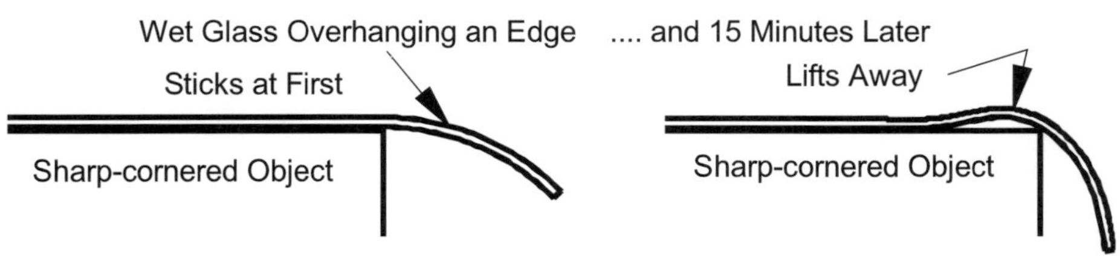

The Corner Droop Problem

10) Peel Ply: If there are ***areas that will be bonded at a later time,*** cover those areas with a layer of peel ply. First, mark the peel ply with a wax crayon or a marker pen so that the marked side is ***away*** from the work. (The wax crayon is preferable because the wax marking does not get dissolved in the epoxy and transferred to the work.) Be sure that the peel ply is truly bonded with the structural layers. If necessary, add a little more epoxy to be sure that the peel ply is wetted, as shown by visible inspection.

> ***Important!*** *Frequently the design will require that several layers of glass be placed in a tapered, or "stair-step" arrangement. The edges of exposed layers may be a little ragged and may have a tendency to curl up slightly, leaving a rough line.* ***Peel ply*** *(marked) should be* ***placed over this transition area*** *and squeegeed down or smoothed with the gloved hand to assure a smooth, snag-free line.*

11) Knife trim: Let the layup cure for three to six hours, to the "green" stage, depending on temperature and the specific epoxy used. When the epoxy has cured to about the ***consistency of warm tar*** or taffy, the layup can be ***trimmed very easily with a utility knife***. Trim the 1/2-inch margin around the layup to a flush edge. The cut may be ***helped by local heating with a heat gun***. If the glass is raised a bit during the trim, just press it back firmly into position.

Special note: Peel ply is close-weaved and has a smooth surface, so it lends itself very well to ***vigorous use of the squeegee***. Step 10, above, should leave some excess epoxy even after completion by an experienced builder. A layer of peel ply can be used as a final layer and squeegeed hard, but WITHOUT adding more epoxy. Excess epoxy is taken up in the peel ply layer. Actually, this is an effective ***"last chance" inspection technique*** because it will show areas which do not produce excess epoxy and therefore might be too dry. More epoxy can be added as needed to assure complete wetting. If you think an area is too dry, it probably is, and must be fixed.

Applying Wetted Glass to Foam

Glass may be wetted before application to a surface, using the "prepreg sandwich" method, described above. In fact, the builder will find that he will use ***wetting under plastic on more occasions than wetting the open glass***, because it is more versatile and allows accurate trimming before placement. Wetting under plastic is somewhat more work and is slightly more wasteful of epoxy than open wetting, but the advantages make it necessary for the builder to be equally good at both methods.

Wetting under plastic will be useful for such applications as placing a very ***large sheet of wetted glass on a large curved surface,*** such as a fuselage. This is especially true when the builder has one or two helpers available and where the piece is so large that several overlapped pieces of glass must be used to fill the required area.

Chapter 6

Flat Tool Vacuum Bagging

Importance of Flat Molding Tools

There is a large engineering literature in the composites field supporting very high performance aerospace applications and very high production automotive applications. Most such applications require tools, materials, and designs far beyond the means or interest of the amateur. This chapter will describe low-cost methods which use materials commonly available to the amateur through the aircraft supply houses or the neighborhood home supply store.

Nearly all of the examples and exercises described in this book will make use of panels or molded structures that are formed against a ***flat mold, or "tool,"*** as briefly described in Chapter 2. In nearly every case, vacuum techniques, as described in this chapter, will be preferred, even though some extra work and materials are required.

Comparison with Commercial Molds

Almost without exception, a composite structure that makes use of vacuum bagging will be formed against some kind of tool. The tool may be a plain flat surface in a homebuilder's shop or can be a complex, multipart, heat-cycled, zero-shrink, ***very*** expensive, tool in the factory of an Air Force contractor.

Commercial molds of whatever shape are polished to a mirror finish, then treated with a mold release agent - usually a wax compound - which does not bond to the layup. The mirror finish does not come cheap. The mold is usually made from an accurate master or "plug" (see Chapter 13), which also has previously had a mirror finish produced at great expense. Kit plane manufacturers are willing to undertake this, because they intend to take many finished parts from a successful mold, but the one-off amateur is rarely willing to go through the full mold-making process. Also, in "real life," a commercial mold may turn out to be unsuccessful at the first attempt and need to be scrapped and repeated.

One reason for the great success of the Rutan designs, and those following, is that the techniques used were "moldless." They were not, however, "finishless." The homebuilder may choose to adapt tools made from available commercial materials which already have a smooth surface.

Conventional mold techniques will be briefly described in Chapter 13.

Flat Tool Techniques

A great many useful aircraft parts can be made easily from a flat tool, using a sheet of smooth plastic, metal, or glass as the tool and using vacuum to complete the operation.

Limitations of an Open Layup

Consider the sketch below, which shows the simplest possible lamination on a plain flat tool with a waxed release surface. Several layers of reinforcing fiber cloth are placed on the mold surface and wet out as described in Chapter 3. The sketch shows an example of rather bad craftsmanship because we exaggerate bubbles and irregularities, considerable excess resin, and poor contact with the tool.

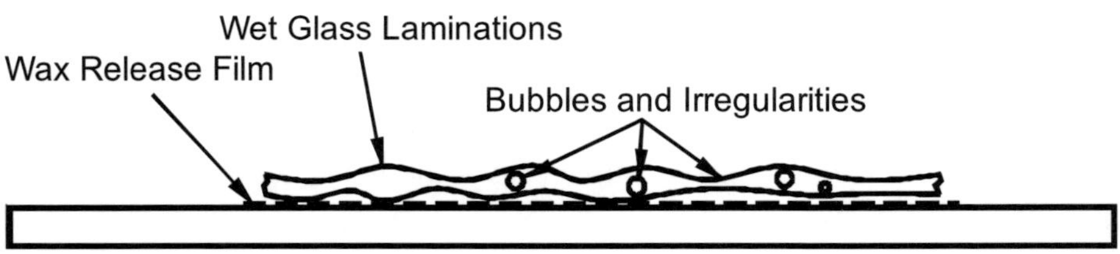

Basic Open Layup (Badly Done)

This very simple system works, but the method has significant limitations for high performance aircraft work. Contact with the mold depends on gravity and the relatively small wetting forces, so significant curves are not practical. Any stiffness in the cloth may lift the cloth out of contact and away from the tool, leaving an unfilled bubble. Enough resin must be applied to assure that there are no voids and that any tiny movement of the fibers will not draw air into the layup. Even so, a simple visual inspection from the working side cannot always assure that there are no bubbles, even if "tinting" has been used (Chapter 3). There will be an appreciable excess of resin, even if the builder is quite skilled.

The open surface is exposed to the air during the cure. Most aircraft epoxies will react with humidity in the air to create a gummy contaminating surface called "amine blush," which must be removed from the finished part. This problem is avoided if air is kept away from the curing resin.

The Basic Vacuum Bag Process

Ideally, we would like to squeeze out any excess resin, remove trapped air, force intimate contact with the tool, and avoid contamination during cure.

Preparation

See the next sketch. Here we have taken the wet layup from the previous section and added three flexible layers:

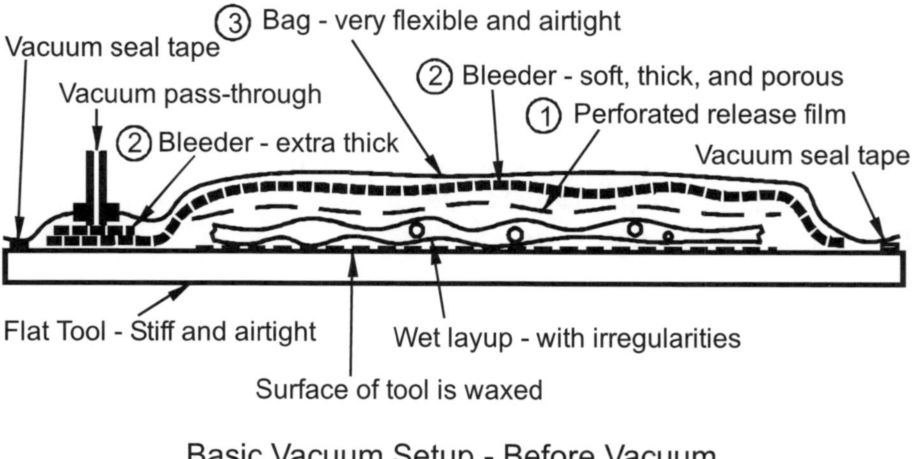

Basic Vacuum Setup - Before Vacuum

1) Perforated Release Film: Thin, non-sticking flexible plastic film, typically 2 to 4 mils in thickness, with many small punched perforations. It is available commercially, or made as needed by the builder from cheap painter's PE plastic, 2- or 4-mil.

As shown, the ***perforated release film*** has been applied smoothly, but not necessarily perfectly. It may be convenient to apply the perf film as a series of strips, with each strip overlapping its neighbor by 1/2- to 1 inch. ***A wrinkle in the perf film will appear as a fine line of excess epoxy on the cured back surface of the layup***, which is usually perfectly acceptable.

2) Bleeder: Ideally, any soft, thick, non-woven fabric such as quilting material. It is sold as a special product for the professional trade, but almost any cheap absorbant porous material from the rag bag will serve. The soft ***bleeder*** is laid over the work in any convenient way and can be made up of separate pieces, which makes it easier to place, the only requirement being that the work area be entirely covered. See that some bleeder material is brought out to the side and that a pad of extra bleeder material is provided under the vacuum pass-through.

Agricultural felt, as sold in garden supply stores for suppressing weeds, is particularly useful as a bleeder where the work is flat or single-curved. The material is very easy to handle and measure, because it is stiffer than cloth or quilting. It has the great advantage of producing an ***excellent bagside surface when used with nearly flat surfaces.*** Where applicable, two layers are usually sufficient. The disadvantage is that it does not conform well to changes in the thickness of the layup.

3) Bag. A good bagging film is any soft, flexible, airtight plastic film. Reliable bagging film is sold for the professional trade and is offered by the aircraft supply houses. Ordinary cheap PE painter's plastic may be used for bagging, but it is ***subject to leaks***, particularly at folds, and may be risky for large, critical layups.

The ***bag*** loosely covers the entire work area and is sealed to the tool with soft plastic sealing tape. The sketch shows the ***usual commercial practice***, where the bag material is simply pressed against the soft sealing tape to make a ***vacuum-tight seal***. In order to make the edges of the loosely fitted bag vacuum-tight, it will be necessary to make short folds, or ***pleats***, to take up the slack in the bag.

The ***vacuum pass-through*** is a simple fitting which seals to the bag and allows connection of a vacuum source. *(It is desireable to place the vacuum pass-through so that it does not contact the uncured work, even* *through a generous pad of bleeder. Vacuum pressure on the hard pass-through may mark the work.)*

Effect of Vacuum

The sketch shows the above setup ***after vacuum has been applied*** and atmospheric pressure has ***reduced the volume*** of the bag and its contents to a minimum. The primary goal has been achieved: the rough layup has been forced tightly against the waxed rigid tool and trapped air and excess resin has been driven through the perforated release film to be trapped in the bleeder. (We assume that the ***bleeder is thick enough to accept all the excess resin*** without saturating.)

The surface of the soft ***bag must wrinkle*** as the volume is reduced. The same effect applies to the bleeder and to the release film. The sketch shows an area on the right side where the bag has been pressed against the tool, causing a ***pinch-off*** - a trapped pocket of air which is not drawn off by the bleeder. As shown, it does not affect the layup.

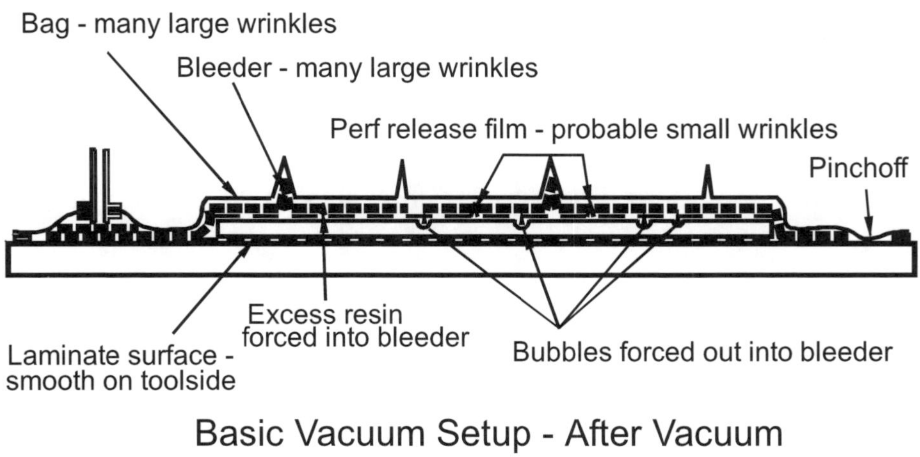

Basic Vacuum Setup - After Vacuum

Bridging

If any layer in a layup cannot conform to the underlying surface when vacuum is applied, a pocket or ***void will be created which will be filled with unreinforced matrix.*** The sketch below shows an exaggerated example. In many situations, a small amount of bridging is quite acceptable.

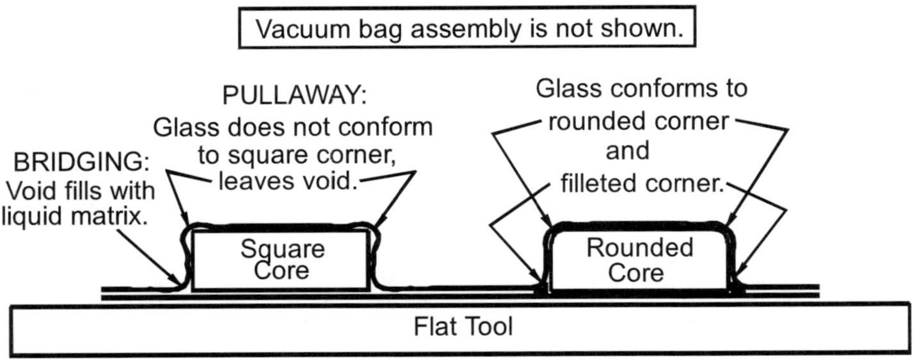

The "Square Corners" Layup Problem

After curing, the vacuum produces an excellent result: a very tight, dense layup, with the excess epoxy removed. The ***toolside*** has a perfect shiny finish reproduced from the smooth waxed tool. As shown, the ***bagside*** of the layup may have some texture from the bleeder or fine lines on the cured surface, which reproduce the wrinkling of the release film. For most applications, these surface irregularities can be ignored.

Note that the vacuum is carried into the work area by the extra bleeder material that connects the vacuum pass-through to the work. This method avoids the risk of having the pass-through hardware make a mark on the work. The extra layer of bleeder assures uniform air flow and reduces the chance of excess resin being carried into the vacuum system.

"Release" Materials

Aircraft supply catalogs list of variety of professional-grade bagging supplies as "release film," "release fabric," "porous ... release fabric," "non-porous ... release fabric," "polyester release ply," etc., and confusion is definitely possible. The only thing these materials have in common is that they do not become a permanent part of the final structure.

For the purposes of this book, we will define three kinds of "release" materials, depending on the application:

1) Plain Release Film: This is any smooth, ***non-porous***, flexible, plastic film which ***does not stick*** to cured resin and is used where the builder desires to remove the formed and cured part from some support, usually a tool. ***Most bagging films can be used for release*** purposes, and special release films are sold specifically for this purpose. Cheap polyethylene sheet plastic as sold for painter's drop cloths and covering woodpiles is an excellent release fiilm, but it is subject to creases.

> ***Warning: Some bagging films are NOT release films, especially the soft, extra-stretchy formulations. Be sure that a film intended to provide a release function actually works before committing to a large layup.***

2) Porous (Perforated) Release Film: This is the same type of smooth, flexible, non-stick, film described above, except that it contains a great many very small perforations so that air and resin - but not glass - can pass through it. The material can be purchased ready-made or the builder can quickly perforate plain release film himself with a ***home-made perforator*** (Appendix 1) or a commercial ***spike roller tool.*** When the perforations are made by piercing the material, as with a nailboard, the holes leave tiny torn edges so that each hole acts as a tiny check valve. After curing, the layup will ***strip freely from the bleeder,*** leaving a smooth surface, except for a very small "pip" at each perforation. The perforated material can usually be reused several times.

3) Peel Ply: In this book, we will take "peel ply" to always mean ***finely woven polyester cloth*** which has the property of ***not sticking to cured resin.*** (The material sold in fabric stores as 100% polyester "liner" or "taffeta" works well.) As used, it may be laid up and wetted out with the regular structural laminations. After cure, it can be stripped off, leaving a clean, slightly textured surface, that is excellent for subsequent secondary bonding operations. In shop practice, ***peel ply is usually left in place to protect bonding surfaces from damage until the actual bonding*** is to be done. Being woven cloth, it is porous and can serve the same purpose as porous release film, as described above, and can be substituted for it - except that it does ***not*** strip freely. In commercial practice, finished work is usually shipped with the peel ply ***and the bleeder*** still attached, to be stripped off when the builder is ready to assemble the part. The author prefers to use perf film to separate the bleeder, just to make things tidier, even if peel ply is to be used. Also, it may be convenient to ***reuse the bleeder*** for another layup, if the used bleeder is not too full of hardened epoxy.

Practical Techniques

Vacuum Capacity

If vacuum bagging systems could be perfect, a very small vacuum source could be used for a very large layup, limited only by the patience of the builder. Real life bagging systems are not perfect and, in fact, perfect vacuum is not required.

The boat building industry, which necessarily must produce very large layups, has made recommendations for pump capacities needed for various layup sizes. Essentially, the boat people choose to live with big leaks and do what it takes to keep the pressure down. They use multiple vacuum sources, high capacity pumps, and large diameter hoses. Fortunately, the aircraft builder typically makes much smaller layups, at a higher level of craftsmanship, with smaller leaks, so a single small "oilless" pump of about 1/3 hp is usually sufficient.

The author strongly recommends the use of a true, purpose-built, commercial vacuum pump for all advanced composites work. Such a pump will be used a great deal and the builder will not wish to be side-tracked with pump problems when he has real work to do. Avoid temporary half-measures like refrigerator conversions and venturis running from shop air.

If the reader completes the Exercises described in Part 2 of this book, he will get a feeling for how sensitive the system is to practical small leaks.

Vacuum Plumbing System

The sketch on the following page shows a complete vacuum system for a basic flat-tool layup, as previously described. The pump may be a standard 1/3 horsepower reciprocating piston pump. The pump pulsations will damage the vacuum gauges, so an external "smoothing" chamber will be necessary. An easy way to make a smoothing chamber is to use a piece of 4-inch PVC pipe, 7 inches long, with end caps. All fittings are standard brass hardware store 1/4-inch pipe fittings, incuding tees, nipples, and barb fittings. The hose is 1/4-inch plastic fuel line. ***No couplings are required***; the system is assembled by simply pushing the tubing over the barb fittings.

The author's setup is to have the pump and smoothing chamber bolted to a 10x 11 inch laminate base. Rubber suction cup feet isolate vibration. The smoothing chamber consists of two stacked pieces of 3-inch PVC pipe, interconnected with a short piece of hose. Two chambers is probably overkill, but the top chamber makes a convenient carrying handle for the assembly.

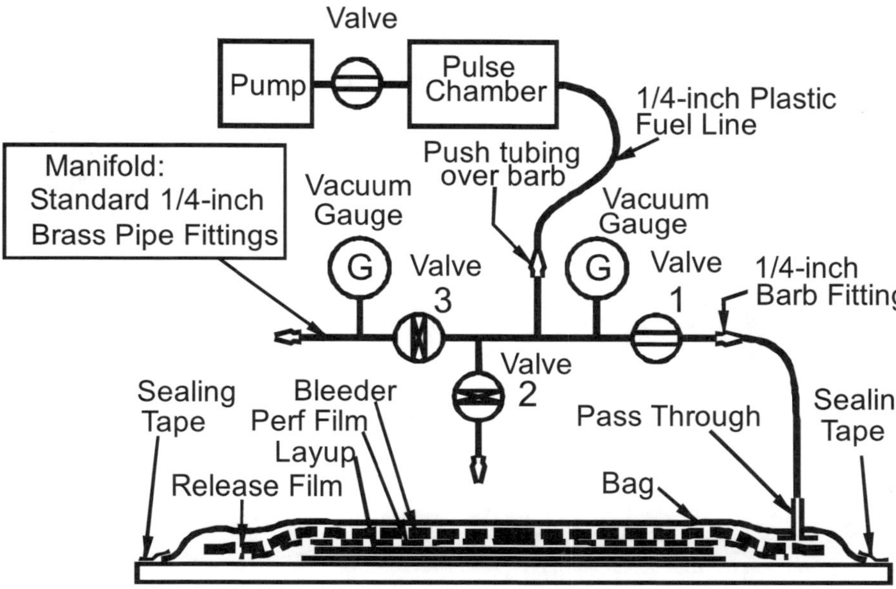

Vacuum System - Basic Vacuum Bag

The sketch shows the left hand side of the gauge manifold as unused, with valves #2 and #3 closed. This part of the system will be used for "double bagging," to be described later.

Vacuum Pass-through "Valve"

The aircraft supply houses and industrial suppliers that cater to the professional composites business offer a full line of vacuum hardware, mostly directed toward applications which assume that the equipment will be used in a high temperature curing oven. Among these is the vacuum pass-through, often called a "valve" even though there is no flow control mechanism. The common commercial product is moderately expensive and does not provide a barb fitting for convenient push-on connection of fuel line hose.

The author has used the simple system shown below for several years. The builder can make up half a dozen of these devices in a single work session and will find that the results will last for years. The O-ring is held in place with automotive windshield sealer. After the flox has hardened, the center hole may be drilled out with a 3/16- inch drill to clean it up and widen the vacuum path. If it should become contaminated with epoxy in use, it can be drilled out again.

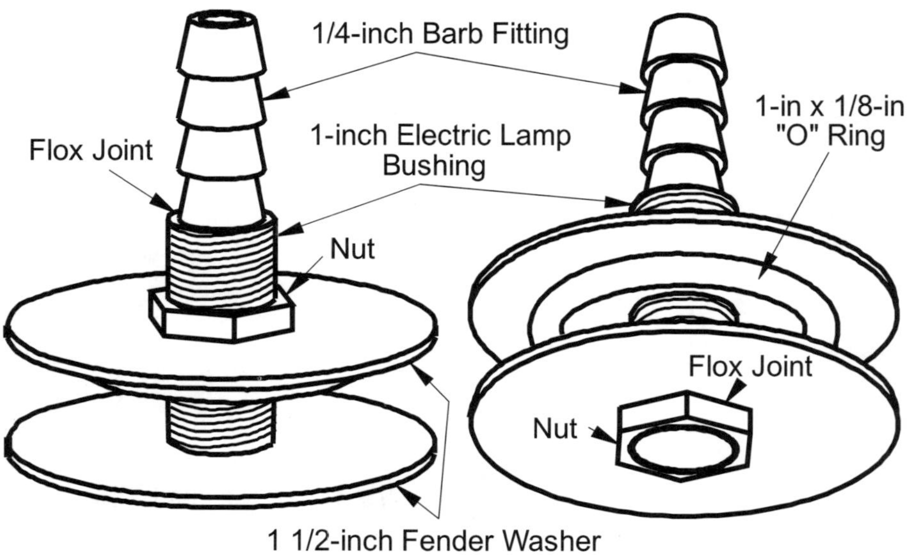

Vacuum Pass-through "Valve"

Alternative Pass-through

The author has also had success with a variation where the soft plastic suction cups sold for hanging kitchen items on refrigerators are substituted for the top fender washer, above.

The center of the suction cup is drilled out to make a snug force fit over the 3/8-inch electrical bushing. To use, the builder simply pushes the soft cup over the bushing and forces it tightly against the bag. Vacuum pressure holds the assembly together. (It is helpful to chill the soft plastic in a freezer to facilitate drilling out the hole for the bushing.) The positive clamping used in the previous example may give the builder comfort, but both schemes work equally well. The required hole in the bag need not be anything more complex than a simple "X" cut with a razor blade.

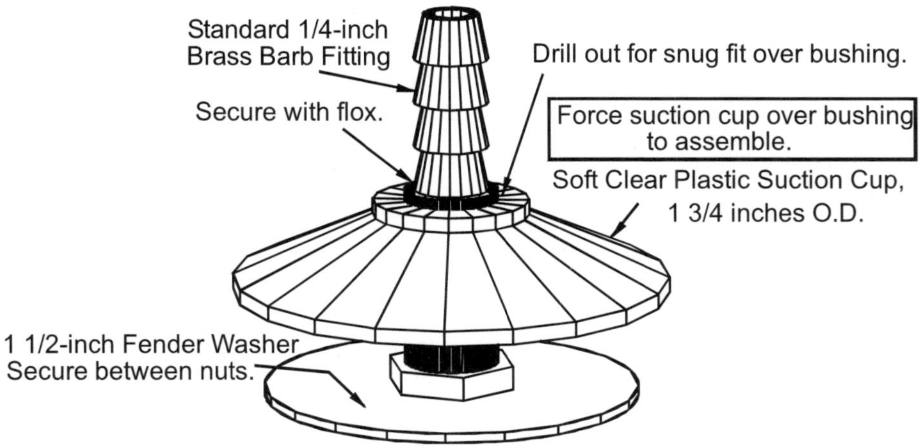

Alternate Vacuum Pass-through

Vacuum Control and Measurement

A vacuum gauge should be included in the vacuum line near where the line plugs into the vacuum pass-through. Most authorities recommend that at least one additional vacuum gauge should be placed on the opposite side of the work, with even more for larger layups. The author has found that the ***second gauge is rarely necessary*** if bleeder is used generously and if the bag is tight. Plenty of bleeder keeps the pressure equal over even a large layup. If the system is leaking, the problem will be obvious at the line gauge and a second gauge will not help much.

> *The author strongly recommends that the builder* ***assemble a gauge manifold,*** *as shown in the sketch on page 90. The* ***gauge manifold puts the valves, hose connections, and gauges that will be used all in one place.*** *The complete assembly can be kept in a drawer and brought out in the usual hurry when needed. It is then only necessary to push the hoses onto the barb fittings as required for the job.*

Bag Material

After the builder has committed to a large layup, such as a wing skin, and epoxy is hardening, the builder must then rig the bagging setup and pump it down. This may require a bag of very big area. This is not a time for a leak-hunting operation. ***Custom composite shops*** that serve aircraft homebuilders ***always*** use commercial grade bagging material, as sold by aircraft suppliers, to minimize the risk of bag flaws. ***The most commonly-used material is pale green nylon 6***, available in wide rolls

(54 inches). It is fairly expensive, but is tough, stretchy, and usually can be reused many times. (It is also a good release film.) As used commercially, the material must be able to stand high-temperature oven curing, which doesn't apply to the homebuilder. ***The amateur will use it for leak-free reliability.***

The homebuilder will often need to make small area layups, where leak control is much easier. In this common case, cheap 4-mil polyethylene (PE) drop cloth plastic, as sold in paint stores, works well enough. The great advantage is that it is very cheap and is always available in the shop for "sandwich wetting" glass and other general purposes. The disadvantages are that the material is somewhat less flexible than the "good stuff" and is subject to leaks, especially at folds and creases.

Bag Sealing

Normal commercial practice for bag sealing is to simply press the bag against the sticky surface of soft, commercially available ***sealant tape***. The sealant tape is first pressed on the tool, all around the work area. Adjustment of the bag is easy: the bag is simply lifted off of the tape, moved to the new position, and pressed down again. The necessary bag pleats are sealed by folding the pleats around short pieces of sealant putty. Often, the sealant putty can be recovered and reused, but is messy to use and sometimes hard to peel off of the original roll, even though a paper separator is used. Commercial sealant putty is also designed for use in a hot curing oven, which the amateur does not require.

Automotive strip calking compound, used for auto body sealing, is actually easier to use as sealant tape than the standard product. It can be used in exactly the same manner but, of course, is not designed for use in an oven. The material comes as short sticks, rather than as a roll, which is actually a convenience.

The author has found that ***2-inch brown mailing tape*** works very well indeed as a bag sealer. It is very quick to apply, but cannot be reused. Situations will occur where strip calking will be needed, in combination with mailing tape, but simply taping the edge of the bag works in the great majority of cases. It is nearly impossible to avoid creases and tucks when using brown mailing tape as a bag sealer, but firm pressure on the area, application of heat with a hair dryer, and sometimes an additional patch of tape will quickly correct a leak. ***The author uses mail-***

ing tape for sealing almost exclusively. (Unfortunately, office supply stores now most often sell clear mailing tape and the once-familiar brown tape may be hard to find. Clear tape is just as effective, but is more of a nuisance to use. It may be well worth it for the builder to ask his local office supply store to special-order brown tape.)

Leak Detection, Location and Repair

Most potential leaks are obvious upon inspection and are quickly corrected. Usually, the really obvious leaks are so bad that only a slight vacuum can be pulled. More subtle leaks are found after a partial vacuum of only a few inches of mercury is measured.

Leaks may usually be found very quickly and easily by using a battery-powered ***electronic stethoscope or microphone***, such as those sold for less than $30 for automotive mechanics. This is an extremely useful tool and will pay for itself with a single rescued layup. To use, the builder simply runs the sensitive probe over suspected areas and listens for the hiss of moving air, which will usually be heard above the sound of the vacuum pump. It may be necessary to turn off the noisy mechanical vacuum pump during the leak hunt; if so, some means of temporary vacuum storage must be provided. (The author uses a vacuum tank made from a 36-inch piece of 4-inch PVC pipe, with the ends capped, and two shutoff valves.) One alternative is to use a long section of vacuum hose and move the noisy pump outside of the shop. With the pump gone, it is easy to hear even a very small leak. Once the leak is found, the line valve can be closed, the long line can be disconnected, the short line reconnected, and the work can be pumped down again.

Nearly all leaks are found at the bag seal and most are fixed by pressing firmly on the seal. Sealing is improved by gentle application of heat from a heat gun. Stubborn leaks usually respond to a patch of brown mailing tape. Creases and folds in the bag are frequent causes of leaks and it is sometimes difficult to get coverage with brown mailing tape, because the tape tends to bridge the crease and not fill it. In these cases, a ***pea-sized piece of windshield putty*** will usually fix it. ***If a PE bag is being used, there is risk of leakage anywhere on the bag surface, especially at a fold.*** In this case, the builder can "sniff out" the leak with the electronic stethoscope, mark it accurately with a felt-tipped pen, and seal the area with a patch of brown mailing tape. Desperate emergencies are not supposed to occur, but they sometimes do, and the situation can sometimes be saved with ordinary ***silicone bathtub calking*** liberally applied with a calking gun. The cleanup will be messy, but worth it if it saves the layup.

Envelope Bagging

The previous discussion has assumed that the vacuum bag will be sealed directly to a rigid, airtight tool. It often happens that it is not practical to make such a seal because the tool cannot be made airtight or because adequate sealing areas cannot be provided. For these cases, the ***entire tool or mold must be encased in bagging material*** and the bag is ***sealed to itself*** to form a soft ***envelope***.

A very good example of envelope bagging is shown in the sketch. Reliable bagging film is first placed on the worktable, then the tool and layup work is added, including the bag itself, and the envelope is closed by taping the top bagging film to the bottom bagging film. Under vacuum, the bag edges are squeezed flat and there is no disadvantage to the fact that the edge is not supported. With envelope bagging, any convenient flat surface, such as a sheet of chipboard, can be used as a support for the vacuum setup. Plain, uncoated Masonite is very good as a tool, if release film is used, because the slight porosity of the Masonite allows the release film to lie flat and form an excellent tool surface.

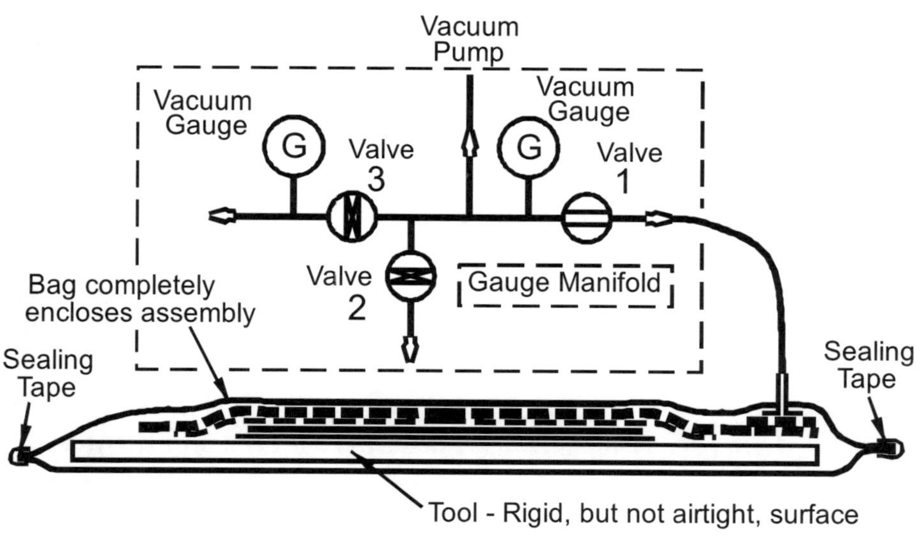

Envelope Bagging

A very useful variation of the envelope bag is to use a ***semi-rigid sheet of airtight material as a sealing layer under the tool.*** Such a sheet is ***much easier to seal*** to than flexible bagging film. The author uses the ***back*** of a 4 x 8-foot sheet of ***FRP white wall liner panel,*** 0.09 inch thick. The intended front has a pebble grain which is hard to seal, but the ***back*** has a very slight matte texture that forms an excellent seal with brown mailing tape and can be used many times, (see Exercise 13).

Flat Tool Mold-Release Techniques

As previously mentioned, commercial composite shops make use of expensive, highly-polished, molds which usually form complex curves. Special mold release agents are used to separate finished parts from these molds. (We will return to this topic in Chapter 13.) These same mold-release materials may be used with flat tools. The very thin film of wax is fragile and must be applied and polished out with care. The amateur builder might be a little too enthusiastic in driving out air from the layup and might penetrate the thin wax film. If damaged in handling, the surface might fail to release and might allow bonding between the part and the mold - damaging both. Mold release materials may contaminate the surface if another bonding operation is planned.

Flat or gently-curved molds avoid many of the problems of mold release compounds by ***allowing the use of a separate release film*** between the tool and the wet layup. Release film gives a positive separation, protects the tool surface from rough handling, and covers possible small irregularities in the tool.

Commercial release films intended for use in a curing oven are available, but have the disadvantage of being expensive. The amateur can use cheaper commercial bagging film as release film, but can also use really cheap PE film in most cases.

Placing Tool-Release Film

When placing tool-release film, the builder must be aware that ***"What You See Is What You Get - WYSIWYG."*** The film must lie smoothly against the tool without creases. If there is slack or creases in the release film, ***it will NOT flatten out under vacuum; the excess slack will appear as a wrinkle on the toolside*** which will reproduce on the part.

Small pieces of brown mailing tape can be used around the edge of the sheet to gently stretch and draw the release film flat. Patience to do a smooth job will be rewarded with a smooth part and the builder can think about how much easier it is to get the release film flat than it is to polish and buff a delicate mold or to fill the imprint of an unnoticed wrinkle.

If the release area is very large, as might be the case with a wing skin, a fuselage panel, or a batch of wing ribs being made as a single layup, there is risk of trapping air between the tool and non-perforated release film (pinchoff). This risk can be greatly reduced by working a ***squeegee*** from the center toward the edges to ***chase out the trapped air***, much as is done with excess epoxy.

PE film always has creases, as delivered on the roll, and it may be impossible to find a large enough section that is free of creases. In this case, a heat gun may be used to ***gently*** warm creased areas while the crease is pulled out. (Remember WYSIWYG: If the sheet of film is not perfectly flat, even after warming, do not use it.)

> ***Tip:*** *It is sometimes difficult to pull cheap PE release film into position when using brown tape; the force required may cause the brown tape to release. In this case,* ***duct tape,*** *which sticks more firmly, will allow quite strong force to be used.*

"Belt and Suspenders" Tool-Release Film

In applications like making a batch of wing ribs, the builder may not care about a perfectly cosmetic smooth toolside surface and can accept the tiny "pip" marks left by the perforations in perforated release film. In this case, the builder can apply ***mold release compound*** to the tool (the "***belt***") and ***also*** use ***perforated release film*** over the tool (the "***suspenders***"). Use of the perfed release film fully protects the tool against hard handling. Any epoxy that works through the perforations and contacts the tool can be cleaned off from the mold release compound after cure, or, in most cases, simply ignored. Toolside trapped air is removed through the work by the vacuum, like any other air trapped in the layup.

Fortunately, brown mailing tape will stick to a tool surface that has been previously coated with mold release, so it can be used to secure the release film. (It doesn't stick very well, but it is sufficient. ***Duct tape*** may stick a little better.)

Note that this system is not applicable in cases where toolside trapped air cannot exit through the layup.

Toolside Bleeding

A very good alternative tool release technique for large areas is to cover the tool area smoothly with ***dry peel ply*** and secure it with brown mailing tape. Leave at least a ***one-inch margin*** of exposed peel ply all around the work area. Place release film over the peel ply, leaving the 1-inch margin exposed, and secure with brown tape. Make and bag the layup in the usual way, but be sure that the bleeder covers the exposed margin of peel ply.

In this application, the ***finely-woven peel ply serves to bleed trapped air*** to the edge of the release film, allowing conformation to the tool. ***Peel ply is not a very good bleeder, but it is good enough*** for the sizes of parts typical of aircraft practice and a smooth surface results. (Actually, the fine weave of the peel ply may produce a ***very slight texture*** in the toolside surface. This texture can usually be ignored when the part is finished.) Note that the peel ply bleeder never gets wetted, so can be reused indefinitely.

An alternative is to use ***agricultural felt*** ("weed felt") as a toolside bleeder. Agricultural felt is a better bleeder than peel ply, but the finished surface is not as good, partly because agricultural felt varies in thickness and texture. It has the advantage that it will ***acommodate a moderate double curve*** better than peel ply because it will pull down to fit a slight double curve.

Another compromise is to use ***PE perf film over a toolside bleeder***, taking care to place the little "pips" in the plastic toward the work. The WYSIWYG rule applies, so it will be necessary to assure that the perf film is pulled quite snugly before the layup is pumped down. If the "pips" are toward the work, the finish will require very little cleanup.

Plain 1/8-inch unsurfaced Masonite wallboard is an excellent tool support surface because it is very slightly porous and does not require a separate toolside bleeder. Because of the porosity, ***Masonite cannot be sealed*** and must be used in an enclosing envelope.

Double Bagging and Joggles

The use of release film and toolside bleeder makes it particularly easy and convenient to include accurate joggles in flat tool work. Cheap Masonite can be used as the tool. A marker pen can be used to define work areas, either directly or on paper patterns. The joggle thickness can be cut out of cardboard and taped to the work area with brown mailing tape. Thin toolside bleeder, such as peel ply, if used, is then laid over the work area, pulled taught, and secured with brown mailing tape.

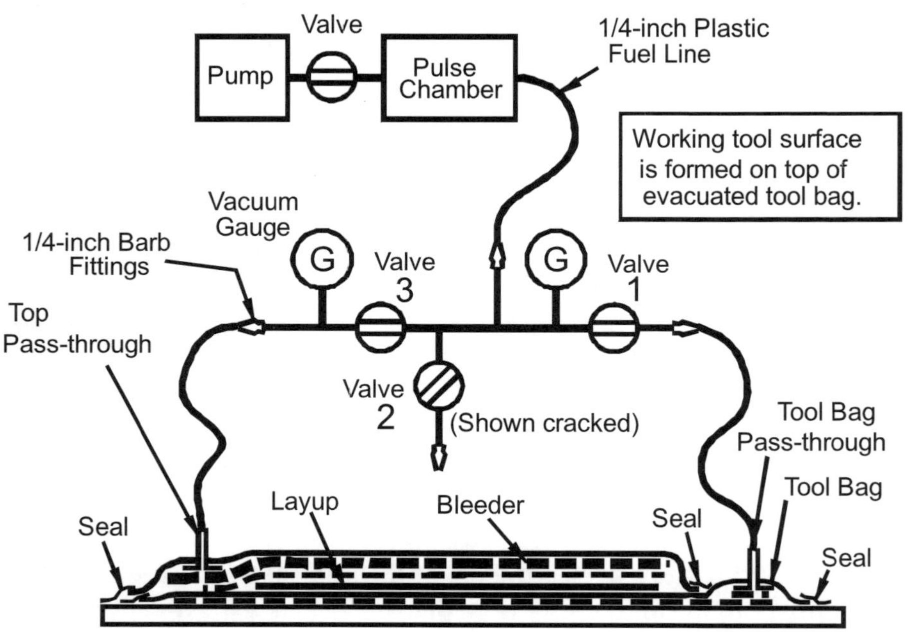

"Double Bagging": Tool Formed Under Vacuum

A vacuum bag is then rigged over the tool, the joggles, and the toolside bleeder. Valve #1 is opened. ***Vacuum is applied very gently*** by using a bypass valve (Valve #2) to outside air to limit vacuum (see the figure). The builder then uses a squeegee to smooth out the bag and assure that the bag lies smoothly against the bleeder. The outline of the joggles will be clear under the vacuum. Full vacuum is applied with the outside bypass valve (#3) fully closed. ***The builder must verify that the bag is quite tight, close to the best vacuum available from the builder's pump.*** Sniff around with the electronic stethoscope to be sure that all leaks are fixed, so that available pump capacity is not spent in maintaining the tool. The drawn-down bagging film can now be used as a new release surface and can be marked for a new layup.

> ***The builder now has a nearly-perfect smooth, clearly-marked, release surface with joggles in it, which remains effective as long as the vacuum pump is running.***

A layup can be made on this new tool surface and rigged for bagging in the usual way. The second bag is sealed to the first bag. (The builder should anticipate this need and should allow plenty of margin around the tool for sealing the second bag.) The second bag will need to be sealed over the inevitable wrinkles from the first bag. This may require both brown mailing tape and windshield sealer. A ***separate vacuum gauge*** will needed, (which is included on the gauge manifold).

To bring in the second bag, the builder must open the connection valve (#3) ***very*** slowly, while observing the vacuum in the first (tool) bag. ***The vacuum in the first bag should not be allowed to drop below about 6 inches while observing the second bag.*** As the second bag evacuates, the vacuum in the second bag should increase. Eventually, the connection valve (3) should be wide open and both bags will have full available vacuum. Note that the pressure difference between the two chambers is now zero, but that full available atmospheric pressure is applied to the entire assembly, and the layup is pressed firmly against the tool.

> *Note that the second bag acts as a cover and seal for the first bag, so any small leaks that might exist in the working surface will not matter, because they will be covered. The risk in proceeding with a job when the tool vacuum is not quite acceptable is due to the fact that most leaks come from the edge seal, not the main surface. A false assumption that a surface leak can be ignored because it will be covered can lead to great disappointment.*

Bagside Surface Finish

Recall that the vacuum process works by reducing the *volume* of the chamber, ***not by pressing "down."*** Irregularities in placement of the perforated release film and/or areas where there is a considerable excess of resin may produce ridges or a pebbled cured surface in the bagside, which will be particularly visible if the resin has been tinted. Usually, the builder will not care, as long as the mold surface is good, the assumption being that the "pebbled" area will not be used for bonding to another part. ***The pebbled effect will not occur where peel ply has been used.***

Good bagside finish is helped by being sure that the perforated release film is smoothly applied. There is no requirement that the perf release film be used as a single piece. With a large layup, it will often be desireable to apply perf film in a series of large strips, with a non-critical overlap of about one inch between strips. The perf film should be rolled out with the grooved rolling tool and squeegeed out thoroughly on the wetted surface before adding the bleeder and bag.

Effect of Thick Cores

It is structurally efficient to use lightweight cores with the maximum convenient thickness. (All of the Exercises in this book make use of 1/2-inch thick cores with a "standard" bevel.) Thick cores may cause difficulty in obtaining a smooth interior finish with either glass or perf film. General-purpose satin-weave BID cloth such as 7725 will usually deform enough to fit smoothly over the bevel and over the corner. Satin weave 7781 will conform quite well. Finer weave crowfoot BID, such as Style 120, may require a razor cut at the corner in order to get a smooth fit. The cut material is used to make a small overlap.

> *Perforated release film cannot deform sufficiently to fit smoothly over the bevel corner on a thick core. If this error in fit is ignored, a slight ridge of cured epoxy will be left along the corner of the bevel. This minor cosmetic problem is avoided by using small strips of perf film to make the necessary transition, overlapping as necessary to get complete coverage. Most builders will ignore it.*

Caul Plates

In certain cases, such as for bolting or mounting surfaces, the builder may require positive control of the bagside finish. A ***caul plate*** is a ***rigid sheet which is placed in the layup behind a release film*** so that the ***smooth surface of the caul is impressed on the work.*** For aircraft work, cauls are usually small and non-critical and can be small plates or disks of plastic or metal which are placed over the desired areas. Obviously, there can be ***no bleeder effect through a solid plate***, but, for small plates, excess resin works out to the side of the plate and is removed in the adjoining bleeder.

Some applications, such as Exercise 15, require a ***flexible caul.*** In Exercise 15, it is necessary to follow the line of the upper surface of a wing section. ***Thin aluminum, plastic, or Masonite*** can be used here.

A single layer of stiff poster cardboard may be a sufficient caul (see Exercise 15). A "perforated caul plate" may be an oxymoron, but thin poster cardboard can be perforated with the perfing tool and used ***over peel ply*** to produce an excellent bagside surface while allowing removal of excess epoxy. (The torn "pips" from the perforating tool should face the bagside to assure a perfectly smooth result.) The peel ply should be slightly wider than the caul. After cure, the cardboard is stripped off with the peel ply, leaving an excellent bonding surface.

For some applications, particularly larger areas, a very good compromise is to use one or two layers of ***agricultural weed felt*** instead of a rigid caul plate. Regular soft bleeder is then placed over the felt in the usual manner. The felt will take only a slight double curve, so this method will be limited to sheet molds.

Industrial Practice

The sketch on the next page is a reproduction of a "full-dress" vacuum stack as described in industrial catalogs, showing all the standard components, including a caul plate. Note that it differs from the processes previously described in several details, including the use of release film.

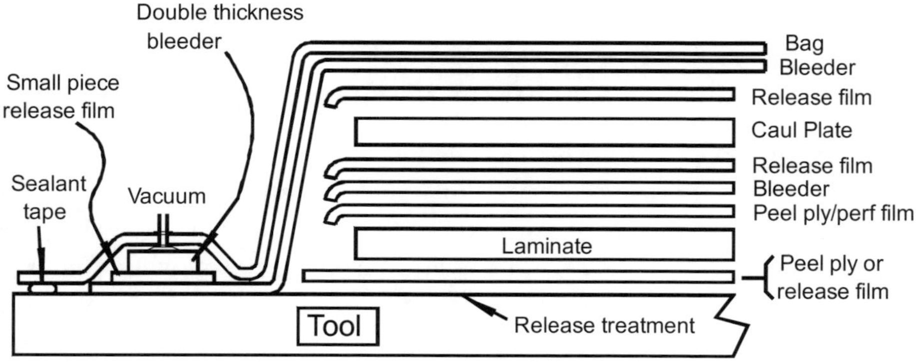

Typical Industrial Vacuum Bagging Stack

Chapter 7

Joints and Adhesives

Introduction

In this chapter, and in following chapters, we will consider the several ways and techniques by which we make connections between parts of structures.

A ***joint*** is the means used to permanently attach two or more separate structures or parts together so that a force may be passed between them. Aluminum airplanes have a great many "joints," usually made with rivets. One of the great advantages of composite construction is that the typical small aircraft airframe will be assembled from only a few large structures and will need very few joints.

An ***adhesive*** is a liquid material which is placed between two structures, and ***wets*** them, to form a ***joint*** when the adhesive hardens to a solid. (Water, frozen to ice, is a good adhesive in certain circumstances, but the temperature limitations of the resulting joint are usually unacceptable.)

An ***attachment*** is a joint which passes loads between low-stress areas, like a large composite panel, and a high-stress area like a steel bolt. Usually, some kind of ***transition structure*** is required to gather and distribute loads (see Chapter 8).

Trimming Cured Skins

After cure, a composite layup always produces a rough and ragged margin of excess material, which must be trimmed to the final dimensions before it can be joined to another structure.

"Green" Trimming of Glass Skins

The matrix materials used in composite construction all go through an intermediate "green," or "B," stage during cure, where the ***partially***

cured material may have the consistency of taffy candy or poster cardboard. The cure time to the green stage varies with the specific matrix used, but is typically around five to ten hours for most aircraft epoxies. Fortunately, the "green trim" or "knife trim" is by far the easiest way to trim a layup. The builder observes the process of curing, testing from time to time, until the matrix has the "right" consistency. He then ***uses a common utility knife to cut off the excess margin.*** Typically, this operation takes only seconds. If a slight delamination takes place, the builder simply pushes the partially cured material back into position. Warming with a heat gun will facilitate rebonding.

The "knife trim" technique is ***limited to open layups.*** Layups which cure under vacuum cannot be disturbed while curing.

Knife Trim of Cured Skin with Local Heat

Cured glass laminates which are not over about four layers can be cut with a standard utility knife, using a principle similar to the "green" trim, softening the laminate by local heating. The cured glass is placed on a cutting surface such as particle board and ***heated in the area to be cut*** with a standard 1200 watt ***heat gun.***

The heated air should be ***applied directly to the cutting knife***, with the area to be cut heated by overflow. This technique will require the builder to wear a leather glove.

The laminate softens in about 15 seconds under the heat and is easily cut with a standard utility knife, leaving a very slightly rough edge. The rough edge can be sanded with 200 grit sandpaper. Cuts as long as 24 inches are easily made with a single application of heat to the cutting area. It takes a bit of practice for the builder to get the feel for how much heat to apply and how large an area can be cut at one time, so ***testing on scraps of cured glass is important.*** Fortunately, the cured glass retains heat fairly well and the process is quite tolerant.

Any doubt about the strength of the glass in the uncut area near the heated area can be avoided by ***protecting the uncut area*** from the heat with a piece of particle board set about 1/2-inch away from the cut.

Foam cores must be protected from high heat, so ***this technique cannot be used for foam-cored sandwiches.***

Router-like Tools

The most general purpose cutting tools for cured skins are essentially small hand-held routers that use the side of a narrow bit to cut. (See Appendix 1 for a more detailed discussion of rotary tools.) Rotary tools can be used in almost every situation, including cored sandwich panels, but they are slow.

Band Saws

Fine-toothed bandsaws, preferably with a 1/4-inch blade, are excellent for fast, safe, fairly straight cuts for all composites, including cored sandwich panels (see Appendix 1).

The Nature Of Adhesives

The Basic Sheared Adhesive Joint

Consider the sketch below, where two sheets of a perfectly flat rigid material are bonded with an ***adhesive*** material. A force is applied which tends to pull the two sheets apart, subjecting the bonded area to a shear stress.

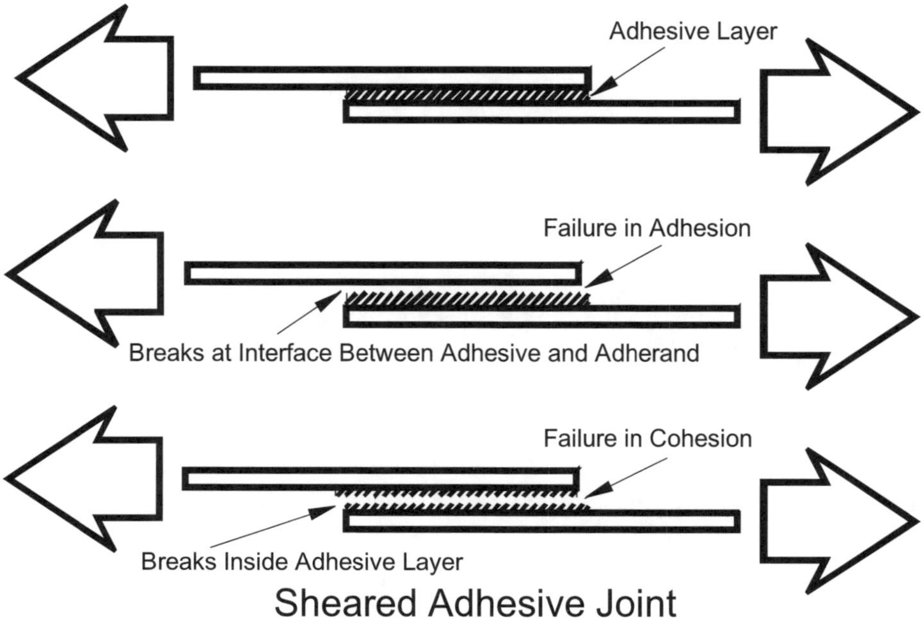

Sheared Adhesive Joint

When loaded to failure, if the interface between the adhesive and the adherend fails, engineers call it failure in ***adhesion.*** If the film of adhesive material fails internally, it is failure in ***cohesion.***

In simple design formulas, the shear stress is just taken as whatever the force is, divided by the total area of the joint, suggesting that the larger the glue area, the stronger the joint. Unfortunately, real life does not work this way.

The Best and Worst Joints

Best: Assume, for a "thought experiment," that the two sheets below are very stiff indeed - let's use diamond. Apply the load, which places a pure shear stress on the adhesive. The sheets of stiff "diamond" will have almost no strain under the load; all the atoms in the sheet of "diamond" will move only a very tiny amount under the load. Effectively, ***all of the strain in the assembly must be in the thin film of adhesive and nowhere else.*** For this ideal case, the reaction force of the simple structure will be the shear stress in the adhesive and that stress will be uniform across the joint. Here, the longer the overlap, the greater the load it will carry, in proportion to the area of the overlap. ***It does not matter what adhesive we use,*** as long as the stress stays below the maximum shear stress that the adhesive can carry.

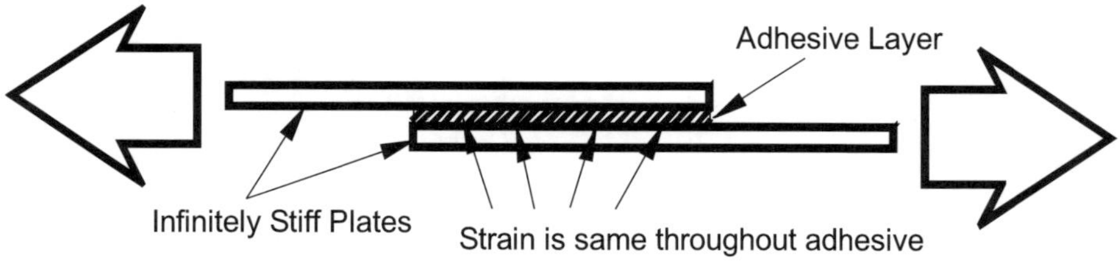

Ideal Adhesive Joint: Load Is Proportional to Bonded Area

Worst: Now consider that the two sheets are made of a ***non-rigid material,*** like tire-tread rubber, and the "adhesive" is a stiff material, like steel. Apply the load. See that the rubber plates stretch the steel "adhesive" interface, but the rubber plates cannot stretch at all ***beyond*** that interface because the stiff "adhesive" does not stretch to transfer the load.

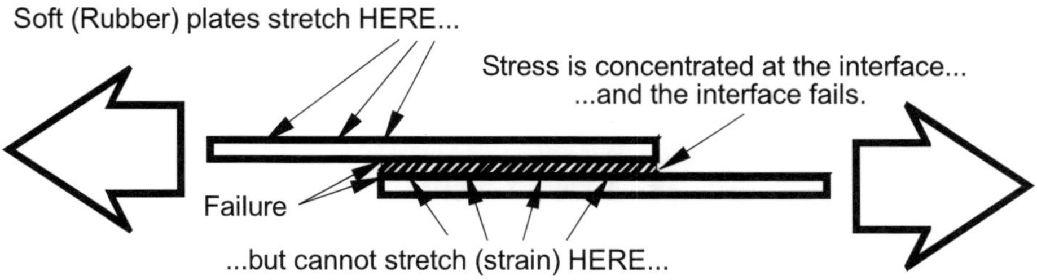

Worst Adhesive Joint: Adhesive Harder than Adherend

All of the load is concentrated at the interface, so the rubber sheets stretch and the adhesive does not; the adhesion fails at the interface and the joint strips out immediately. Again, ***it does not matter what adhesive we use, or how large the joint is, because any practical adhesive will not stand the stress concentration.***

Real Adhesive Joints

Real life lap joints with real materials are hard to design for theoretical excellence. The "ideal" real joint would have a uniform safe shear strain throughout the bonded area and would not have any stress concentrations, as discussed above. If the adhesive has uniform characteristics, this leaves only the bonded structure itself to be manipulated for a better joint design. In practice, designers use various combinations of tapering and/or doubling the bonded structures, plus ***generous safety factors.*** Engineering handbooks discuss tapered lap joints, but our composite skins of woven cloth cannot be tapered. We can only choose to add or subtract a discrete cloth layer and/or use a zig-zag edge. We can approximate a true taper by using several layers of different widths (see below). Simplicity and good appearance are also valued, so there are many compromises to be made.

A practical lap joint between uniform, non-tapered plates will have the adhesive stresses concentrated near the edges of the joint, distributed approximately as shown in the sketch below.

See that there is ***little advantage to making the lap area very long***, because the middle of the lap joint carries very little load.

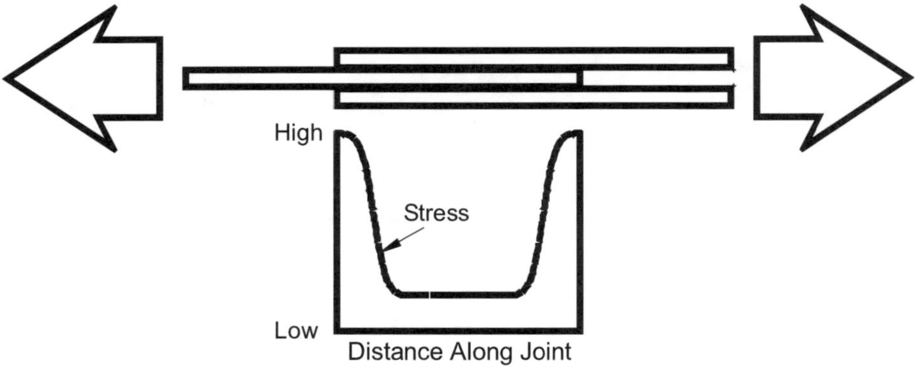

Very Approximate Distribution of Stress in an Adhesive Bond

If we are to make bonded joints between cured composite plates or panels, the "rule of thumb" conclusion is that ***the adhesive should be chosen to be significantly less stiff than the plates to be bonded.*** (This is similar to the "rule" that a material to be used as a matrix in a composite must be less stiff than the fibers in the composite.) The materials sold as two-part ***structural adhesives*** meet this stiffness requirement. ***The matrix epoxies we use in making composite structures do not.***

Rigid structures to be joined will not be perfectly smooth and perfectly matching. The thickness of the adhesive layer must vary with the irregularities between the faces. In fact, ***some adhesives require a minimum thickness*** and include some small solid particles (like glass spheres) so that the gap cannot be made too small (see below, methacrylates).

Considerations of building methods, handling, and alignment of parts generally require aircraft builders to bond surfaces that are flexible and very thin compared to the bonded area. All our practical bond areas are "long," many times longer than the width of the finished bond, so only the ***ends of the bond area will receive a significant stress.***

Practical Adhesives

Every large home supply store displays a huge number of glues, sealants, and adhesives, ***none*** of which are directly applicable to composite aircraft stuctures. In this book, we will choose to limit the discussion to only those few adhesive systems that meet aircraft standards for bonding non-porous surfaces. This restriction ***rules out any "one-part" system that requires a solvent to evaporate or requires a chemical*** reaction with oxygen or moisture in the air, to harden; the liquid ***adhesive material must be self-contained*** and with very small shrinkage when hardening. An acceptable adhesive must also have the ***right stiffness range.***

Aircraft supply catalogs offer a wide range of one-part and two-part adhesives which are in general use by aircraft builders. The one-part adhesives are applicable to builders who wish to bond wood parts and must ***depend on the porosity*** of wood to allow completion of the hardening (literally drying) reaction. Several chemical systems are used, but most of the two-part adhesives are derived from the epoxy family of systems.

Systems labeled ***structural adhesives*** have the desired qualities of surface adhesion and intermediate stiffness. A specific choice will depend on ambient temperature, pot life, curing time, and maximum allowed temperature of the cured joint. In general, the ***best advice is to use the adhesive that a successful designer tested and recommended.***

Matrix Material as Adhesive

Builders sometimes use the same matrix material they use in making primary composite structures as an adhesive to join those same structures ***after cure***. They usually "get away" with it because the stress at the joint may be sufficiently low and the safety factor may be large. The problem is that unreinforced matrix is too stiff to spread a shear load over a large bond area and a stress concentration results. (Recall the "worst joint" example above.) In general, a ***true structural adhesive is preferred*** when bonding cured surfaces.

Methacrylate Adhesives

The automotive industry, sports equipment, the marine industry, and such semi-custom applications as Recreational Vehicles, make wide use of the ***methacrylate*** family of adhesives. These materials have not found regular use in General Aviation and are not presently offered in aircraft supplier catalogs for Experimental Aviation. The author is informed that manufacturers have not chosen to undertake the special testing and record-keeping that is required for materials that are intended for certified aircraft. The builder of non-certified Experimental Category aircraft is free to use such materials on his own responsibility.

The characteristics of the methacrylate adhesives are quite attractive for the home builder, so they will be described here.

Methacrylates are offered in a wide range of formulations, varying in pot life, viscosity, stiffness when cured, fixture time, and working temperature. All are ***two-part systems***, mostly mixed at a 1:1 ratio. Very ***simple and ingenious tools are available*** for measuring, mixing, and dispensing the materials. With these tools, it is convenient to dispense and locate accurately exactly the quantity required. Correct mixing is easy, especially with the tools provided, however the mixing ratio is not critical as it is with most epoxies.

Snap Cure

Methacrylates have the very unusual property of ***snap cure***. After mixing, the materials remain at nearly the same viscosity for the entire pot life (typically 4 to 15 minutes), permitting adjustments of the surfaces to be bonded (see below).

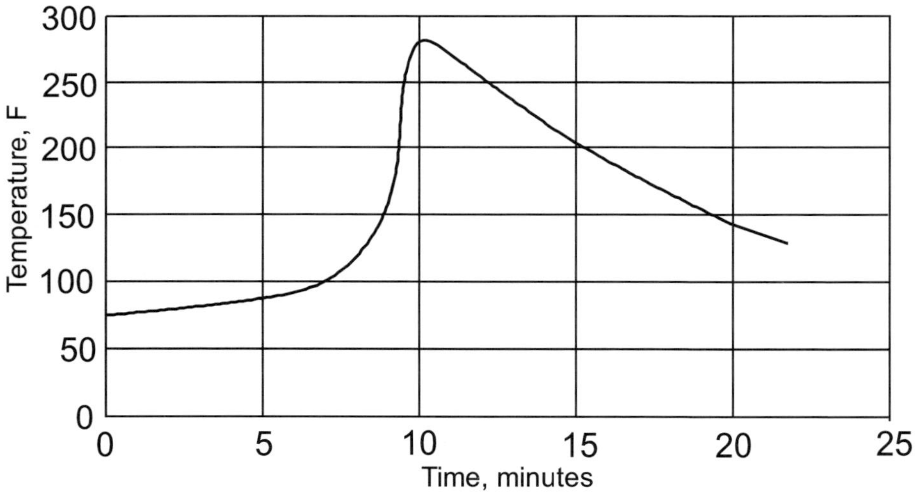

Typical Exotherm for a "Fast" Methacrylate Adhesive

At the end of the pot life, the adhesive heats up (exotherms) and ***cures quite suddenly*** ("snaps"), usually in well under one minute. The assembly can be removed from whatever holding jig was used (fixture time) in typically 12 to 35 minutes. Essentially full cure is complete in less than one hour. A typical cure curve for a fast-acting formulation is shown. Note that the bond line can get quite hot at the "snap," so that it may not be suitable for a joint where foam is immediately adjacent to the bond line. A similar formulation is available which has a lower peak temperature, but requires about twice the cure time. Cured glass/epoxy will generally tolerate the brief period of high temperature.

The temperature curve above should be taken as very approximate for practical thin skins which are typical in aircraft practice. If the bond area is well ventilated, the bond can dissipate heat quickly enough to avoid the highest temperatures which might cause damage.

Minimum Bond-Line Thickness

A second unusual characteristic is that ***a minimum bond-line thickness is required***. The cure ***time***, but not the cure ***quality***, of the mixed adhesive depends on the thickness of the film. Methacrylates are mass dependent. A bond thickness of .030 inch is generally recommended, but a ***thickness of .010 inch is satisfactory***, the only cost being cure time. The cure time is typically of the order of minutes, so the added cure time is not a significant disadvantage. If a bond line less than .01 inch thick is used, the adhesive may not cure completely.

The minimum bond line thickness can be assured very easily by sprinkling ***commercially available .01 inch glass beads*** over the open adhesive surface. The beads prevent the builder from closing the gap beyond the minimum. Where appropriate, ***bare copper electrical wire*** can be used as an alternative spacer.

A relatively thick adhesive joint, of course, must also weigh more than a corresponding thin joint made to aircraft standards, but this may not be practically significant. The ***extra joint thickness must be allowed for in forming a joggle***. A methacrylate joint will weigh slightly less than a corresponding glass-epoxy joint made by the "wet tape" method (see below) because the material is less dense and does not add a glass layer.

Small Boat Application

Methacrylate adhesive has been spectacularly successful in the large-scale manufacture of small polyester composite boats, where the application is in joining deck moldings to hull moldings. Typically, the tolerance in the gap to be closed by the adhesive is quite large, by aircraft standards. Operators of fleets of small boats, such as sailing schools, have found that the performance of the bonded joints has been extremely reliable in the face of very hard use in a marine environment. The system has also been used very successfully in large racing speedboats and yachts to 120 feet.

The Secondary Bond - A Compromise

Current composite aircraft designs very frequently require that the builder make a ***wet layup*** over a similar ***already-cured*** composite surface - a ***secondary bond***. In this case, the new layup is seen as a continuation of the previous structure, except that the composite fibers ***do not extend through the joint***. There is no separate "adhesive film" because the old and new surfaces must fit perfectly and the surfaces have identical characteristics.

We use woven glass cloth in composite construction, instead of loose fibers, because it is practical to handle the material. We accept the fact that the finished structure is a little weaker between layers of the cloth (delamination). In aircraft practice, we make every effort to make a large layup all at one time to minimize the risk of delamination. The secondary bond is seen to be a ***compromise - not as good a structure as a continuous layup, but generally better than a true adhesive joint***.

Peeling

A peeling force is any component of the force on a joint which tends to pull the faces apart — a vertical component. The aircraft adhesives we use have specified peel test ratings, as measured by idealized comparison tests, but whatever they are, the peel ratings are always small compared with their shear strength. The amateur builder should assume that the ***peel strength of the adhesive is practically zero*** and design for only pure shear loads in the bonded area.

Bonded Joints

Single Lap Joint

The single lap joint has the great advantage of simplicity. An important disadvantage is that there is a stress concentration at each end of the bonded area. See the sketch.

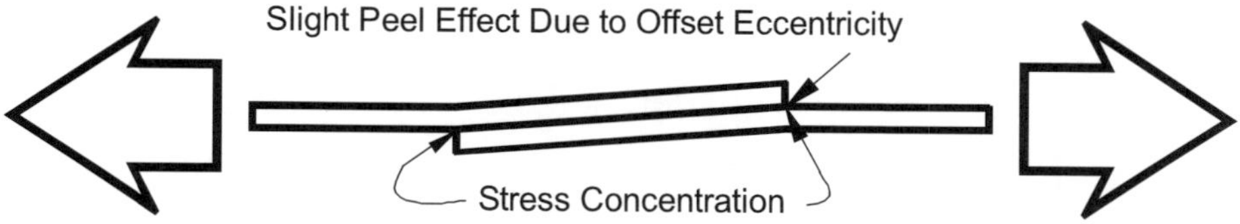

Simple Single Lap Joint

Whatever the thickness of the bonded pieces, the structure in the region of the bond will be thicker than either skin alone, therefore much stiffer. When a tensile load is applied, the necessary ***bending is concentrated at the edge of the bond*** and a peeling force is induced in the joint, exactly where the ***highest shear stresses*** are located.

Single Lap Joint with Joggle

A better joint results if one of the bonded sheets has a small displacement equal to its own thickness, plus an allowance for the thickness of the adhesive layer, as in the sketch below. The appearance of the joint is greatly improved and the peel stress is reduced.

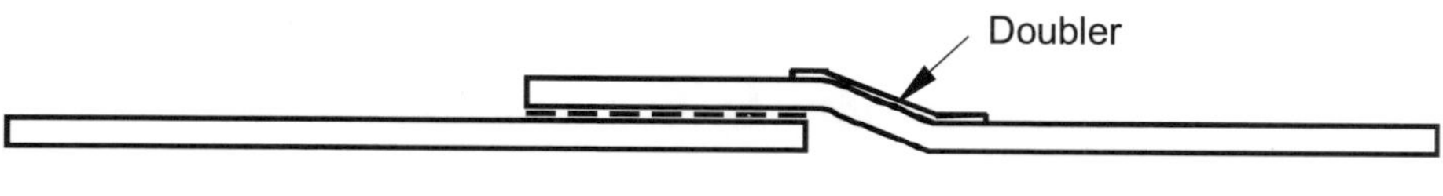

Reducing Stress at a Joggle

The following sketch shows the example, widely used in the boat industry, in which the adhesive is applied as a strip of glass tape which has been soaked with resin.

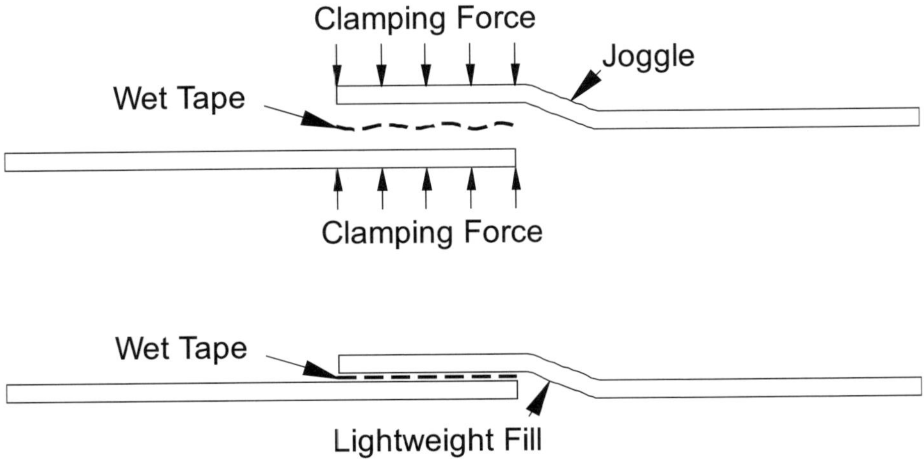

Wet Tape Joggle Joint

The weave of the glass cloth provides resin-filled voids which allow ***considerable variation in spacing*** between the bonded surfaces while still providing a fiber-reinforced material in the gap. Each face of the joint is a secondary bond. This is usually an excellent bond, but may be more work and weight than using structural adhesive.

The problem of the stress concentration at the edge of the joggle remains. Stress at the bend of the joggle is reduced by adding a layer of glass, as shown, either applied with the original layup or added later with a secondary bond.

Strap Joints

A strap joint is formed by placing the sheets to be bonded in opposition and then bonding a separate sheet, the "strap," across the junction, as shown below. A joint with two straps is balanced so that there is no peeling force on the bond. The sketch following shows an example of practical strap joints which might be used in small aircraft construction.

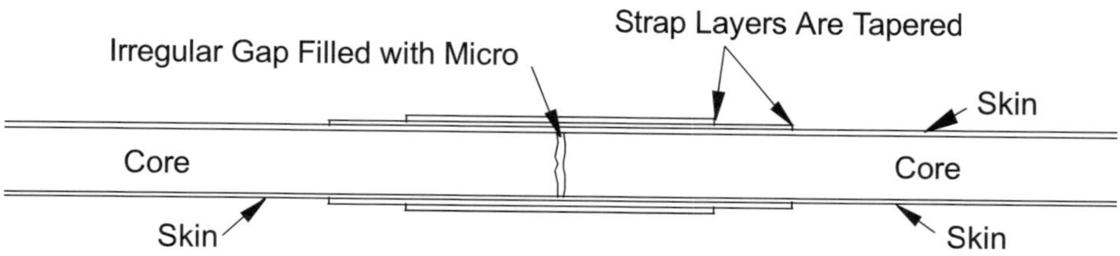

Butt Strap Joint - Cored Sandwich Panels

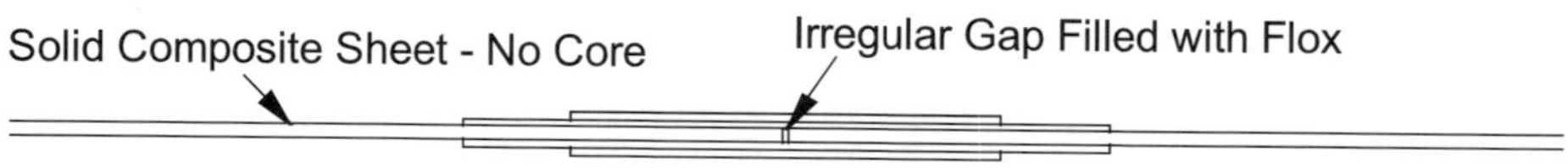

Butt Strap Joint - Solid Sheets

Vibration Considerations

We want to use simple joints in homebuilt designs. Any practical design we choose will add weight and stiffness to the joint area. If the joint area is heavy and stiff and the area just beyond the bond area is unsupported flexible skin, ***bending stress will be concentrated where the flexible thin skin meets the stiff thick joint***. A small aircraft is a high vibration environment and the structure is subject to resonances, so vibration fatigue is an important consideration. A few million cycles of a resonant "buzz" at the sharp edge of a joint could put the structure at risk of failure due to fatigue. Care should be taken to avoid sharp discontinutities of stiffness by ***blending the edges of joints by tapering layers***, using "zig-zag" edges as cut with dressmaker's "pinking" shears, varying the overlap, or all of the above.

Edges

The Filled Edge

Situations will occur where the edge of a composite sandwich exposes the core and the edge is not closed by attachment to some other structure. When the core is plywood, the designer may choose to leave the exposed edge as cut. When the exposed core is foam, however, the exposed edge must be protected. Bumps and bruises in the shop can cause nicking and high stresses at the edge which can start debonding. One solution is to ***dig out foam*** to a depth about equal to the thickness of the core and then to ***fill the volume with "dry" micro***. The exposed inside surface of the skin should be buffed and painted with epoxy before adding the micro. ***Care should be taken not to make matters worse by starting to debond the skin while cleaning up the inside surface.*** The filled edge is often more work than it is worth and its use should be avoided.

The Wrapped Edge

An exposed core edge can be closed by wrapping directly with light bias-cut glass, such as 3-ounce Style 120, and ***not*** digging out and replacing the foam, as above. It may be necessary to touch up the exposed core with a little dry micro before wrapping. The sharp edge of the skin remains, so this is an ***"outside corner" situation***, as described below, so the builder must be aware that the ***cloth will tend to bend away*** from the sharp edge. The cloth can be smoothed down with the gloved hand after the epoxy has begun to set. Usually, there is a better solution to a situation where a sandwich edge is unsupported.

The Glass Edge

A ***glass edge*** is formed when the two skins of a cored sandwich are brought together over a tapered core, forming a primary or secondary bond, as shownbelow. Usually, the ***margin of the glass edge is used as a bonding surface*** for connection to some other structure. The glass edge is also used at the edge of open access ports in bulkheads or in any other situation where no corner will be used.

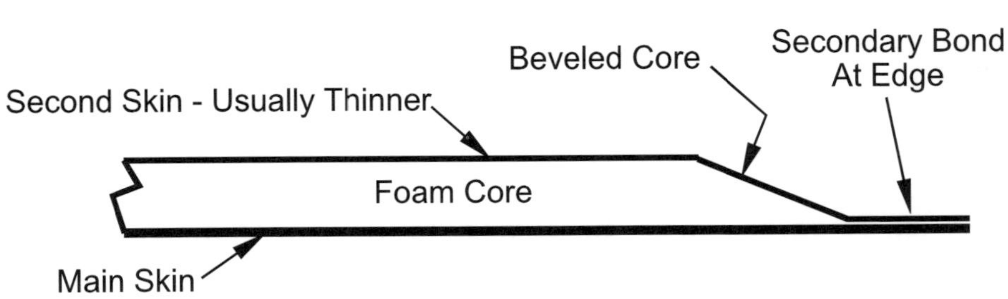

Glass Edge

Recall that the composite sandwich gets its considerable strength in compression because the thick core stabilizes the thin skin against buckling, much as the spreader on a sailboat mast prevents a mast from buckling. In the glass edge, the core is tapered and gently rounded, so that compressive forces do not see a discontinuity which could start buckling.

We will make considerable use of the glass edge in this book, so we will assume that all the bevels are made with the ***beveling tool*** described in Appendix 1. The beveling tool makes an accurate angle between the surfaces, as shown. The builder also should make a ***light, non-critical, dressing cut with sandpaper to round off the angle.*** This "dressing" can be done easily with a couple of light strokes of sandpaper.

For skeptics: A rather convincing experiment is to make two light test panels with a glass edge all around (exactly as described in Exercise 4,) but to ***use soft, non-structural, heat insulation foam*** as the core for one panel and ***rigid structural foam*** in the other, keeping everything else the same. For the experiment, round off about 1/3 of one edge fairly generously, 1/3 with only a bit of rounding, and 1/3 left "sharp." Bend the test panels over the knee, first so that the stabilizing skin is in tension, second so that the stabilizing skin is in compression. Both panels feel very stiff in tension. When bent so that the "inside" stabilizing skin is in ***compression***, the panel with the soft core will fail almost immediately by buckling close behind the bevel and will ***buckle first at the sharp edge of the bevel.*** The "good core" panel will fail at the same points, but at a much higher bending load. See also that the ***part of the edge that has not yet failed will delay the failure of a neighboring section.***

Tacking Tape

It will frequently be necessary to attach parts temporarily - "tacking." One application is the "lost Bondo" corner described in Chapter 8, where the parts to be joined are accurately positioned by a temporary joint which is ground away and replaced by a permanent joint. Another example is making a temporary jig from PVC pipe, where very thin strong joints are required at arbitrary angles. The method will also be useful in mold-making.

Basically, tacking tape is made by wetting out a single layer of fiber in the usual plastic film sandwich and cutting small strips of wet fiber from the sandwich as needed. The gloved hand is used to smooth the wet tape into position. The epoxy used should have a ***generous quantity of fused silica*** added to prevent running and a ***small amount of flox*** to greatly improve "gumminess."

The disadvantage of standard epoxy is that the builder must wait for the cure of the epoxy to be complete before work can continue. If the builder expects to do quite a bit of tacking, it becomes worth while to stock a faster-setting resin just for that purpose. Where only one or two "tacks" are required, 5-minute epoxy can be used for the matrix.

Another alternative is the use of ***methacrylate adhesive***, such as Plexus MA300, in the same manner as with 5-minute epoxy. If only a very small amount is needed, the two parts can be mixed on a disposable mixing surface, avoiding the waste associated with the disposable mixing tube (Appendix 1). This material is so tolerant of variations in film thickness that may be possible to avoid the use of a binding fiber altogether.

Style 7715 UNI is extremely strong in the direction of the fibers. Tacking tape made with UNI and standard epoxy produces a very strong joint with a single layer, which makes it very suitable for jig-making. An advantage of using UNI for the purpose is that it is easy to keep the fiber orientation along the load line and to make a thin, smooth joint.

Style 120 BID glass is very thin (.004 inch) and makes excellent smooth tack joints, but is somewhat harder to handle with sticky gloved hands than the heavier materials. A good alternative for the thin Style 120 is to ***paint small strips of the dry cloth into place***, using a small paint brush, and finish with a wipe of the gloved hand. A bit of trial makes the best technique obvious.

Cotton cloth, such as is used for bed sheets, makes excellent tacking tape when used with standard epoxy, 5-minute epoxy, or water-catalyzed carpenter's glue, because it fits a sharp corner easier than glass.

Alignment, Clamping, and Inspection

One of the great advantages of wet fiberglass is that the builder can ***see through several layers,*** even if the matrix has been tinted. She can then ***see clearly if a dark bonding adhesive has been spread evenly over the bonding area.*** A strong auxiliary light may be necessary, but inspection is still possible. Examples will be given in following chapters and in the Exercises.

When bonding very large parts, such as a fuselage turtleback to a fuselage tub, it will be necessary to very accurately position the parts (fixture), then clamp the joint over the entire length while the adhesive hardens. In contrast with typical boat parts that are to be joined, such as a deck or cabin tub to a hull, ***typical aircraft parts will be quite flexible.*** To assure complete bonding, it is necessary to ***clamp the joint at close intervals*** all along the joint. (Exercise 14 demonstrates an example.)

Trial Assembly

In most cases, the builder will not have access to both sides of the joint, so normal shop clamps cannot be used. Temporary ***"pull-up" screws serve as clamps*** and permit full inspection. The procedure is as follows:

(1) Temporarily fit the parts to be bonded. Mark any margins that need to be trimmed. Clamp as needed. (Often, the ends of the parts will be accessable and can be held with soft-pad shop clamps.) Adjust alignment as needed. When alignment is "perfect," place a very few ***alignment screws*** through the joint to fix the final position. Ordinary round head sheet metal screws can be used, but the builder will find that the special screws sold as ***#8 Washer Head Self Drilling*** Self Countersinking Carpentry Screws for #2 Phillips drive are particularly quick and convenient. Inspect the trial fitting. Mark the ***alignment screws*** by ***circling*** with a marker pen.

(2) Drill holes at 4- to 5-inch intervals for "pull-up" screws all along the bond line, but do not mount the screws. ***Mark each hole with an "X,"*** using a marker pen. (The builder will need to locate the correct holes quickly when bonding the joint!)

(3) Disassemble the trial fit. Clean up the drilling shavings. Use a second drill to open up the many pull-up holes (NOT the few alignment holes) in the surface toward the drill by ***one drill size.*** (A thin glass skin will strip very easily so it is necessary for the pull-up screws to slide freely in the surface toward the screw head.) Trim the margins of the bond area as needed.

Bonding

(4) Be sure you have sufficient help if you are going to need extra hands! ***Prepare the adhesive of choice.*** If a paste structural adhesive is used, apply it evenly on both surfaces to be bonded. Follow the manufacturer's recommendations. (It may be desirable to use a ***fine-toothed linoleum adhesive spreader*** for this operation.) If the wet tape method is used, place segments of wet tape on one surface so that there is a small overlap between segments, just enough to assure coverage.

(5) Assemble, using the previously marked alignment screws to fix the alignment. (It will be convenient to have two or three tapered spike tools available to assist in quickly adjusting alignment.) Replace the alignment screws first and verify that no mistakes have been made.

(6) Place pull-up screws in the previously marked holes. This operation goes very quickly with an electric drill driver. Inspect the bond area as pull-up screws are installed and verify that the bond area is fully covered with adhesive. If coverage is adequate in an area, do not install a pull-up screw. If additional clamping is necessary in an area, drill and install an appropriate screw. Inspect the entire bond line and verify that sufficient adhesive has been successfully spread between the surfaces. Allow the work to ***cure to firmness.*** (If the work is allowed to cure completely, the only disadvantage is that removal of the screws is somewhat more difficult.)

(7) Remove the pull-up and alignment screws. The holes left by the clamping screws will be filled at the final finishing operation.

Chapter 8

Taping and Basic Corners

We will use the term ***taping*** to mean the placement of strips of wetted glass cloth to form a secondary bond joint between two ***previously cured*** composite parts. Taping is one of the ***basic building skills***, second only to wetting out, and the builder will be called upon to use the skill constantly. She will, in fact, use the techniques to be described so much that she will get quite good at it.

Basic Tape-making

First, it is necessary to estimate how many running feet of tape will be required during a work session. Allow an extra 20% for overlaps, rejections, and surprises. With experience, the builder will develop her own version of this basic skill. One procedure is as follows:

1) Work Area: The surface of the work area should be smooth and covered with PE plastic, preferably 3- or 4-mil. Spills will happen.

2) Wetting plastic: Obtain a piece of 3-mil PE plastic somewhat larger than twice the area of glass required. Locate the center, and ***fold the plastic double.*** Use a marking pen to mark the fold line and a sufficient number of strips to meet requirements. Mark the ***outside***, because epoxy dissolves the marking pen ink. (It may be convenient to define a work area about 2 by 4 feet. A 2-inch wide ruler is handy to mark the strips.)

3) Perforation: (This step is optional.) Identify the side of the wetting plastic ***opposite*** to the marked area and ***perforate that side only*** with the perforating tool (Appendix 1). Place the wetting plastic on the worktable, ***perforated side down***, and open the fold.

3***) Bias-cut the glass*** to be used, sufficient to cover the work area, allowing some excess for trimming. Place over the wetting out area. A single layer of medium-weight glass, such as 7725 BID, is preferred, to save weight, make application easier, and allow tapered edges if more layers are required. It is almost always permissable to use glass scraps to make tape. Overlap adjacent pieces about 1/2-inch. In general, do not try to use more than two layers of glass at one time.

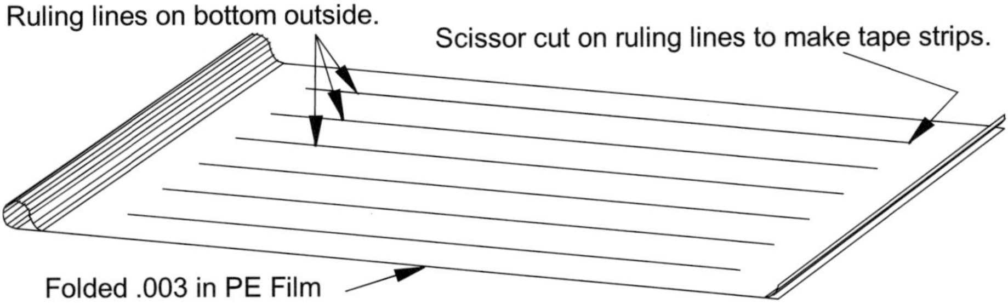

Wetting Tape Under Plastic

4) Mix and tint sufficient epoxy to wet out the glass. At least some of the tape may be required to be applied in a vertical position, thus subject to running while the epoxy is liquid, so ***add about 1/3 volume of fumed silica, to prevent running. Add about 1/10 volume of flox to improve "stickiness."*** Place the glass in the wetting out area.

5) Wet out. First: Pour the mixed epoxy on the dry glass and distribute fairly evenly with a squeegee. Close the folded plastic. Second: ***Turn the sandwich over so that the perforated side is up.*** Squeegee and roll with the grooved roller so that the glass is thoroughly wetted. Use the heat gun to speed the process. Turn over and inspect the bottom side for full wetting. Turn over again so that the perforations are up and roll thoroughly with the grooved roller. ***Excess epoxy will be driven up through the perforations.*** (See that this process is somewhat like vacuum-bagging. Epoxy is forced through the work, assuring full wetting, and excess is removed, reducing weight.)

6) Cut out the tape strips along the marked lines with scissors.

Corner Bonding

Corner bonding will be necessary wherever two nearly-flat structural surfaces meet at a sharp angle - usually 90 degrees. Good corner design requires that any deflection of a part of the aircraft under design load cannot cause a local stress concentration large enough to cause local failure. In general, this means that ***stress concentrations due to sharp radius turns must be avoided.***

Simple Outside Corners

Cloth cannot be laid to follow a perfectly sharp edge. The inherent stiffness of the cloth will make it lift away from the surface to be covered, leaving a void on both faces. The heavier the cloth, the greater the required radius. A good rule is to allow at least 3/16-inch radius for medium-weight cloth, but more is definitely better. Unless there is some special reason, it is never desireable to try to use a sharp radius. Cloth that is to be laid over an edge should be cut on the 45-degree bias, which allows a much sharper radius than a 90-degree crossing. ***Every builder, as part of his education, must once experience the frustration of making a "perfect" layup over a corner and finding - after it has cured - that the cloth has slowly oozed up out of the viscous epoxy and left air-filled voids all around the corner. Use an adequate radius.***

Inside Corners

Fillets

Inside corners occur very frequently. (See the sketch). A fillet of some kind is required to form a controlled radius and stabilize the load-bearing skin, similar to the core in a flat sandwich. The fillet also occupies what would otherwise be a void under the tape.

Normally, a small amount of "dry" micro is placed in the joint, formed into the desired smooth curve, and is covered with wet glass tape in a single work session; i.e., the experienced builder proceeds with placing the glass without waiting for the fillet to harden. The beginner may find that it is a bit difficult to handle squirmy lengths of wet tape and place them accurately without disturbing the wet fillet. The cautious approach is to ***place the fillets needed for the entire work session, in-***

spect them, cleaning up as necessary, and allow the fillets to cure. The glass tape is then added in another work session.

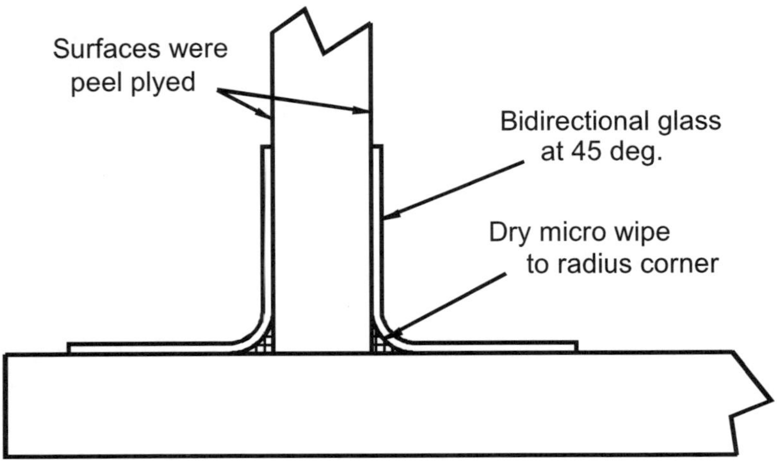

The Inside Corner

It is often difficult to get into tight places to apply the micro fillet and micro will get on the faces where the tape will be bonded. Use masking tape in these cases. Care should be taken to wipe off the excess micro and clean the bonding face with alcohol.

The "Pastry Bag" Tool

When there is quite a bit of corner taping to be done, the ***"pastry bag," sold in kitchen supply stores for decorating cakes, is extremely convenient.*** A small round nozzle is selected **(#3 or #4)**, the bag is filled with dry micro, and a very neat bead of micro can be laid in the joint without getting excess on the bond area. The gloved finger tip can then be used to round out the bead. "Dry" micro does not pour by definition, so it must be scraped into the pastry bag with a trowel or squeegee. The partially full bag may then be laid on the worktable and the micro can be driven toward the nozzle with a squeegee. This operation takes only seconds. This method may be rather wasteful of micro by leaving excess in the bag, but saves so much work that it is worth it.

Applying Tape

The perforated layer of plastic is removed and the exposed wet side is pressed into position with the gloved hand. If the corner line is curved, as with a wing rib, the builder will find that the ***bias-cut cloth will smooth into the corner very nicely, but that the plastic carrier film will not***, so the wet cloth must be gently coaxed from the plastic carrier and smoothed into position. A little experience will quickly teach the required technique. The beginner should start with strips of tape about 8 inches long, or even 4 inches for curves. The experienced builder can handle 12-inch strips for fairly straight runs.

A good technique for placing sections of wet tape is to ***fold the tape section over the edge of a plastic squeegee,*** force the edge of the squeegee into the corner, and smooth out the glass. After a fair amount of wet glass has contacted the joint, the ***glass will adhere well enough to permit removal of the remaining plastic*** without too much disturbance. The glass can then be smoothed into final position with the gloved hand. A small square ***plastic pot scraper***, with rounded corners, as sold in kitchen stores, is an excellent placement tool for confined spaces.

Usually, inside corners get only minimal stress, so resist the urge to "make it strong." In most cases, a single layer of medium weight glass is sufficient.

> *Care must be taken to assure that the wet tape is fully pressed into the corner, making* ***full contact with the fillet,*** *and that the glass has not left an unsupported void by bridging between the two surfaces. If the tint used is "opposite" to the tint used in the main structure, any voids are easily seen and removed.*

In almost every case, ***place a strip of dry peel ply over the outer edge of the wet cloth tape.*** This will assure a ***smooth blended surface*** inside the structure and avoid little sharp glass snags and "glove busters" that tend to occur. The builder should keep a bag of short strips of peel ply handy.

The "Lost Bondo" Corner

This technique produces an easy to make, very strong, radiused corner, used to connect the ***glass edges*** of two ***beveled*** panels, meeting with a radius of about 1/2-inch. It is suitable for connecting flat surfaces of almost any size, where the builder can have access to the ***inside*** of the joint. The method has the considerable advantage of being nearly self-jigging. ***(See Exercise 9, the Mailbox, for a specific example.)*** The steps described below will cover the case of surfaces meeting at a 90-degree angle, but it will work as well for an oblique angle up to 120 degrees.

"Lost Bondo" Sequence

1) Jig up the surfaces to be bonded, as appropriate. Use dabs of Bondo to hold the work to the worktable and brace with wood scraps and/or pieces of cheap thinwall PVC water pipe. Hold the faces in place with small pieces of duct tape or mailing tape so that a "best" fit is found. Adjust and reposition as needed.

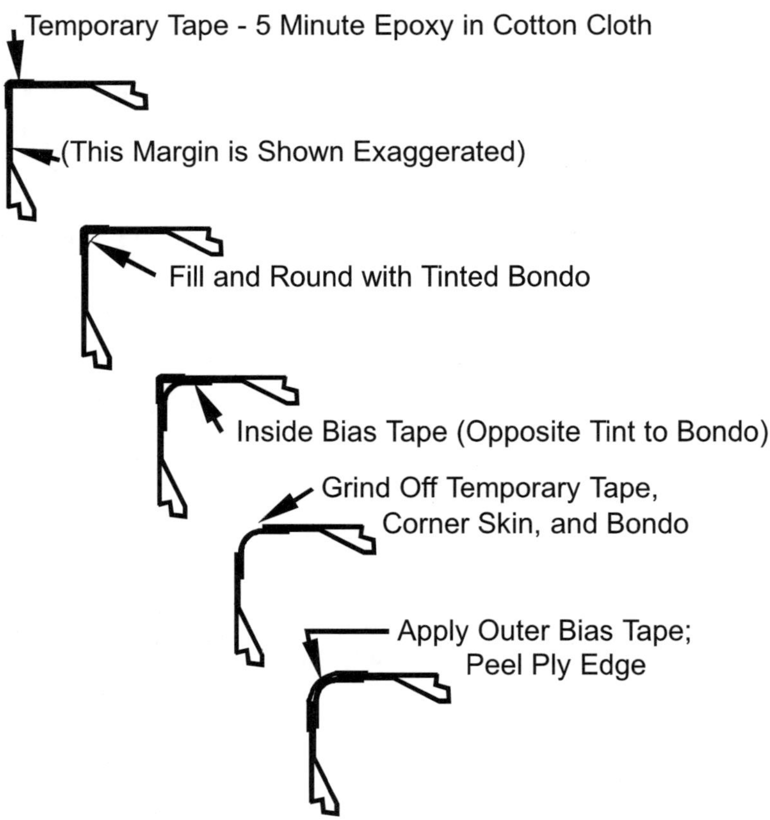

Sequence of "Lost Bondo" Corner

2) "Tack" bonding: Find a smooth cotton rag similar to a standard bed sheet. Use the plastic sandwich wetting method and ***5-minute epoxy*** to wet out a piece of rag about 6 inches square. Cut out patches of wetted rag about 1-inch by 1 1/4-inch. Remove the plastic from one side and, with the gloved hand, smooth these patches at about 6-inch intervals along the junction. Allow to harden to ***fix the final position of the parts to be bonded*** .

3) Remove the temporarily-bonded workpiece from the table and use polyester body filler (Bondo) to ***make a radiused fill all along the inside of the joint.*** After mixing the colored catalyst into the Bondo, tint the Bondo with the ***same color*** of carpenter's chalk so that the red or blue color of the Bondo hardener is clearly enhanced. Smooth the Bondo out with the curved end of a standard tongue depressor or the finger of the gloved hand. ***Work fast, because the Bondo will set quickly.***

> *It is a good idea to use masking tape to protect the inside surface of the joint where Bondo will NOT be used. The masking tape should be set back about a half-inch from the corner, leaving the corner open to be filled with Bondo. Placing Bondo is a messy business, driven by the quick setting time of the material, and it is difficult to get the soft Bondo smoothed into place in time without getting some on the bonding surface.*

4) Wet a single layer of medium BID glass, like 7725, under plastic, marked for 1 3/4 inch bias-cut strips. (Tint with the ***opposite*** color to whatever color was used with the Bondo - red, if blue was used, etc.) Cut into strips, as marked, and ***place the wet tape strips over the Bondo***, exactly as with a regular tape joint. Cure.

5) The cured assembly will be quite stiff indeed after the inside surface is formed. Use an electric belt sander with a 100 grit belt to ***grind off the corner so that the tinted Bondo is just removed*** - the "lost Bondo". This is a very fast operation, generates a lot of glass dust, and is best done out of doors with a dust mask.

6) Wet more glass and ***place wet glass tape on the outside of the joint area.*** Peel ply the edge. Cure. The result is an excellent, stiff, radiused joint.

The Pipe-formed Corner

This technique is a simple variation of the "Lost Bondo" corner for joining beveled core panels, which may be more suitable for larger structures, such as assembling the fuselage for the *Crosswind*. The method produces a joint with a ***larger radius than the "Lost Bondo" corner***, so it produces a corner with lower stress concentrations. With this system, the builder is ***not constrained by the short working time of Bondo*** and can take whatever time is needed to jig and align the parts to be joined. The best application is the joining of large prefabricated panels, such as flat or very gently curved fuselage sides.

> *Note that this corner technique is very general and* ***does not require that the joint be straight****, as with a folded joint. Folded joints will be discussed in detail in Chapter 8.*

One very important advantage of this technique is that the panels to be joined can be vacuum-formed on a large flat or curved tool so that the finishing work is greatly reduced.

The PVC Pipe Jig

The basic principle is to form the smoothly-radiused corner over a simple one-time jig made of cheap, lightweight PVC water pipe.

The PVC water pipe determines the radius of the corner. Brown mailing tape adheres well to PVC pipe, so it can be used as a reliable release film. The parts to be joined are then temporarily attached to the jig with more brown mailing tape and/or small self-tapping screws. A smooth glass tape joint is then made over the pipe. After cure, the glass tape is temporarily fastened to the jig and the tape joint is completed from the inside. After the second cure, the jig can be removed. See the following figure.

Standard "Schedule 40" PVC pipe should be used to make the jig. The technique for assembling the jig is almost exactly the one used by most readers of this book to make model airplane fuselages.

Pipe-formed Corner Procedure

The procedure is as follows, as shown in the figure.

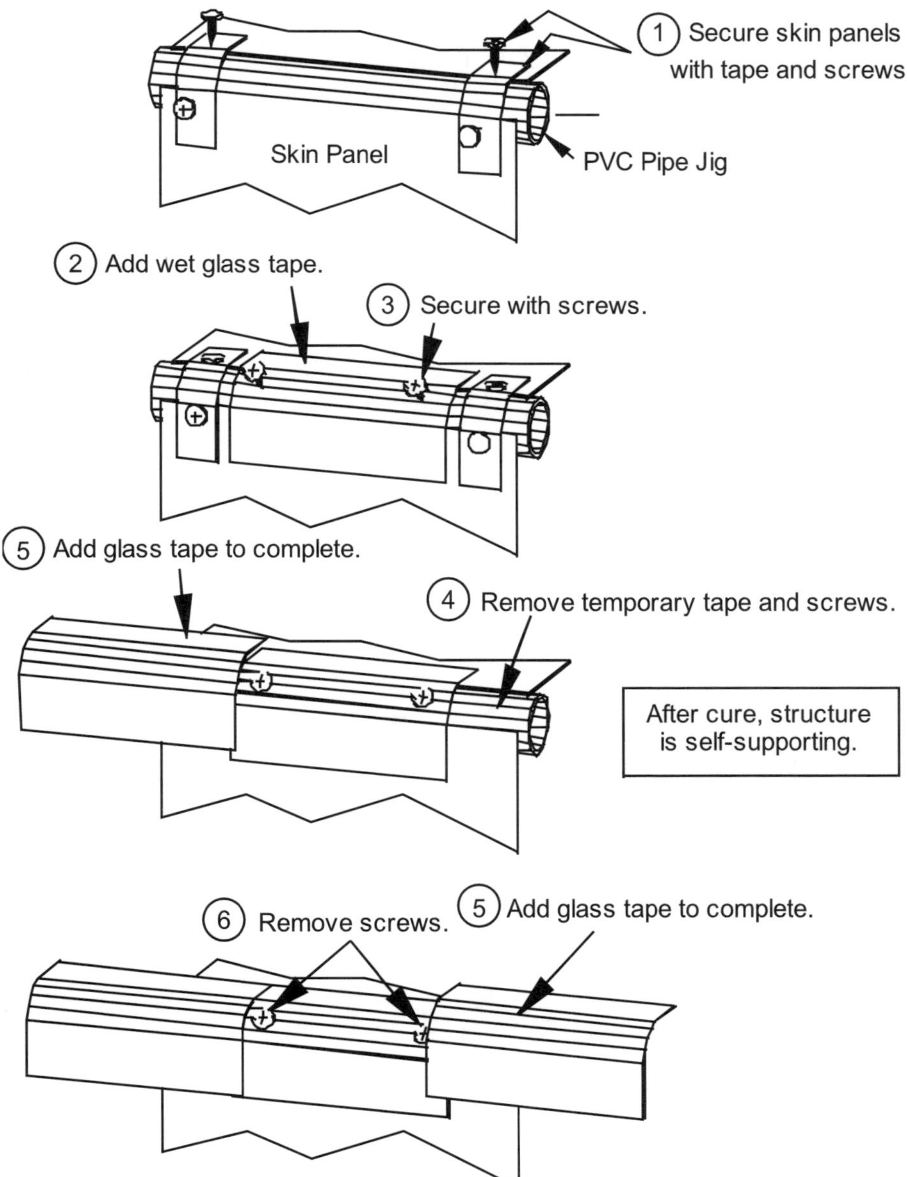

Pipe-formed Corner Sequence

1) Set up an appropriate jig, using standard PVC water pipe (7/8-inch or 1 1/16-inch O.D.). Bond the jig to the worktable or to an appropriate strongback. (Lightweight steel studding material is particularly convenient for the strongback.)

2) Cover the exposed surface of the water-pipe jig with brown mailing tape wherever the jig will contact the parts to be joined.

3) Make a preliminary assembly of the parts to be joined by attaching the parts to the pipe jig with brown mailing tape or duct tape. Adjust and move as needed. Tack temporarily, as needed, with #8 Washer Head Self Countersinking Carpentry Screws (similar to sheet metal screws).

4) Make an accurate assembly of the parts to be joined, using mailing tape or Washer Head screws, as needed, at approximately 6-to 12-inch intervals. Double check the accuracy and adjust as needed.

5) In the open intervals BETWEEN the temporary bonds, ***make a standard BID tape joint*** connecting the parts to be joined, so that the tape overlaps the parts to be joined by at least an inch. Peel ply the free edge. Cure.

6) Temporarily attach the jig to the cured joint by drilling through the cured tape as needed and attaching with Washer Head screws as needed (see the figure). Now, REMOVE the original brown mailing tape and/or Washer Head screws.

7) Clean up as needed. ***Make a tape joint over the open spaces*** left by the original Washer Head screws. Peel ply the edges. Cure.

8) Remove the second group of Washer Head screws and ***remove the jig. The structure is now free-standing, independent of the jig, and very firmly located by the full-length skin of cured tape.*** Make a second layer tape joint over the entire joint area, overlapping the first layer by about a half-inch. Peel ply. Cure.

9) Place a tape joint on the INSIDE of the new radiused joint. Peel ply the edge. Cure.

Oblique Corners: The "Barrel Joint"

Approximating a Complex Curve

A typical round wooden barrel is made up planks (staves), each of which can be either bent or twisted, but not both at the same time. The round barrel is actually a polygon shape where the interface between adjacent staves is very carefully bevelled and fitted so that the barrel will not leak. ***The interface line between staves is a curve***, held under compression by steel bands (in tension) around the barrel.

While it is true that the ***ideal*** structural and aerodynamic shape for an aircraft surface is usually a complex double curve, designing and molding such an ideal surface is a lot of work. The one-off homebuilder will often want to avoid this work by approximating the ideal with a set of intersecting single-curved surfaces that are almost as good aerodynamically and a lot less work. Classic fabric-covered aircraft like the *Stinson Reliant* do this when fabric is stretched over light, non-structural, "false work" to approximate an ideal faired surface.

> *A great many successful boat hulls have been made primarily from four long sheets of plywood. To make a practical boat hull, two side pieces and two bottom pieces are bent over a very simple form, bringing the edges together at an oblique angle, forming the "chine" and the "keel". The joint is usually secured with screws into a wooden strip, and the joint is strapped together with overlapping layers of glass. The method is, actually, making "barrel joints", except that tension loads across the joint are carried by the bonding skin rather than by an external steel band.*

Applying "Planks"

It is very easy for the builder to quickly make large quantities of high quality, ***flat stock, without a core,*** on a very simple flat tool. A single glass outer skin is laid up on the tool, cured, and cut to shape as needed to make "barrel stave" strips. Without a core or a complimentary sandwich layer, the thin strips will be quite flexible, but still stiff enough for easy handling. The sketch on the next page shows a typical case.

The procedure for making a useful form with oblique corners, such as a fuselage tub, boat hull, or large removeable inspection panel, is described below.

1) Prepare a set of formers. Cut out plywood or chipboard formers to ***define accurately the outside surface*** of the glass skin. Secure the formers to a frame or "strongback" made of wood or light steel studding. Usually, the position of the formers will correspond to the location of bulkheads and four to six formers will be required for adequate definition of a surface as large as a fuselage tub.

2) Prepare patterns for the "planks." Plain, unsurfaced, ***1/8-inch Masonite*** is a good compromise for low cost, weight, and ease of handling. Temporarily secure the patterns to the formers with 1/2-inch #8 Washerhead screws. Fit the patterns as needed by trimming with a spiral saw or a small battery-powered circular saw. (The small circular saw will cut a moderate curve if the wider carbide-tooth saw blade is used.) ***An approximate fit is sufficient.*** Mark the location of any bulkheads. Remove the fitted patterns. If appearance is important, ***identify all joggles that will be used and place the joggle spacers on the tool support.*** Make all necessary marks on the tool support to define trim lines.

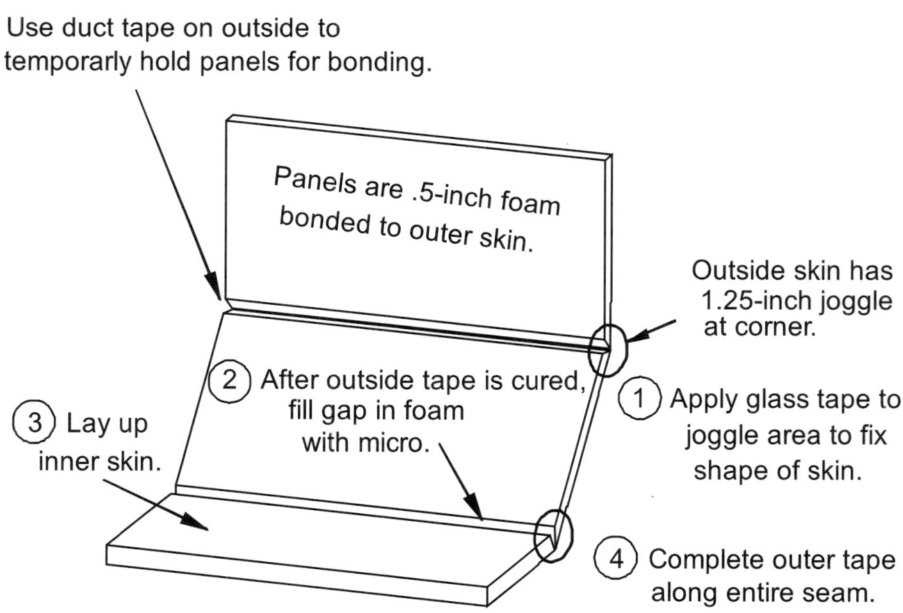

135-Degree Butt Joint for Single Skin Panels

3) Lay up the outer skin. As previously described, prepare a flat tool surface with any joggles. Place ***peel ply*** over the bonding area and lay up the outer skin over the peel ply as required by the design. Note that ***wet peel ply will not pass air easily*** and that significant work can be saved by ***perforating the peel ply*** before making the layup. (In many aircraft applications and for light-weight boat applications, such as a high performance kayak, a single layer of 7725 BID may be adequate for an outer skin.)

4) Add perforated foam core panels as required. Use slurry as previously described. Vacuum bag as usual for a flat tool. Cure. The result is a ***large flat area of single-skinned foam - not yet a sandwich.***

5) Cut out the planks. Use the plank patterns to mark the cutting outlines. Cut out with a spiral saw or bandsaw. The small battery-powered circular saw is very fast for nearly straight cuts.

6) Fit the planks. Trial fit the planks to their required positions. Use a "cheese grater" tool and/or a sanding block to adjust the fit. The exact fit is not critical becase any small gap will be bridged by the tape.

7) Secure the planks in final position. Use #8 Washerhead screws or hot melt glue to secure the planks to the forms. Inspect the result. Verify that the planks are positioned exactly as required. If some supplementary support is required, this is the time to rig it.

8) Tape the joint lines. In the usual manner, wet out and apply enough 2-inch wide tape material to cover all of the joint lines. ***Leave the screws that attach the planks to the forms exposed.*** If it is necessary to tape over the screws, use a sharp tool like an awl or a nail point to "tease" the glass fabric away from the screw heads to expose them. Peel ply the edges of the tape and smooth out. Cure. ***Leave the attachment screws in place.***

9) Apply additional core panels as needed. Perforate, cut, and bevel core panels as needed. Slot the cores where required, as described in Chapter 4. Mix slurry and place in the core panels in position. Depending on the degree of curve in the skin, sandbags can be used to hold the panels while curing. See that slurry is forced up through the perforations to verify that full contact without voids is achieved. Clean up any excess slurry that escapes past the edges of the panels. Cure.

10) Lay up the inner skin. Add the inner skin as an ***open layup*** (not vacuum bagged), to the extent permitted by whatever attachment screws are needed to assure accurate contact with the forms. If necessary, any open screw holes will be patched later after the surface shape of the skin is accurately determined. Cure. Remove the attachment screws.

11) Attach the bulkheads, as needed. Clean up and sand the bonding areas. Insert the bulkheads and tape in place. Verify that alignment is exactly as required. (The addition of the bulkheads will make the structure quite stiff and adjustments will not be possible after the bulkheads are installed.) Cure.

12) Finish outside tape and cleanup. Invert the structure, remove the outside peel ply, and tape the outside of the joints and clean up and add any supplemental tape that may be needed, including patching holes made by the attachment screws. Some applications may require a second layer of tape over the outside joints.

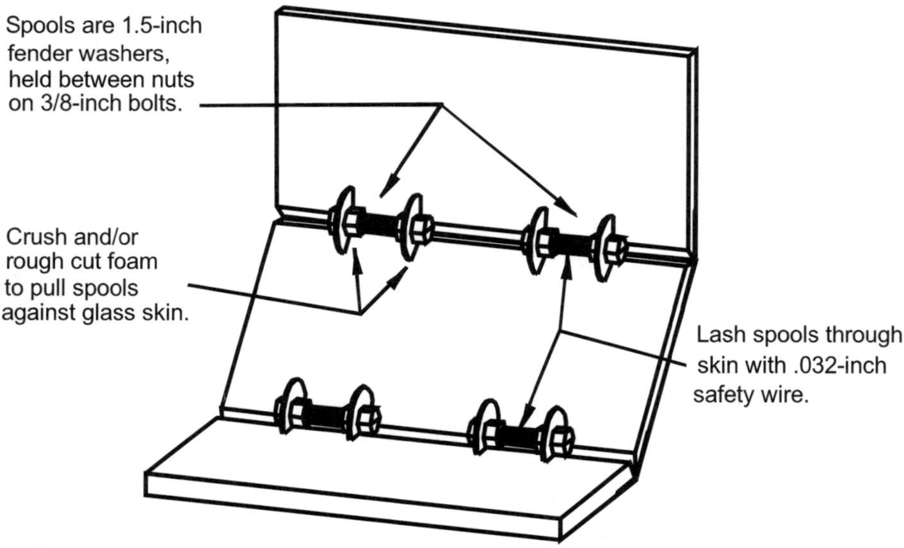

Jigging Single-skin Planks with Forming Spools

The "Forming Spool" Self-jigging Trick

The technique described above requires that the large panels to be joined must be held in position by some kind of simple form or jig. It may be possible to simplify the form and get a self-jigging effect by using the "forming spool" to align the edges, as shown in the sketch above. The simple tool is easily made with 1/4-inch bolts and 1 1/2-inch fender washers. The edges to be joined are lashed together with aircraft safety wire, as shown. See that the edges of the skins will be held flush by the forming spool.

Short pieces of PVC pipe may also be used as forming spools.

A Hybrid Variation

A few "hybrid" airplanes have been built which use a steel frame or "cage" as the primary load-carrying structure for heavy, centrally located components such as the engine, landing gear, main payload, and wing attachments, but the remainder of the structure is composite. (The ***GlaStar*** is a current example.) The designer might choose to use all composite flight surfaces and tail cone, then use a non-load-bearing composite shell to cover the steel cage cabin area.

This raises the interesting ***possibility of making the non-load-bearing shell so that it could be conveniently removed in large sections,*** similar to the way many engine cowls and some agricultural aircraft are designed, using Dzus or Camloc fasteners.

The builder would make curved sandwich panels from flat glass stock, as described above, ***without the foam core***. The "planks" would be cut and fitted as above, working around the steel frame as necessary and fastened with temporary fasterners while the edges were taped. The single layer "planks" will be much too flimsy without a core and a second skin layer, but can serve to define the surface. After the joining tape cures, the surface is determined and the attachment points are accurately located. Foam core material can then be applied in selected critical areas and held in place with sandbags until the slurry cures. The inside layer of glass is then added over the selected areas. After the inside glass cures, the shell will be quite rigid, even though the entire surface is not yet completed. The partially complete shell can now be removed from the cage, the remaining foam core added, and the inside glass layer completed.

The temporary fasteners can now be drilled out and replaced with Dzus or other appropriate removable fasteners.

Boat-building Considerations

The general method described in the preceding section is particularly applicable to building a high performance ocean kayak. The author has corresponded with a builder whose application is summer kayak touring in the remote high Arctic. This builder requires aircraft-level reliability and weight control, but also requires that the very light boat stand up to repeatedly being beached on gravel bars. This requirement was successfully met by using a ***partial bottom skin of thin Kevlar***. Where the Kevlar outer skin is used, it is not necessary to use additional glass tape over the joints.

The builder might also note that the gaps in the foam between adjacent planks might offer the opportunity to embed carbon graphite yarn in flox in the open slot. The desired effect would be to make an internal frame of very stiff carbon at a very small added cost in money, complexity, and weight, which might seem very attractive at first consideration. The consequence is that the ***very stiff carbon becomes the primary load path*** in the immediate area of the carbon and the nearby glass receives no strain and becomes redundant. To be useful, the ***load carried by the thin carbon must go out into the nearby structure as shear***, and these local shear stresses may be very high. This may be understood by imagining an infinitely strong and thin "wire" running the length of the structure. If the structure is loaded in bending, so that the wire is in tension, the adjacent real-life structure cannot support the infinite shear forces necessary to attach to the "wire" and ***failure of the bonding*** to the adjacent structure - not the wire - will occur. This differential stiffness problem can be avoided by ***using sufficient unidirectional glass tape instead of carbon*** to produce additional bending stiffness. As a "thought experiment", the designer might ask herself what would happen if the strong wire had been greased, so that it could not carry shear.

Chapter 9

Formed Curves and Corners

Bending Single-skin Foam

The "tension cracking" problem is avoided if ***one surface of the core is skinned previous to the bending operation.*** (This skin must be thin enough to accept the required bending angle, of course). The cured sheet can then be pressed into the form or jig so that the ***skin takes the tension*** load, there is ***no tension strain in the foam,*** and ***all*** of the compression strain goes to the foam. This process obviously has a limit because the necessary displacement in the foam must go *somewhere.* Excessive crushing will impair the integrity of the foam cells. Fortunately, the ***compression problem occurs at the non-skinned surface,*** where patient application of a heat gun may help the bending process considerably.

This "compression only - no tension" process has a ***practical limit for bare cores of about 1/4-inch thickness.***

The "bending skin" technique can be combined with slotting, which is particularly effective when using the more brittle urethane foam. Here, the slab of flat foam is ***skinned on one side with a light skin, such as a single layer of Style 120, 0.004-inch glass,*** before the slotting operation. Situations may occur where the builder may wish to put a "bending skin" on a large piece of foam, such as a fuselage side, ***only in the specific areas where the bend is required.***

This method is limited to single-curved applications, such as seat backs, because straight lines are preserved in bending a flat surface.

Folded Corners

It is possible to make excellent finished structures from flat sections, using a flat tool, where flat surfaces are completed as rigid two-sided sandwiches. Surfaces which will require curvature are ***formed in the same layup as the main surfaces, but with the area to be curved left***

unskinned on the inside. The builder bends ("folds") the large sheet into the final shape using a very simple bending form.The builder then adds the remaining inside skin to the curved fold area to complete the part.

The folded corner results when ***two stiff flat panels are joined by a section of flexible skin, then folded along the joint line to make a straight line joint.*** The joint is then completed by adding whatever skin or core material is needed inside the joint. The fact that ***the folded corner must make a straight line may be a compromise with the ideal, but it is often a compromise very much worth making.***

The sketch below shows the simple case where two cored panels have been formed on a flat tool, leaving a common area without a core. The common area can then be bent (folded) as required. Additional skin layers, and/or core and skin, can be added after the bend is formed (see also Exercise 9, the Mailbox).

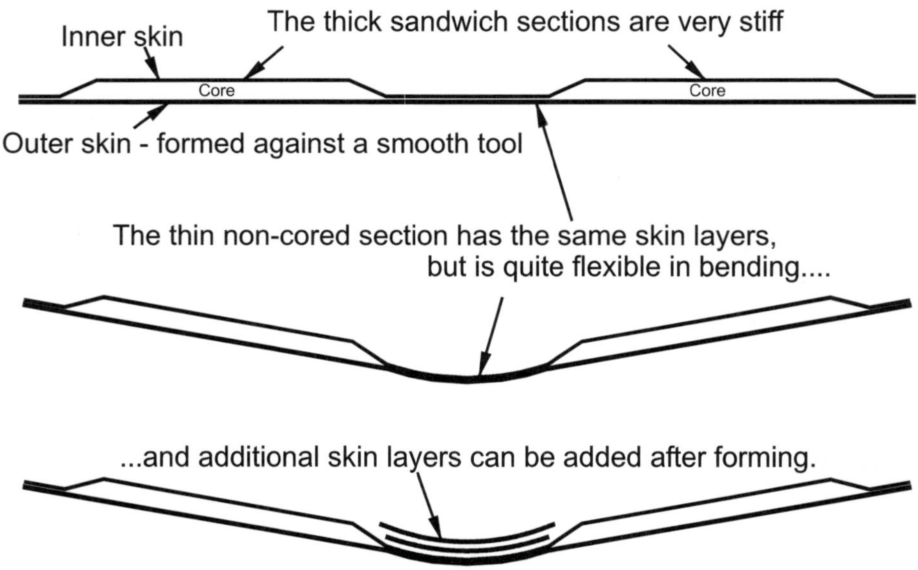

Forming with a Non-cored Bending Section

The folded corner has several ***very important advantages:***

1) If the panels to be joined can be vacuum-formed as a single operation on a flat tool, the ***entire joint has a perfect tool finish.***

2) The "folding" operation is ***extremely simple*** and can be done using the simplest of bending jigs, often simple wooden forms without the complexity of an accurate forming radius.

3) Where the flat panels must be formed as separate pieces, the ***bending area can be added later***, often without the need of a tool surface.

4) Several flat forms can be stacked together for shipping, greatly reducing shipping costs for kit parts.

Bending Slotted Sheets

Where the core thickness is too large for the required bend, it will be necessary to cut slots ***so that the remaining web is made thin***, of the order of 1/16- to 1/8-inch. The builder must then fill the slots with soft micro before placing the sheet in the final bending form. Excess micro will ooze out of the slots and can be removed with a squeegee (see the following sketch). The builder can proceed with laying up the final skin before the micro has hardened (see also Exercise 10, the Mailbox).

The "single skin" technique can be combined with slotting, which is particularly effective when using the more brittle urethane foam. Here, the slab of flat foam is ***skinned on one side with a light skin, such as a single layer of Style 120, 0.004-inch glass,*** before the slotting operation. Situations may occur where the builder may wish to put a "bending skin" on a large piece of foam, such as a fuselage side, ***only in the specific areas where the bend is required.*** See the following figure.

This method is limited to single-curved applications, such as seat backs, because straight lines are preserved in bending.

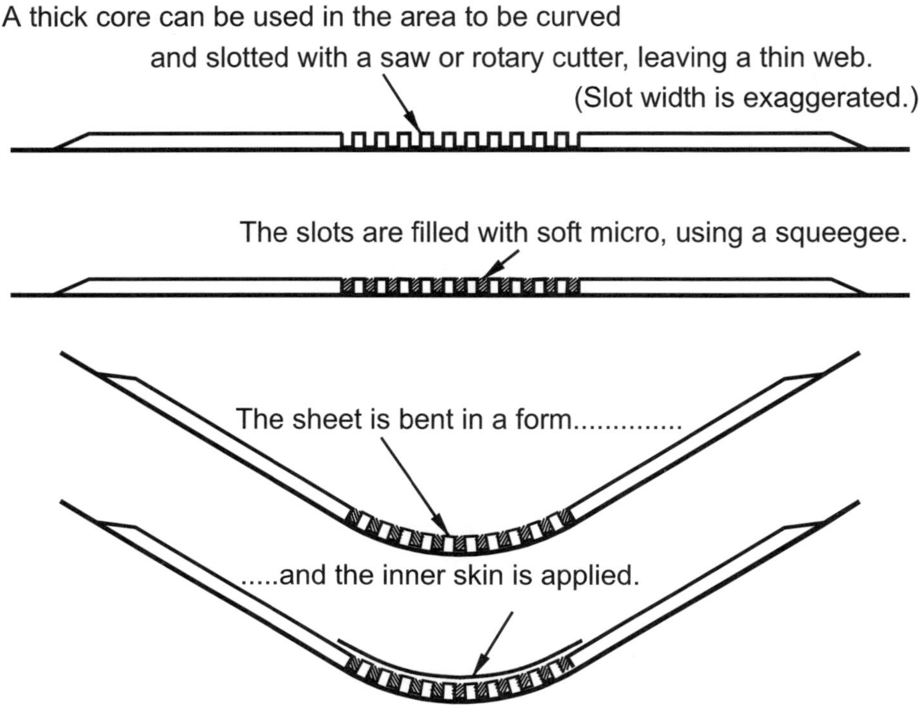

Use of Slotted Core at a Bend

The Short-radius Folded Joint

It is often impractical for the home builder to provide enough tool area so that several panels and their common folding areas can be laid up at one time. In that case, the separate flat panels must have the ***folding areas attached before bending.***

The basic process is: (1) make a ***temporary flat*** flexible joint, (2) bend the temporary joint to the required form, (3) complete the joint from the ***inside***, then (4) remove the temporary joint, leaving a final joint with an excellent surface finish.

The sketch following shows one practical example of the process for a joint about 1.25 inches wide. The temporary flexible bending joint is provided by ***two layers of peel ply*** over a third layer of .004 inch Style 120 glass.

The three layers are wetted out under plastic in the usual manner, trimmed to size, then placed between the flat sections to be joined, as shown, using a flat tool. Notice that this places a ***single layer*** of Style 120 glass on the ***outside*** of the joint.

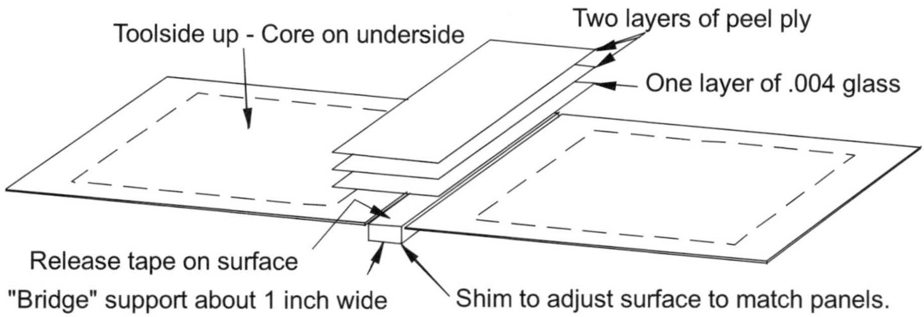

Temporary Bonding of Two Flat Cored Panels

The sketch above shows the use of a "bridge" tool covered with release tape. The bridge is necessary to avoid sagging of the wet tape before it has cured. Note that the width is not critical. The three layers of wet fabric will span a gap of up to about half an inch, but will require support for longer spans. The tape-coated bridge may be reusable.

> *The builder must be careful to place the **glass** layer against the work. Wet peel ply and wet Style 120 glass look almost identical!*

After curing, the three layers are sufficiently strong to allow handling, but flexible enough to allow the required fold. The ***bending jig can be extremely simple***. In fact, the builder will probably find it sufficient to make a bending jig by attaching two pieces of wood together to form a right angle and ***not bother with a forming radius at all.***

The sketch following shows the attached panels in a simple right-angle forming jig. (In practice, the work may be held in the jig with sandbags or fastened temporarily with small washer-headed screws). As described, the joint area will bend easily to a smooth radius.

See that the squared-off edges of the sections to be joined form a discontinuity at the junction with the very thin outer skin. (The skin thickness shown in the sketch is exaggerated). This discontinuity should be ***blended in as shown with a small amount of flox and smoothed out with the gloved hand.*** The two load-bearing skins are shown as 7725 BID. (Ideally, the overlapping edges should be staggered slightly, as shown, to avoid a stress concentration at the transition. The disadvantage is that overlapping edges requires two separate applications of tape).

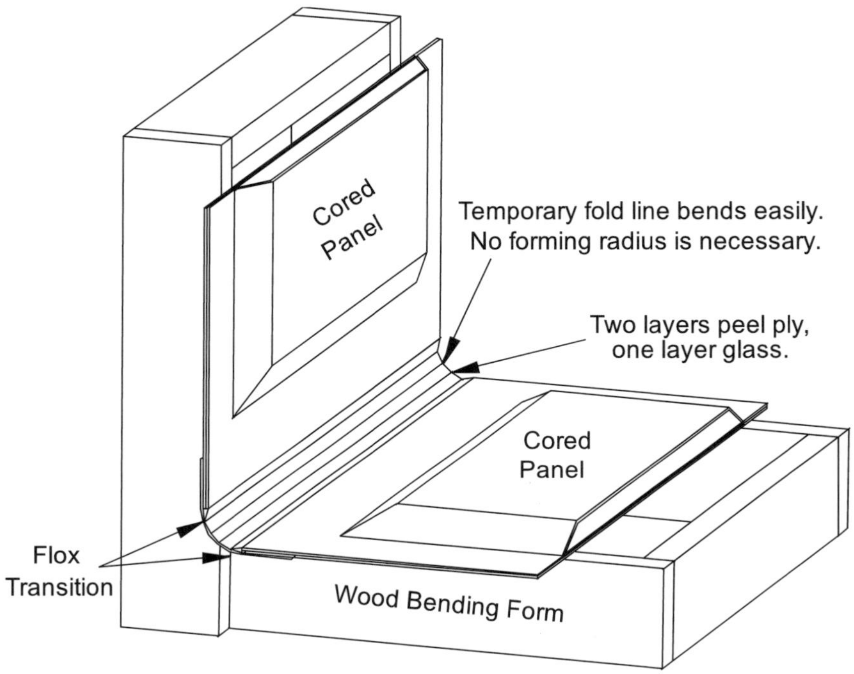

Folded Joint - Previously Cured Panels

Inside Layers

The sketch following shows the completed joint with an additional two layers of glass on the inside.

After cure, the ***peel ply is removed*** from the outside of the corner. This operation leaves a very small discontinuity formed by the thickness of the Style 120 glass and the slight surface texturing formed by the weave of the peel ply.

In principle, the temporary bending joint could be provided using three layers of peel ply and no outer layer of glass at all. The thin outer glass layer makes a more conservative joint, ***greatly reducing the risk of delamination***, because it ***protects the transition*** between the main skins being attached and the strong glass layers used inside the joint.

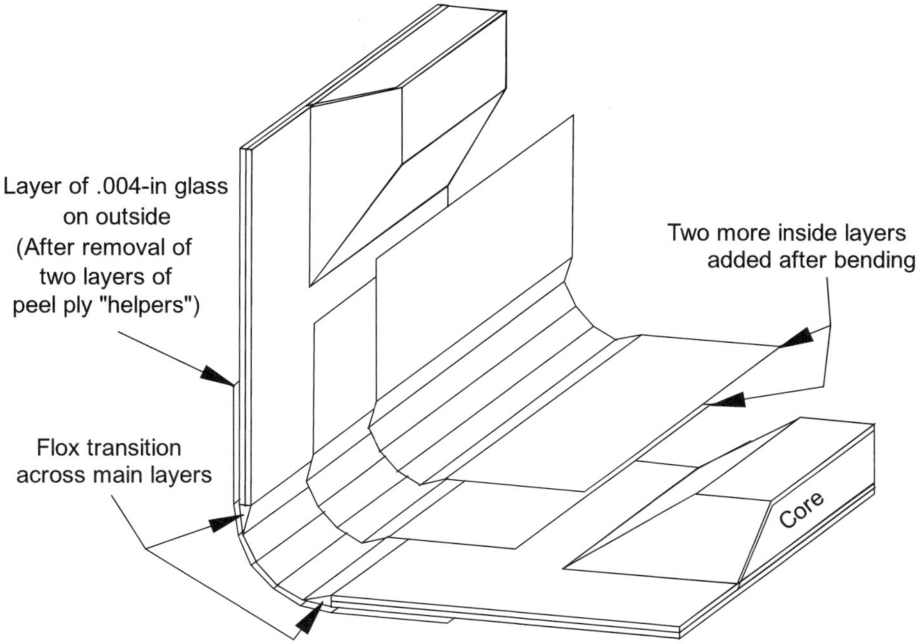

Flox Transition and Added Inside Layers

The Cored Corner

The "Cored Corner" technique is a variation on the basic folded joint described above that produces a very efficient corner. The technique has two advantages: (1) The joint has an efficient thick cross section that may save one layer of cloth, at about the same total weight as a conventional joint, and, (2) the joint allows easier access, which will be welcome in the tight quarters of an aircraft fuselage.

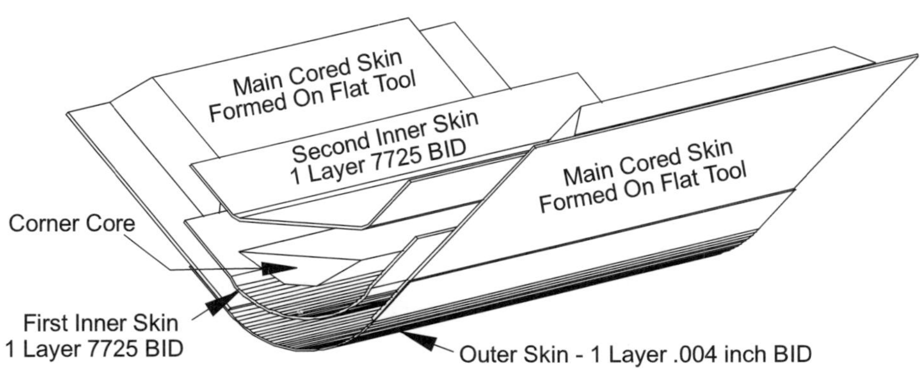

Folded Cored Corner - Layers Separated

The sketch above shows the process. The temporary flexible bending joint is the same as above, provided by ***two layers of peel ply*** over a third layer of .004-inch Style 120 glass.

The ***core is imbedded*** in "peanut butter" micro. The sketches show the case where the core is cut using the "standard" all-purpose beveling tool used thoughout this book to cut and bevel sheet cores. (Ideally, the builder could shape the core with a sanding block to make an accurate fit to the radius, which would save some micro and a very little weight, but would be more work).

The next sketch is a cross section view of the cored corner.

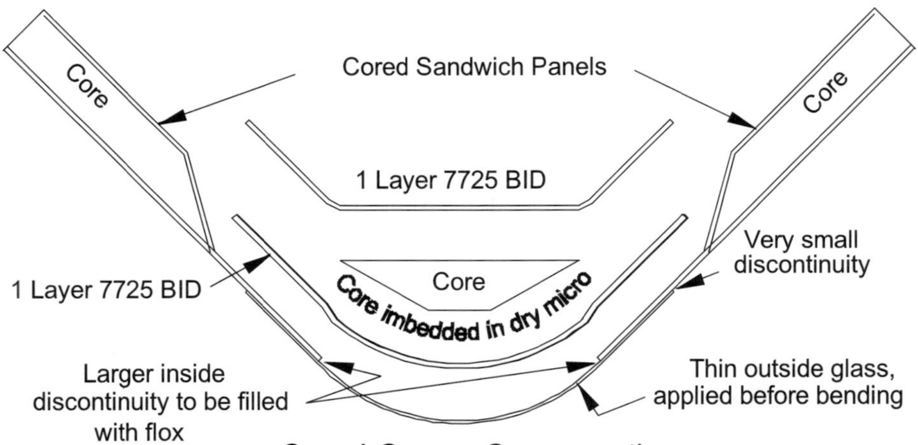

Cored Corner Cross-section

Shallow Bending Angles

The following sketch shows a simple variation for shallow bending angles, which may be applicable in some cases, particularly boat building. Assuming that the main panels use 1/2-inch cores, and the joint area uses 1/4-inch cores, the builder will find that the main panels will be very much stiffer than the joint area and that the bend can be controlled quite easily.

There is a practical limit to the allowable bending radius of the skin and core combination and the builder will need to experiment for each specific case. (The core must compress to some extent at the inside surface). For such applications as a seat or turtleback, core thickness will probably be limited to a quarter-inch.

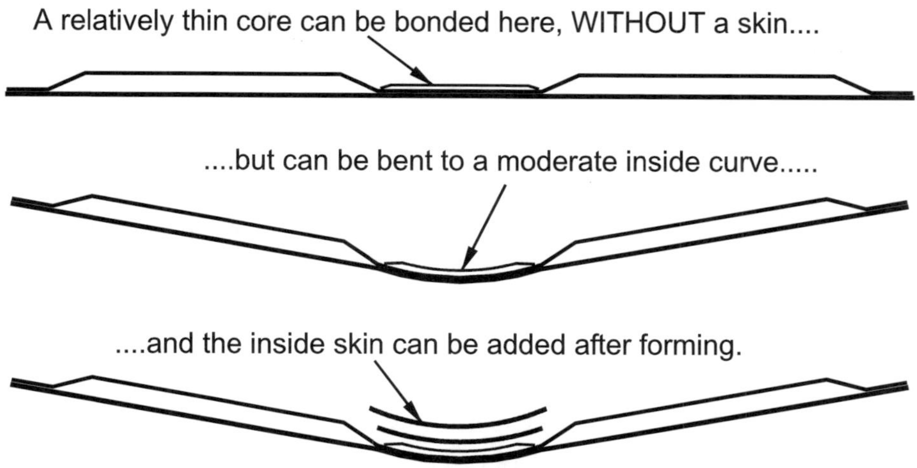

Bending Section with a Thin Core

Joggles In Folded Joints

The very slight discontinuity caused by the outside layer of .004-inch glass, discussed above, may not be acceptable in some cases or the designer ***may wish to use a much thicker layer of glass on the outside*** (boatbuilding may be an example). In this case, the builder must include an appropriate joggle when laying up the panels to be attached. A single layer of 7725 BID and a single layer of peel ply will allow a small bending radius. (Of course, the builder must try a test section for each specific case to verify that a good choice of bending area has been made. The suggestions above are typical).

"Green" Folding

The builder may find it convenient to do the folding operation before the joint area between panels is entirely cured. An example would be where the builder has completed a flat joint between panels in the evening, then returned the following morning, while the epoxy is still soft, to make the fold. The fold will be extremely easy, expecially if aided by local warming with a heat gun. Timing a "green" operation may take some careful watching, but the "time window" of around 12 hours is usually close enough. The use of "green" folding may allow the builder to use two layers of 7725 across a bending joint area where it otherwise would not be practical to bend two fully-cured layers.

The Vision Aircraft Example

The sketch below is similar to the lower fuselage skin of the *Vision* aircraft, the plane on the cover of this book. The graceful shape is formed by flat surfaces, cylindrical curves, or sections of cones; i.e., ***the surface is formed entirely from straight lines.***

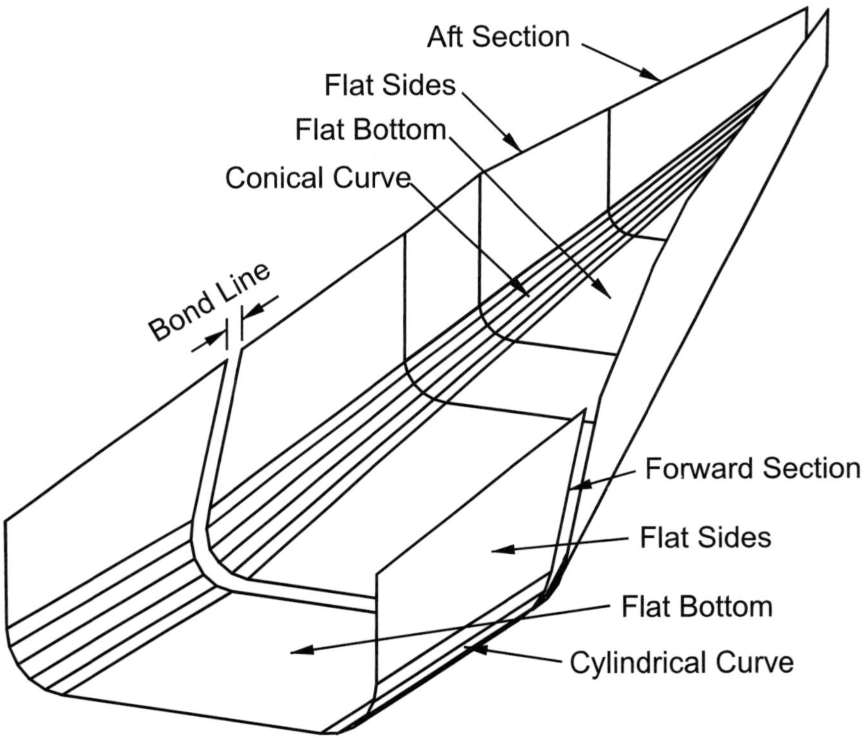

Fuselage Tub for the Vision Aircraft

The center and lower front of the fuselage is a straight-sided open trough shape. The aft lower fuselage is a straight-sided tapered trough with cones at the corners. The fuselage turtledeck is essentially a straight half cone with flat margins. ***These few components are bonded together to form the final fuselage shape.***

In principle, this structure could be assembled from only two flat-molded parts, a front section which is essentially a cylindrical trough, and a rear section which is a cone with flattened sides and bottom. (In practice, the sides of the actual aircraft have a slight twist to allow for the widening at the upper edge, but this slight twist is easily accommodated by the skin panels).

The sketch following shows the surface of the fuselage skin "unfolded" into sets of separate flat surfaces: One set forms the trough-like front section, the second set forms the cone-like aft section. Flat surfaces are thick-cored sandwich panels, formed as described in Chapter 6. Areas where bulkheads are to be attached are shown. Surfaces which are to be curved are layed up at the same time, but are left unskinned on the inside. Each of the large flat surfaces (in principle) could be formed in a single work session on a smooth flat tool.

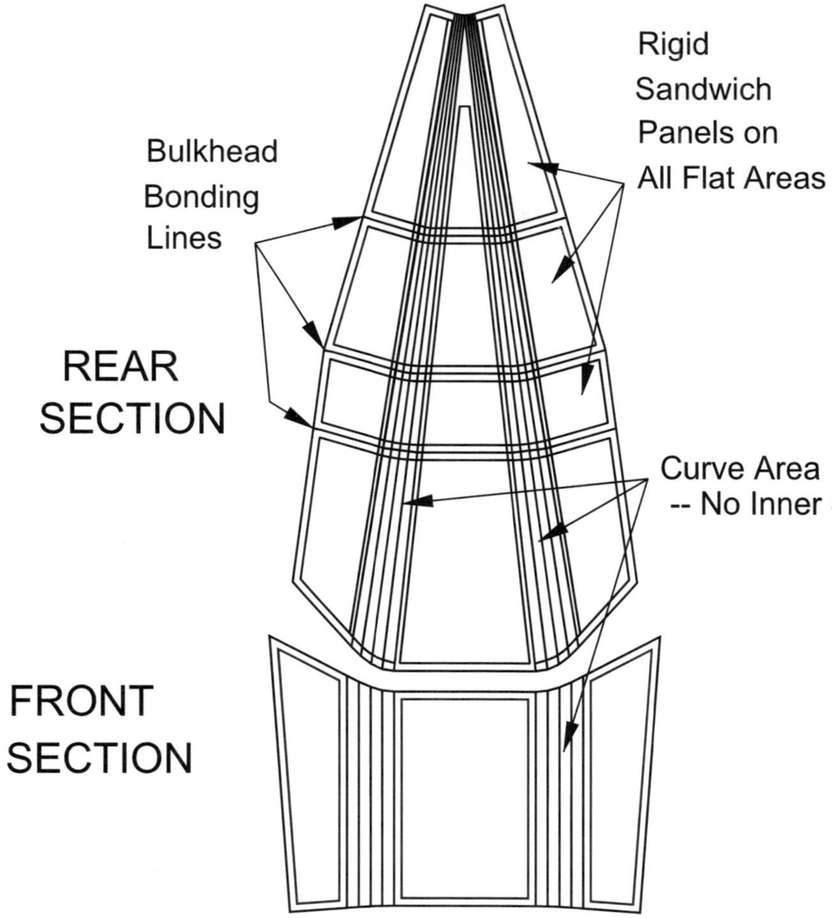

Fuselage Tub for the Vision Aircraft - Flat Projection

Practical Flat Sections

The front trough and rear cone sections shown above are too large to be practical for the homebuilder working alone with a tool size limited to 4 x 8 feet. The sketch following shows the fuselage skin divided into ***six smaller sections*** with the splice areas shown. The builder then makes the flat splices as required before making the final assembly of the front and rear sections.

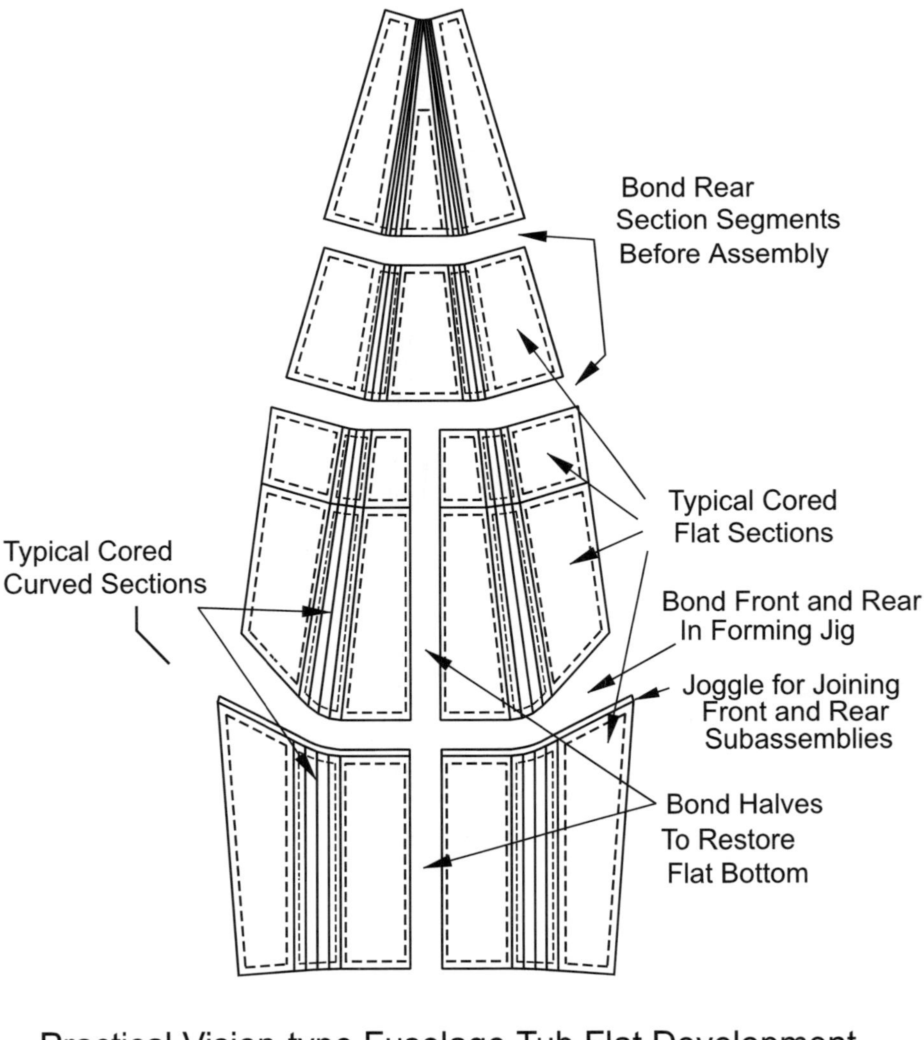

Practical Vision-type Fuselage Tub Flat Development

The "payoff" comes from the very simple final assembly. For the *Vision* example, ***as few as three very simple plywood or chipboard bending forms are sufficient to define the final shape.*** (See the sketch below.) It may be convenient to mount the bending forms on a light weight frame of steel studding material, though equally good alternatives will occur to the builder. ***The inner skin is added to the curved areas to complete the rigid structure.***

Forming from Flat Sections

The sketch below shows the *Vision* fuselage example.

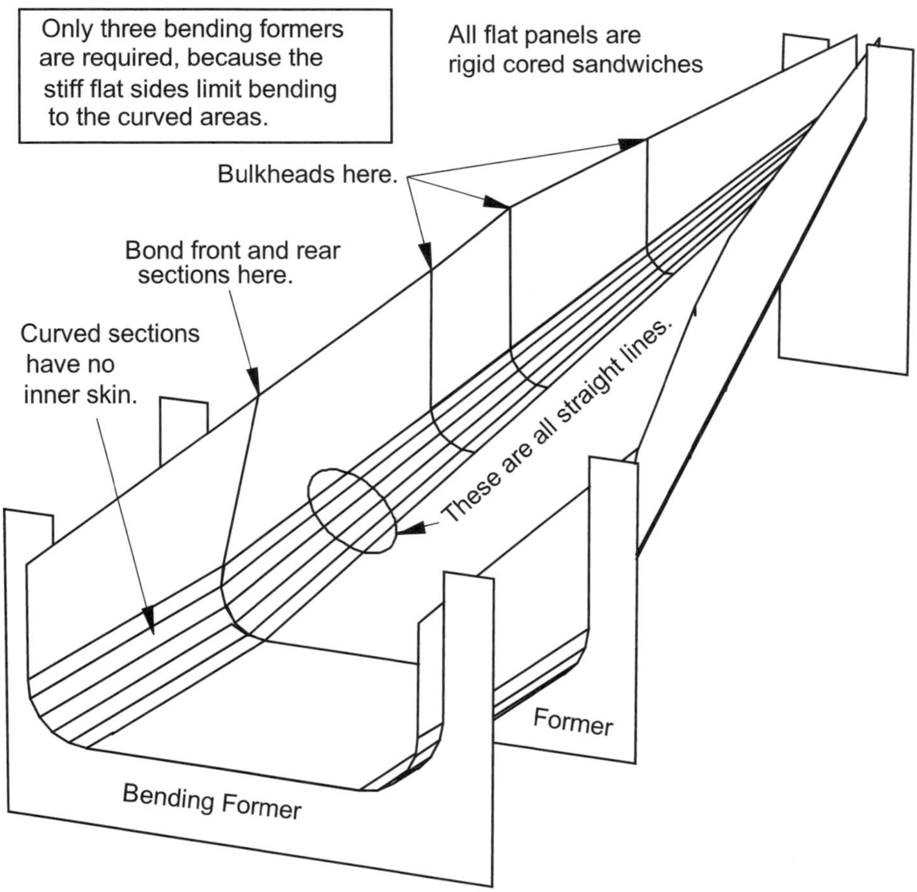

Vision Fuselage Tub - Forming and Assembly

The flat sections discussed above have been "folded" into the ***three very simple forming jigs*** shown. The rigid flat panels will force the bending to be limited to the intended curved areas, where the inner skin has not yet been applied. The front and rear subassemblies are bonded in the jig to establish the final form of the skin. The stiff structure is completed when the bulkheads are bonded on the lines shown.

Comparison with Conventional Methods

It is true that a single part made by the above procedure is more work and only a little bit heavier than a similar part taken from a high-production industrial tool. The builder should note that, when amortization of a very expensive tool and shipping of a large and fragile part is considered, the ***flat sheet method is much cheaper***.

The method has proven to be quite successful. American Affordable Aircraft has developed a variety of *Vision* components using these concepts, under the name ***"FoldaPlane."*** The builder receives the considerable advantages of having the critical flat layup work done by an expert, leaving less critical tasks to the builder, all without the costs of shipping very large, fragile components. Vacuum-bagging techniques (Chapter 6) can be used by the manufacturer, ***but the builder is not required to have vacuum capability to complete the final assembly.***

Chapter 10

Flanges and Closing

The Closing Problem

The preceding chapters have described methods for making and shaping large structures, where the builder has ***access to the inside*** of the structure. To make a practical aircraft, these subassemblies must be accurately assembled and closed. This means that situations will arise where a ***closed structure must be assembled without access to the inside of the structure***. A good example is the addition of a wing skin to a partially assembled wing which already has one skin and the ribs and shear webs in place. It is usually easy to do the necessary assembly and taping operations when there is open access to the working side. Special techniques are required to prepare and locate a bonding surface so that the ***closing surface can be assembled accurately in a single operation.***

Flanges

The ***flange*** is a very important special case of the inside tape corner, which is used when the ***corner must be formed before a later assembly operation can be made.*** One side of the flange is a ***secondary bond***, laid up in place with wet tape. The other side of the flange is an ***adhesive bond*** made to the cured surface of the tape.

A temporary structure, the ***flanging tool,*** will be necessary to form the adhesive surface of the flange. Basically, the builder will place wet tape between the original structure (like a wing rib) and the surface of the flanging tool, exactly as with a standard tape joint. The flanging tool must be covered with a release surface. After curing, when the flanging tool is removed, the exposed flange surface is a very good replica of the tool. Bonding to the face formed by the flanging tool must be by an ***adhesive bond***. See the figure on the next page.

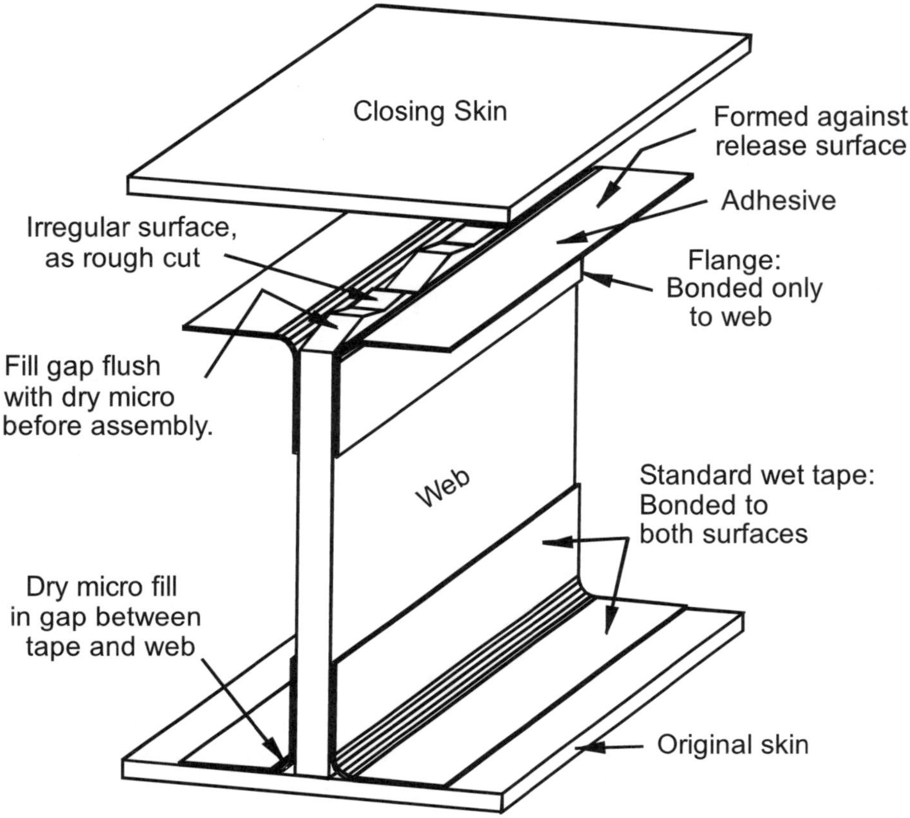

Flange Joint and Tape Joint

Use of the Flange Tool

The flange tool is basically a flexible flat strip, coated with a release surface, which is laid across the panel to be flanged and held in position by some means. The sketch on the following page shows an example in which narrow strips of 1/8-inch Masonite are laid across the flange tool strips and held in place with hot-melt glue. Sandbags, or other convenient weights, are used to hold the tool assembly in position. Corner tape is then applied in the usual way, allowing a radius. The radius will leave a gap, as shown, which will be filled with dry micro when the joint is closed.

The flange tool should be somewhat wider than the intended finish dimension of the flange to allow for trimming to an accurate edge with a spiral saw.

Fiberglass strips cut from two or three layers of 7725 glass and covered with ***clear*** mailing tape are very convenient as flanging tools. The builder can see through the cured glass and verify that the tape is properly applied and all voids have been pressed out.

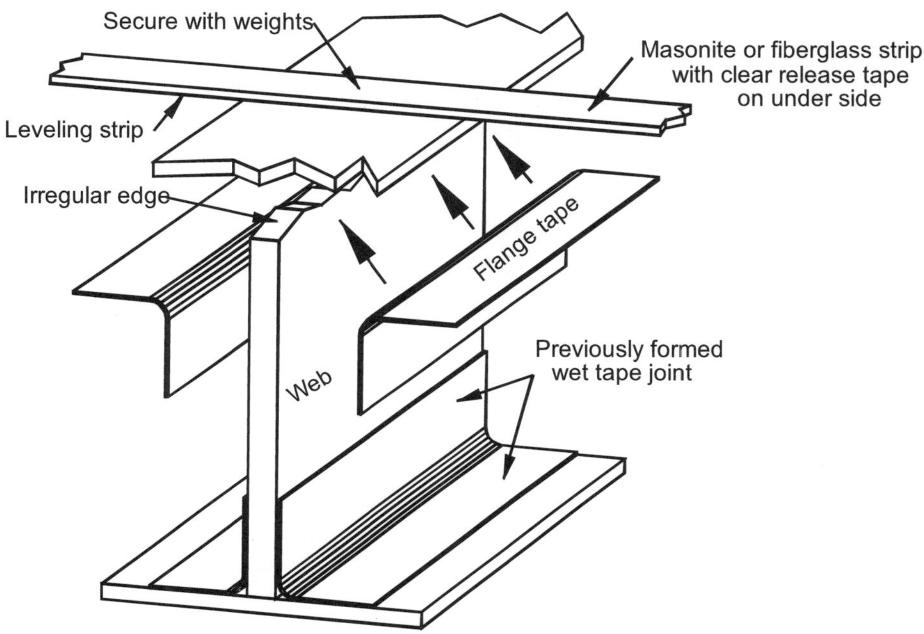

Flange Tool Application

The author has had success by gluing dry peel ply to the face of the flanging tool with artist's spray adhesive or rubber cement, which does not penetrate the peel ply. The peel ply transfers to the flange when the tool is removed. When the peel ply is removed later, after cure, it leaves an excellent, clean, bonding surface on the face of the flange.

Locating the Flange Line

There is almost always a ***significant uncertainty in the location of the surfaces to be bonded with a flange.*** (The two preceding figures show a ragged edge on the surface to be flanged.) For the case of attaching a bottom wing skin to a rib that has already been attached to an upper wing skin, there will be variations in the thickness of both skins and in the shape of the rib itself. ***These variations require that the rib be cut undersize.*** The resulting gap of as great as 1/4-inch will be closed by the flange. See the preceding sketch.

In general, some kind of ***temporary jig*** will be required to define the bonding line accurately.

The outside shape of the finished structure will be defined by a reference form, such as a wing rib master. This master will define the ***ideal outside surface*** of the finished assembly, ***not the line formed by the inside of the skin.***

In the case of a wing, the mold that was used to form the outside (upper) surface of the main wing structure may serve as a reference for the wing rib master. The finish line for the bottom skin is then determined by the master rib.

Transfer Jig

The builder must construct an accurate ***movable transfer jig*** so that the line determined by the master can be transferred to the structure to be closed. In the simplest case, this means that the structure to be closed is held in an assembly jig and the master can be used to transfer the line to the temporary jig. See the figure following and Exercise 15 for a detailed specific example.

One way to make a transfer jig is as follows: Temporarily position the master to the wing mold, or other reference, with Bondo.

> *(Release tape can be used to temporarily protect the surface of the wing mold and the wing master. Bondo, rubber cement, or 3M Super 77 can be used to attach braces of wood scraps.)*

Rough cut a form of 1/8-inch particle board, such as Masonite, as shown in the figure following. Use release tape to protect the edge of the master and the surface of the mold sheet. Locate the Masonite sheet over the reference, as shown, and fill the gap with 5-minute flox, trowelled in and smoothed with the gloved hand. Cure. Separate the Masonite form. The builder now has a ***very accurate reference to the desired closing line.***

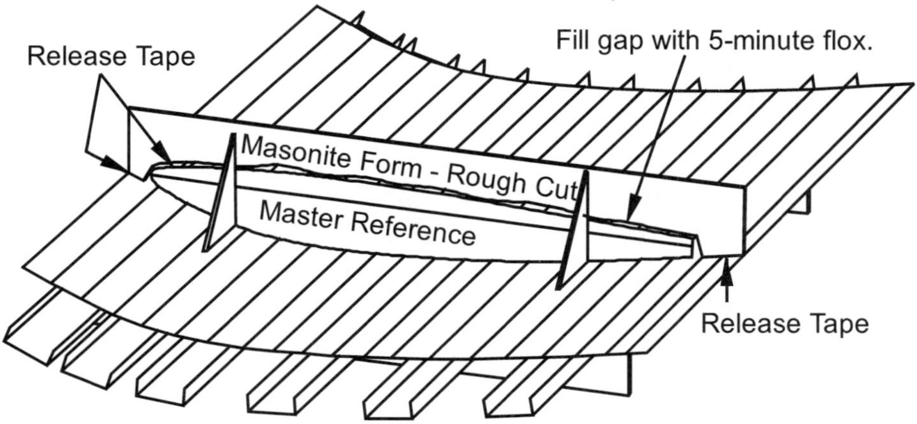

Transferring the Reference Line for Closing

The transfer jig determines the ***ideal outside line*** of the closing skin and ***does not allow for skin thickness.*** The inside skin thickness can be approximated by gluing poster cardboard to the transfer jig. After the ***inside line*** has been determined, cover the Masonite form with brown release tape. The transfer jig now accurately carries the desired location of the flange.

Setting the Flange Line

Make "glass flanging aids": Find a flat scrap of cured BID glass that is thick enough to be reasonably stiff (two or three layers of 7725). Use a bandsaw to trim out strips about a half-inch wide. Cut into sections about 1 inch long. With the transfer jig in place, use 5-minute epoxy to place a line of glass flanging aids between the jig and the surface to be flanged, at intervals of about three to four inches. Check to see that the flanging aids do not creep out of position. Cure. ***This method defines a very accurate and clean bonding line.*** (See Exercise 18 for a detailed example of specific techniques.)

Flush Flanging Tools

The simplest flanging tool for a ***curved surface***, such as a rib, is a smooth strip of bendable material, such as 1/8-inch Masonite or a strip of cured fiberglass, which has had ***mailing tape applied to one side to serve as a release film.*** The completed flange will be used in a secondary bond, so it is desirable to use **peel *ply over the bond area*** that will be formed. Cut the peel ply about 1/4-inch wider than the flanging tool.

(The adhesives hold sufficiently well for the purpose, even though brown tape is a good release surface.) For a ***straight surface***, such as a shear web, almost any straight scrap wood, such as chipboard, will serve.

Assuming that the bonding surface has been defined, as above, the flanging tool is placed over the defined line and held in place with strips of wood or Masonite. Hobbyist's hot-melt glue is an excellent quick-setting adhesive to hold the flanging tool in position. Bondo is also satisfactory. Sandbags or other convenient weights are used to hold the tool in position. Care must be taken to verify that the flanging tool is not tilted from the desired line. See the sketch.

Tint and wet out tape in the usual manner. Place the wet tape between the flanging tool and the existing surface and smooth the edge of the tape with peel ply, as with a regular corner joint. Apply the tape and peel ply over the "glass flanging aids." Cure. Remove the flanging tool and strip off the peel ply. Trim wild edges, as necessary, with a spiral saw or Dremel tool.

> ***See Exercise 18 for a detailed description and example of these techniques applied to close a wing structure.***

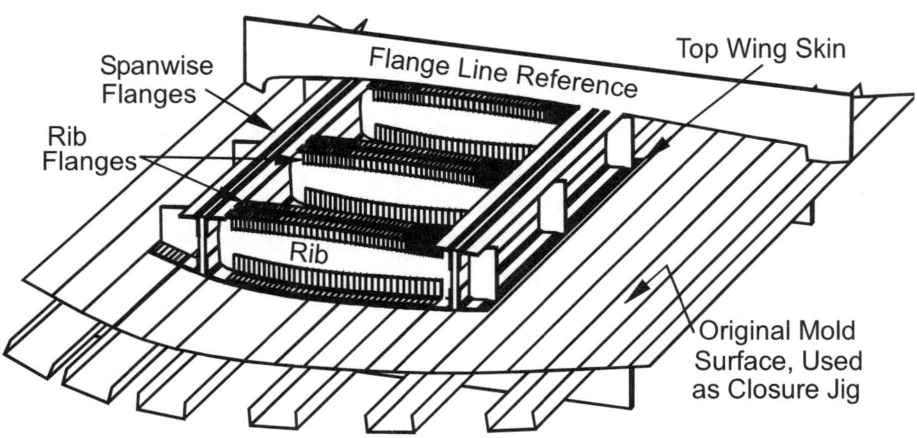

Closure Operations

The figure above is a view which summarizes the techniques that have been described. See that the mold that formed the upper wing skin now serves as an accurate jig to position all the component parts. The Masonite ***flange line reference*** determines the position of the flanges. The ribs are secured to the upper wing skin with glass tape, as previously described.

> *Note that there may be considerable uncertainty in the exact shape of the ribs: It may be difficult to determine exactly how thick the skins will be, especially if the skins taper in thickness and have local variations due to overlapping glass layers during the layup. An accurate fit is not critical, however, because the tapes will carry the skin loads. The builder must be careful not to allow unfilled gaps between the skin and the rib.*

Offset Flanges

It is sometimes necessary to make flanges which are ***offset by a precise distance from the edge of the supporting surface.*** An example is placement of a flange on a wing rib to support a thick skin panel exactly flush with the edge of the rib, so that the form of the rib will be preserved.

There is considerable room for builder's ingenuity in making the flanging tool. Basically, the flanging tool must have a slot which exactly fits the supporting surface and has exactly the depth of the desired offset. The face that forms the flange must have a release film of some kind and that release film must not buckle in compression when the thick flanging tool is curved over a concave support. Our old friend brown mailing tape does not work here, because it buckles in compression, but duct tape is soft and often works well. *(Duct tape adhesive will clog a table saw!)*

Cheap, non-structural, insulating foam is very convenient for making offset flanging tools. A ***table router*** is convenient for cutting the required slot in a single pass. If no router is available, a ***table saw*** can be used for sawing out the slot to the ***desired width in successive passes.*** The strip of foam can then be placed over the rib and held in place by the slot. The soft foam will conform to a fairly short radius, such as the leading edge region of a wing. One alternative is to put the slotted tool in place, secure it in an appropriate manner, then add the release tape.

When working with slotted flanging tools, is very important to assure that the tool is firmly bottomed in the slot. Any error in positioning the tool will produce exactly that error in the offset and probably require scrapping the part.

Another variation is to form the slot by gluing a strip of foam to the backing strip that is exactly the thickness of the offset. This scheme allows greater curvature.

A useful trick: Flanging tool stock can be mass-produced on a table saw by first gluing a sheet of marked peel ply, at 45 degrees to the weave, to a sheet of 3/4-inch ***non-structural foam*** with 3M Super 77 Spray Adhesive. The ***bias-cut*** peel ply will follow an appreciable curve without buckling. (Note that the crayon-marked side of the peel ply must be ***toward the foam***.) A batch of flanging-tool stock can now be quickly cut out on a table saw. The builder should make a test before making too large a committment to this method, because it is essential to get a clean cut at the edge of the peel ply. A fine tooth saw and a slow feed may be required. See that ***release film is not used*** here. The flange layup is made directly against the peel ply layer and incorporates the peel ply. The expendable flanging tool is destroyed after the flange has cured. The marks on the peel ply become visible to remind the builder to remove it, leaving a clean bonding surface.

Prefabricated Flange Stock

Wherever the flanged joint is a ***straight line***, work can sometimes be saved by using prefabricated flange stock. ***Aluminum "ell" stock*** in 1-inch and 3/4-inch sizes can be purchased at a home supply store and ***faced with brown mailing tape***. Peel ply and 7725 bias-cut BID scraps can be laid up over the mailing tape. Allow to cure to the "green" stage and trim along the aluminum edge. The cured glass "ell" will strip freely from the aluminum mold and can be cut to length as needed. *Do not forget to remove the marked peel ply when installing the flange stock.*

*Note that the use of **prefabricated flange stock** requires the use of true **structural adhesive** bonds between the previously cured surfaces.*

Closing Operations

Trial Fitting

It is very important to ***assure that the two surfaces to be bonded are accurately located.*** The emotional shock in discovering that an entire wing skin is permanently assembled with an error of 1/4-inch is unacceptable to most builders. Depending on the application, it may be necessary to make up an assembly jig to measure alignment. In other applications, it may be sufficient to align the parts and mark the final positions with a marking pen. Position the parts as accurately and firmly as possible, using clamps and/or duct tape to hold the position. Choose two to four "registration holes," as suitable, to precisly locate the parts, and fix them with wide-head self-tapping #6 screws. ***Mark the registration holes*** with distinctive circles, on ***both pieces*** to be bonded.

Note that this cored-skin construction method produces very rigid subassemblies, especially if many layers of glass have been used to meet compression and tension specifications for a wing. This is not a problem for the upper wing skin and the associated internal structure, which have been cured in an accurate jig. Presumably, the builder has attached the internal structure using sandbags or other weights to force the components against the mold/jig. The bottom wing skin is often very nearly flat, but rarely exactly flat. The range of bending adjustment when fitting the bottom wing skin will be small, and the builder ***must not assume that a significant curve can be forced into or out of the cured bottom wing skin.***

Bonding the Closure

Replace the pieces to be bonded and drill a line of "pull up" holes for #6-20 x 1/2-inch Washer Head self-tapping screws at intervals of 3 to 4 inches all along the bond line. Mark the "pull up" holes with distinctive "x"s. Separate the pieces, vacuum out the drilling debris, and ***open the registration holes in the part to be bonded by one drill size..***

Estimate the amount of wet tape necessary to cover all of the bond lines in the structure. ***Tint the epoxy with a chalk color opposite to that used***

for the wing skin and wet out a single layer of 7725 BID, as with regular tape. Place the tape.

With assistance, if necessary, place the parts together. Weight the surfaces to engage the parts. ***Inspect the bond lines to see if full wetting has occurred.*** The different-colored bond area can be seen through many layers of cured glass. Wherever there is a gap, insert the #6 s-t screws and pull the surfaces together. (An electric drill driver will be convenient, but be very careful to avoid stripping the fragile holes.) Continue, as necessary, until all of the bonded area shows wetting. If a screw is not necessary in an area, do not use one. Cure until half-hard and remove all the screws.

Fewer "pull-up" screws and some weight may be avoided by using ***structural adhesive paste instead of wet glass tape.*** If structural adhesive is used, it should be applied somewhat generously and smoothed out with a notched spreader as used by linoleum layers. Excess adhesive will be squeezed out of the joint and ignored. The dark colored adhesive may be inspected easily at the time of assembly. The sketch below shows an exploded view of a cored-skin wing section.

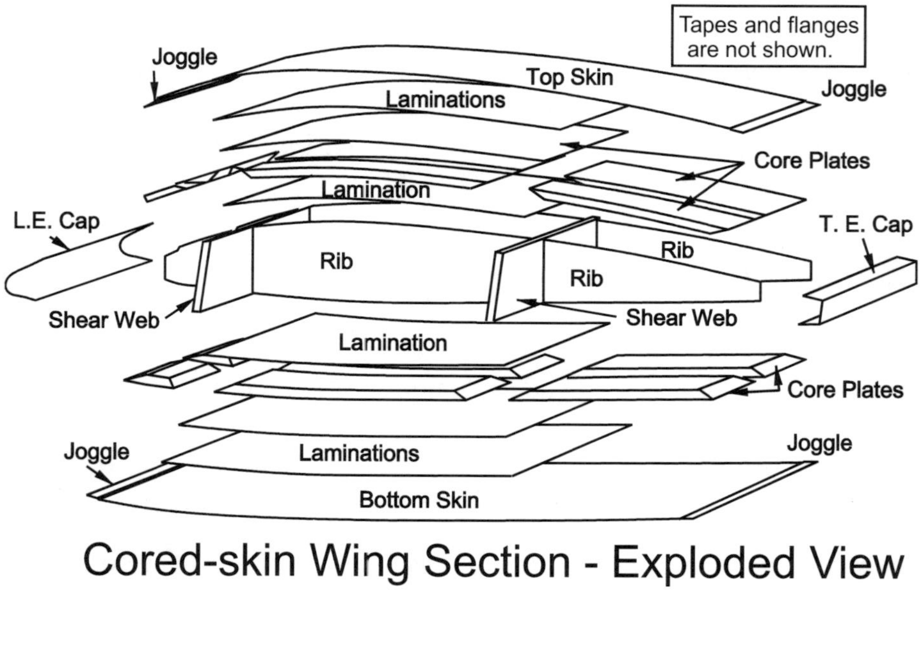

Cored-skin Wing Section - Exploded View

The "wet tape" bonding with the "pull-up screw" method is quick, easy, secure, and allows a long working time. *Note that the working time will be substantially shorter if a typical structural adhesive is used. If you use a structural adhesive, plan the assembly operations carefully so that you will not exceed the allowable time limit.*

Chapter 11

Attachments

Load Transitions

Load Matching

When we hang clothes in a closet, we do not hang the flexible garments from a single point, like a nail. We prefer to reduce the stress by distributing the load with a familiar load-matching structure called the ***common coat hanger.***

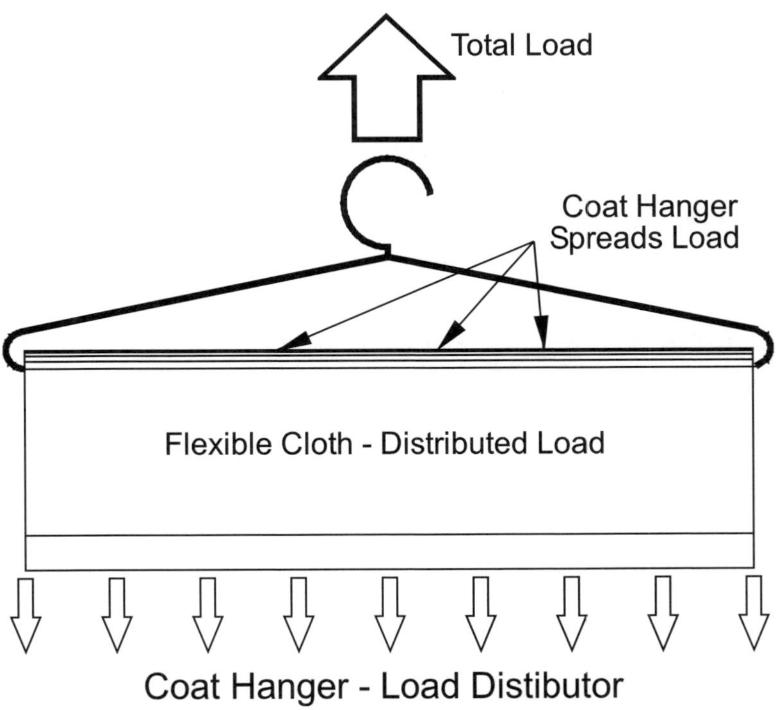

Coat Hanger - Load Distibutor

Assume, for a "thought experiment," that the experimenter wishes to suspend a load of several hundred pounds from a nail and requires that the load be carried through one of the light test panels described in Exercise 3. There is plenty of glass area across the full width of the panel to carry the load. If the experimenter just drills a nail hole in the margin of the panel and applies the load, the thin glass margin will fail from the high local stress, as the steel nail cuts through the glass margin. One solution might be to bond a rigid strip of steel across the edge of the panel, then drill a nail hole in the steel strip. The nail and the glass margin never "see" each other; stress is carried evenly to the edge of the glass, and the load is suspended successfully.

The real problem is the design of an efficient, light-weight, load matcher.

The Problem of the "Perfect" Bolt

Suppose the builder had the "best bolt in the world," of some unnamed exotic material, but extremely light and strong, and she wished to impress her friends by using this wonder as a wing attachment bolt for a composite wing.

How might the builder safely use such an ideal bolt? First, some kind of perfectly fitted, ground and tapered steel bushing might be used to get enough outside surface area to spread an allowable stress to a second, considerably larger, bushing of hard aluminum. The aluminum bushing might then be embedded in a larger pad of a strong core material, perhaps phenolic, or wood, or thick laminated glass. This large area pad is then bonded very securely to a strong skin that spreads and tapers to join the working skin over a wide, low-stress, area. Obviously, the weight of the necessary "transition" stucture must be charged to the total weight of the attachment.

Plans for some current homebuilt aircraft show bolt attachments which go directly through the thin skin and into an embedded bushing or other stress-spreader. The fact that the edge of the bolt is shown contacting the skin over a small area is incidental; ***the real intent is to attach the bolt to the underlying large pad.*** The ***large-area pad*** then spreads the load. In fact, the design would theoretically be slightly improved if a clearance hole were drilled in the skin where the bolt penetrates so that the bolt did not touch the skin at all - only the bushing.

Relation to Standard Practice

The engineering handbooks discuss the standard cases for bolted or riveted joints in more detail than most builders care to deal with. Unfortunately for our problem, in most cases the data apply to the situation where the bolts and the bolted objects are of roughly similar isotropic materials and dimensions - like steel bolts in steel plates - which is *not* our situation. We nearly always need a transition between a low-stress distibuted composite structure and a high-stress attachment device. One exception is the common case in which a light fiberglass wingtip is attached to an aluminum wing with a line of small machine screws. The fiberglass edge is drilled for bolting directly, but the stresses in this case are trivial.

Practical Bolt Attachments

Cores for Load Transitions

The sketch below is a cross section which shows a somewhat idealized scheme for attaching a bolt to a thin-skinned composite sandwich. It requires that a core material be used which can carry the high stress load transmitted by the bolt and distribute that load to a ***sufficient skin area*** to reduce the stress that can be carried by the skins.

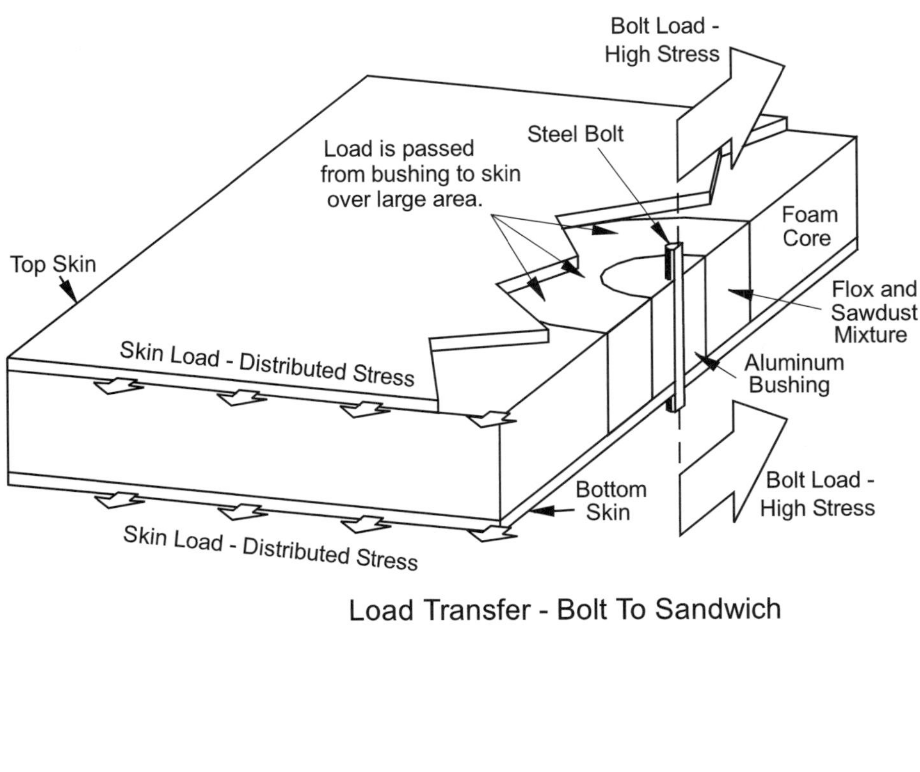

Load Transfer - Bolt To Sandwich

As shown, the required "strong" core area, a large bushing, is fitted into the main body of the core before the skins are applied. The skins are then added in a following operation and form a strong bond over the large area of the bushing. The ***bushing consists of two parts***: (1) an aluminum section, which can accept the high local stresses applied by the bolt, and (2) a second section, of much larger area, which delivers load to the skins. The hole for the steel bolt is drilled out of the aluminum core after the skin and composite core has hardened.

Practical Example

The sketch on the opposite page shows the construction of a practical variation which is very easy to build. The builder simply saws out a hole with a spiral saw (to avoid delamination) in the completed sandwich sufficiently large to accept the bushing, places a core of aluminum in the center of the hole, and "pots" the aluminum core in a mixture of epoxy, flox, and sawdust. The ***"potting" mixture hardens in place to form a pad of large area which joins the rough edge of the core perfectly and bonds very well to the skin.*** The bolt hole is drilled out after the entire assembly has cured, so minor errors of alignment are removed. The aluminum core can be slightly smaller than the thickness of the main foam core, because the entire core can be embedded in the potting mix, which saves the the builder the work necessary to make a perfect fit. As shown, the cutout is in the rough "as-cut" condition, without any special cleanup.

> *The strength of this transition is improved somewhat by placing a small piece of glass cloth in the wet filler before placing the aluminum bushing, which fills the cavity, then placing a second piece of glass over the bushing and cavity before closing.*

One easy way to do the potting is to cover the entire opening of the cutout hole with brown mailing tape, place the work on a supporting surface, pour in the mix, and smooth to level. The exposed sticky surface of the tape can be used to hold the aluminum core in position. After cure, the tape is removed, leaving a perfect fill, ready for skinning.

The sketch shows two layers of glass used to close the attachment, which is somewhat irregular in shape. If multiple layers are necessary, they should be tapered to avoid stress concentration. The irregular edge also helps to avoid stress concentration.

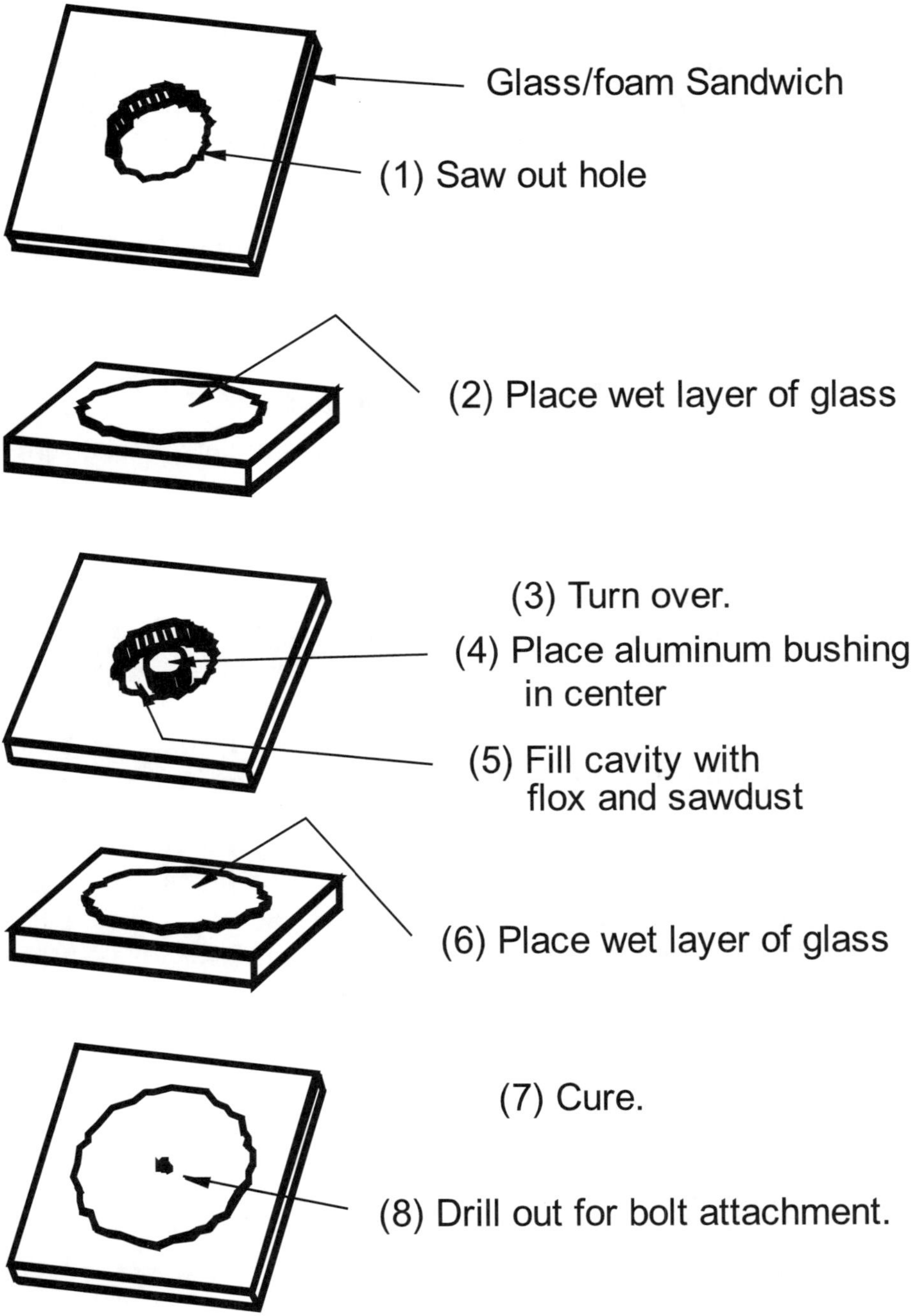

Bolt Attachment Sequence

*The "potting" method of making the outer bushing requires that the uncured potting material be fairly thick, approximately the same thickness as the main core, and considerably thicker than normally encountered in a layup. This introduces the **risk of exotherm** if the builder is working in a particularly warm environment. The author has not encountered exotherm when working with half-inch foam with mixes of about one quarter mixed epoxy and the rest flox and sawdust, but strongly recommends that a simple experiment be made to get a feel for the process.*

The "Skin to Pin" Transition Problem

The *Crosswind* wing described previously makes use of a "torque box" as the main load-carrying structure of the wing, following current design practice for modern jet aircraft. The classic strong spar, as a narrow separate member, is not used. The flight loads are carried by the wing skins in tension and compression and the two shear webs which complete the "box". In the case of the *Crosswind*, this creates the problem of bringing the distributed loads in the skins and shear webs to a single strut attach point near the middle of the wing and two attachment pins at the wing root. The strut attach point must therefore be located *between* the shear webs, not on a spar as in the original *Tailwind*.

Ideally, the strut might be attached to a perfectly rigid rib which then carries the load to the skin area without deflection over its entire circumference. The situation is similar to the close-fitting plastic plugs used at the ends of cardboard mailing tubes. A thick steel plate in the shape of a rib would transmit the load evenly across the skins, but would be prohibitively heavy.

The Stub Web

A more practical alternative for strut attachment is to use a short, stiff, "stub" web which transfers load to the skin across a large span-wise "footprint." The sketch on the next page shows the center section of the *Crosswind* wing, with the bottom wing skin removed. See that the spanwise strap on the steel attachment acts to distribute the load along the stub web and carry the distributed load into the flanges. The stub web, as shown, is also at the most heavily loaded part of the wing and shares part of the total shear load, as well as passing tension and compression forces to the skins through the flange "footprint".

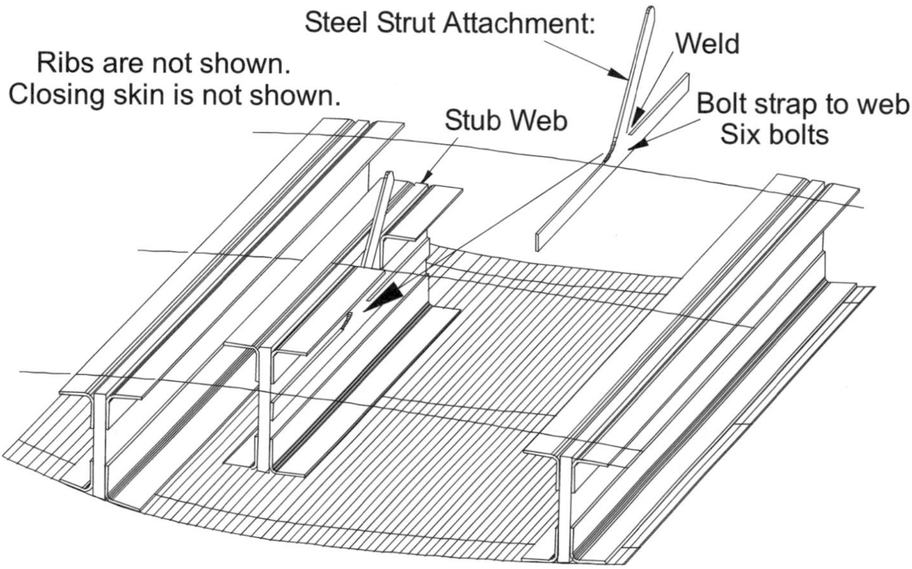

Strut Attachment Example

The wing root attachment problem in the *Crosswind* is similar except that the root rib is small and has two attachment points. Also, the attachment points as originally designed do not correspond to the locations of the new shear webs. The design requirement is to transmit the attachment loads directly to the skins from the small rib. One way to do this would be to use double layers of edge tape between the skin and the root rib and bolt steel plates to the root rib to carry the actual attachments which would be welded to the steel plates.

"Synthetic Spruce"

Many classic aircraft designs, including the *Tailwind*, specify aircraft-grade spruce wood for major structural components, such as the main wing spar. In this application, the spruce is usually a simple beam of rectangular cross section, drilled out for attachments, as required. Load spreading at attachment points is provided by using several bolts through steel fittings on each side of the wooden beam.

Engineering handbooks show the density of aircraft spruce at normal moisture content to be about 20 to 22 pounds per cubic foot. It is easily possible to exceed the performance of a spruce spar by replacing it with a built up composite spar of the same or smaller dimensions. The

Tailwind uses a main spar which is 1 1/4-inches wide, 4 3/4-inches deep, and 120 inches long, tapered in depth at the wing tip and sharply tapered at the wing root. Steel attachments are provided at the root and at a strut position in the middle of the main span. (The *Tailwind* wing also uses a plywood skin which carries significant flight loads in addition to the loads carried by the spar. This example will be confined to replacing only the spruce spar.)

One approach for replacing the spruce main spar in the *Tailwind* is as follows:

1) Cut and assemble a 1-inch thick, 4-inches deep, spar core from 2-pound urethane or polyurethane foam. Taper at the tip and root as called for in the *Tailwind* plans. Butt-join separate pieces as necessary.

2) Identify areas where attachments are required at the root and strut. Dig out the foam in a generous area around the attachments (about 3 by 6 inches, with at least a 1-inch radius on the corners) and fill flush with the surface with a paste mixture of 1/3 micro, 1/3 flox, and 1/3 coarse sawdust. Use peel ply on both sides. Cure.

3) Round off the top and bottom edges of the core to form a semicirclular cross section. Wet and lay a single layer of 7715 UNI glass, 3 inches wide, along both rounded edges. Smooth with peel ply and cure. Remove the peel ply.

4) Cut, wet, and lay two layers of 7725 BID on one side of the core, so that the glass overlaps the rounded edge. At the same time, add two extra layers of 7725 BID to generously cover the attachment area. Smooth the edge with peel ply. Cure. Remove the peel ply. Turn over and repeat for the other side of the spar. Cure. Postcure.

5) Drill for the attachments as called out in the plans. The open layup (not vacuum-bagged) described here will have a more irregular surface than milled spruce, so it will be necessary to embed the attachment hardware in wet flox at the time of assembly.

The result is essentially a box beam with a total of five layers of glass in each spar cap. The necessary shear web is a sandwich between two 2-layer glass surfaces. This composite beam, as described, will outperform the spruce equivalent. See also that some weight may be saved by recognizing that wing loads drop off sharply near the wing tip, so a single layer of shear web glass should be sufficient in the wing tip area.

With composite technology, the opportunity exists to "improve" on the original design in "22-pound" spruce. For example, the largest loads are in the immediate vicinity of the strut, so the builder may wish to add more glass in this critical midsection area and use less glass in the low stress areas, but if a plain spruce beam was proven sufficient for the original design, there is no benefit to overdesigning. And - obviously - a proof test is required after completion.

Chapter 12

Single-Curved Molds

Sheet Molds

The home builder cannot make the large investment in time and money needed to make tooling for the complex double-curved parts typical of commercial high production applications. As discussed in Chapter 6, many practical aircraft structures can be constructed from single-curved surfaces that are ***generated from straight lines, i.e. elements of flats, cylinders, and cones.*** In fact, most commercial aircraft make extensive use of flat-rolled aluminum sheet material that is formed to a single curve.

The previous chapter discussed methods for making flat panels that are very stiff and light and have an excellent surface finish from a smooth tool. This chapter will discuss ***extending flat tool methods*** to form large single-curved rigid structures with very simple tooling. The basic problem is how to produce an accurate, smooth, vacuum-tight mold from a flexible sheet. The main aircraft application will be forming skins for flight surfaces.

Bent Sheet Mold - No Plug

A very good single-curved female mold can be made by forming a thin sheet of smooth, vacuum-tight, flexible material into the desired shape. Aluminum, galvanized roofing steel, plastic shower liner, Formica, or thin fiberglass formed on a flat tool can all be used. The mold surface must support either conventional mold release compounds or release films as described in Chapter 6. The process is: (1) make an accurate bending form, (2) bend the mold sheet exactly to the form and fix the shape in some manner, (3) remove the formed sheet from the bending form.

The tool design problem is how to bend the sheet to an accurate form, support it, and then firmly fix the sheet in position. The builder will find that a short radius curve, such as near the leading edge of a wing, is easy to keep on a straight line, but possibly hard to bend; and a long radius curve, such as near the trailing edge, is easy enough to bend but tends to sag between supports. Selecting and supporting a sheet material will require some ingenuity by the builder, but fortunately the mold will not have to fly and whatever additional reinforcement is required will be much easier than making a conventional mold.

Exercise 14 describes in detail the specific case of bending plastic shower liner material over a rib-shaped form supported by a frame of steel studding material.

Bending Form

The bending form must accurately define the desired curve and be strong enough to carry the forces needed to form the sheet. The builder must work out the details for each case, depending on the material chosen to make the mold, the size, and the amount of curvature required. For a wing skin, the bending form can consist of a series of chipboard rib forms at intervals of about 18 inches. Again, see Exercise 14 for an example.

The sketches on the following page show a simple frame made of steel studding to support four bending forms.

Mold Sheet Material

The hard-pressed composite material sold for ***kitchen counter surfaces,*** under trade names such as ***Formica,*** is an excellent molding sheet. The surface is smooth and the combination of stiffness and flexibility permits simple supports to hold it in the desired shape. The rear surface is textured and ***lends itself very well to secure and accurate bonding.*** Problems with it include that it is quite expensive, brittle, and is not easily available in large sheets.

Thin sheet steel, like the kind sold for roof flashing, has been used. Home supply stores carry the material in limited widths, but wider pieces can be had through heating and air conditioning stores. It may be useful for short radius applications such as leading edges. Bonding to formers may be more difficult than with Formica, but satisfactory adhesives exist.

Sheet aluminum can be used and may have the added advantage of being conveniently available in an aircraft environment.

The author has used ***plain 1/8-inch Masonite*** which has been ***painted with pure epoxy*** and lighly sanded after cure. This material is easy to handle and bond, but the surface finish is only fair and is not entirely reliable for vacuum tightness. One fairly successful variation is to use ***white plastic faced Masonite,*** which has a good surface finish but is not reliable for vacuum, and ***paint the back*** with a penetrating base coat to seal it.

The author has found that an excellent compromise material is ***FRP plastic shower liner***, .090 inch thick, sold in 4 x 8 sheets as better-quality shower liner. The intended front surface is pebbled, so is useless, but the ***rear surface is quite good.*** The material is reliably vacuum-tight. Bonding to the back (normally the pebbled front) is easy if the gloss is removed with an electric sander (see Excercises 9, 13, and 14).

Shaping the Mold Sheet

The mold sheet must be ***forced evenly against the bending form by some straight-line uniform support*** that does not cause sagging between the reference forms. Another way to say this is that ***true straight lines must be preserved***, so that concentrated forces like sandbags cannot be used directly to bend the sheet.

One effective method is to use the ***rolled edges of steel studding material,*** which are very good straight lines, to force the mold sheet into position. Sandbags and other weights can be placed on the back of the steel studs to produce the necessary conforming force. Six steel studs adequately support a typical small wing section by giving twelve defining straight lines. The sketches following show the method, in which four formers are supported on a simple frame made from steel studding.

It is then necessary to secure the forming support to the formed mold sheet. After the form is bent to the desired shape, the "forming means," like steel studs, can be removed one at a time, adhesive can be applied, and the formers returned to position and re-weighted.

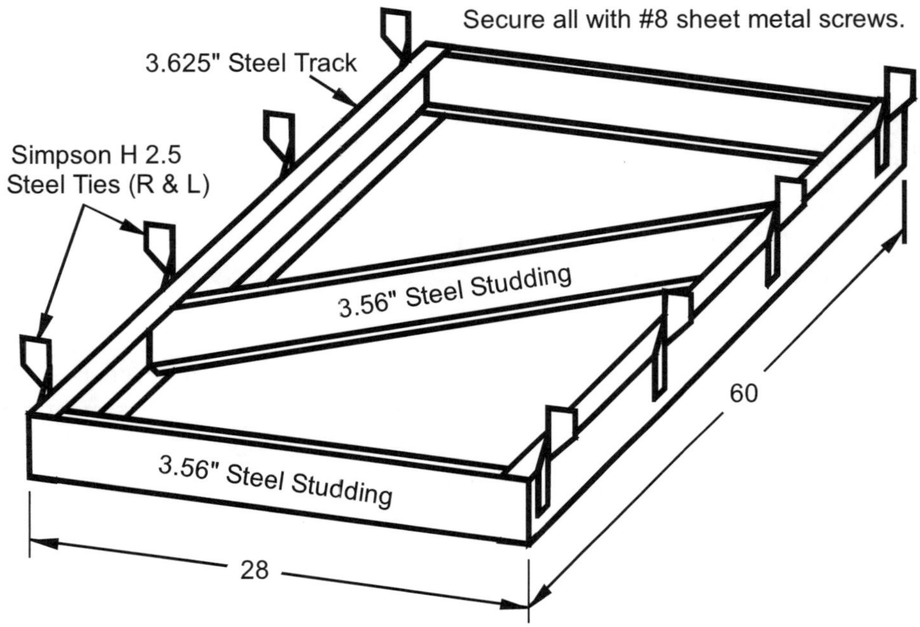

Wingskin Mold Bending Frame

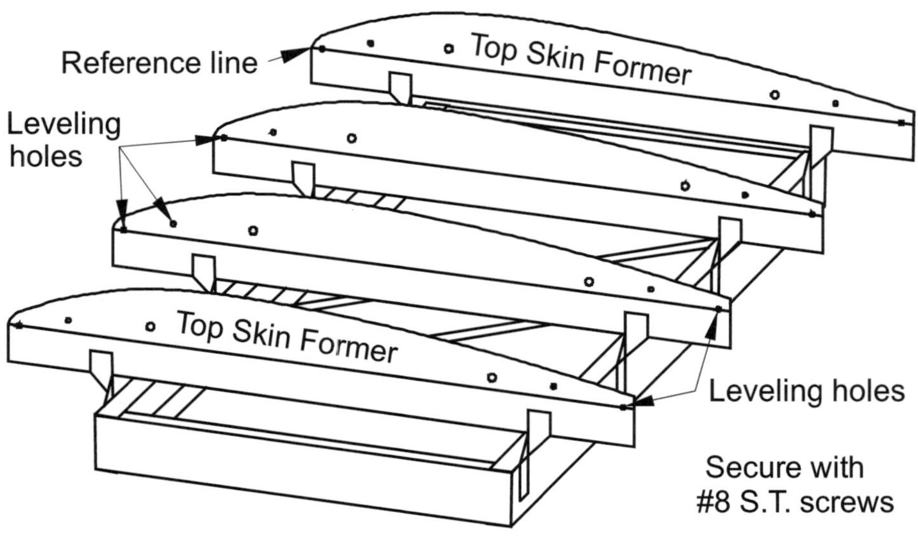

Wingskin Mold Formers Mounted on Frame

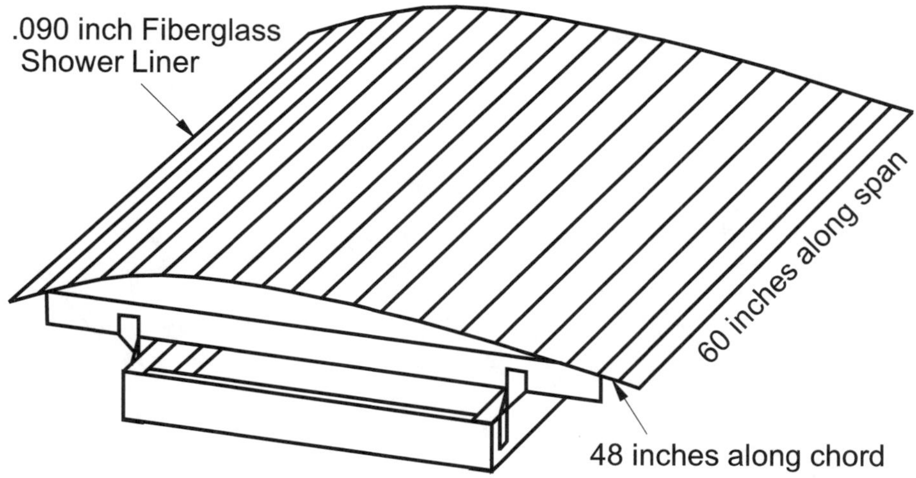

Bending Mold Sheet Over Form

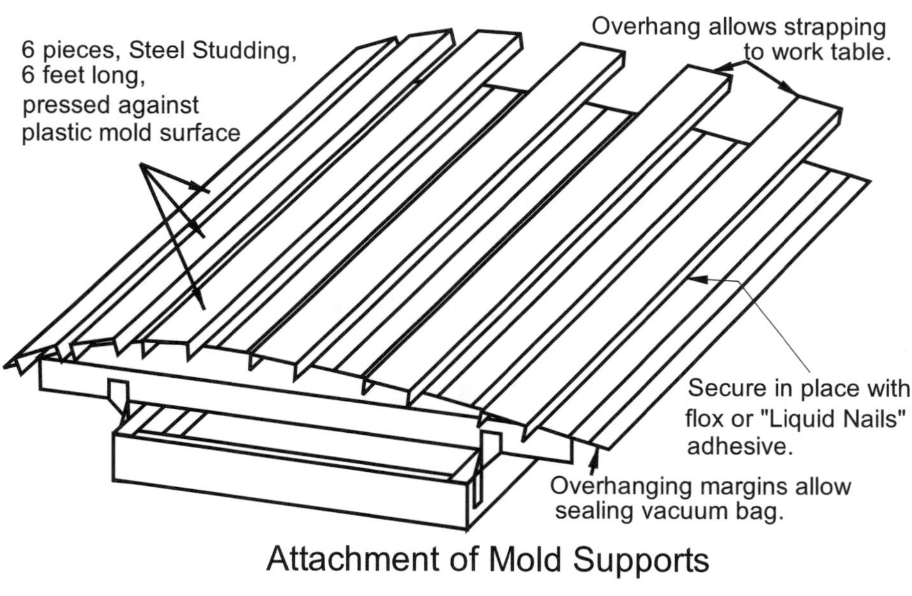

Attachment of Mold Supports

Rough wooden supports can be secured to the forming strips with steel ties to hold the curve. ***Again, see Exercise 14.*** With the supports in place, the ***completed female mold sheet can be lifted from the bending form and transferred to a workbench.*** Note that the sheet mold is very stiff in bending, but is ***not stiff in torsion***. Torsion stiffness must be provided by the supporting workbench. See the following sketch

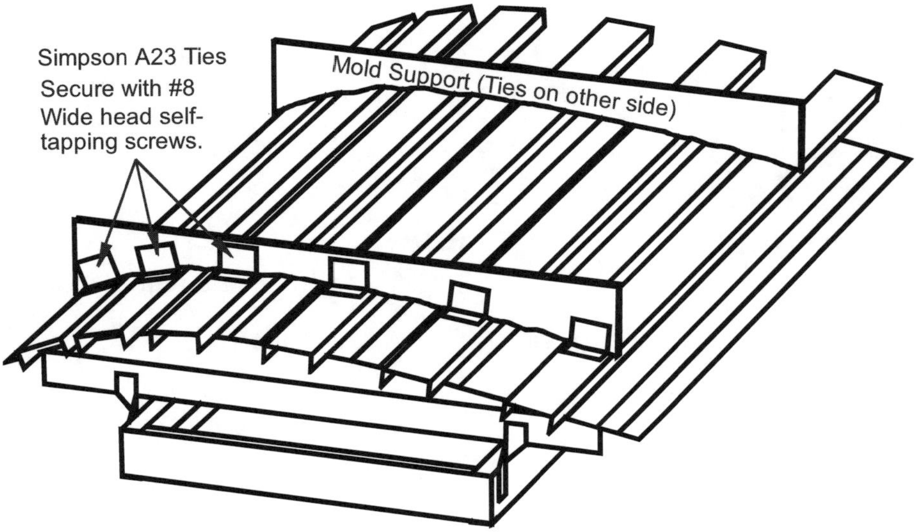

Securing Mold Supports

This method produces an accurate, smooth vacuum-tight female mold, without the necessity of making a perfect plug.

Using a Single-curved Mold

Vacuum-formed Release Film

The basic techniques for using a single-curved female mold are ***essentially the same as Flat Tool Vacuum Bagging,*** described in Chapter 6. One fortunate difference is that the ***concave curve of the mold makes it a little easier to place a smooth vacuum-formed release film.*** The film can be raised up somewhat above the surface of the mold, so there is less tendency to "pinch-off" and form wrinkles when the vacuum is drawn.

Placing Flat Core Panels On A Curved Skin

Flat, perforated, beveled core panels as thick as 1/2-inch can be coated on the bottom with slurry and placed on a moderately-curved surface.

The panels will stick in position sufficiently, even though the straight flat panel does not match the curve exactly.

> *Note that the mold-side (outer) skin must be almost completely cured before the builder attempts to add core panels. This is to prevent a small, but sharp, bending discontinuity in the uncovered skin, just before it is covered by the edge of the core panel at the edge of the panel.* ***This discontinuty will greatly weaken the structure.*** *For a small job, the builder may be tempted to do an entire layup, outer skin, core, and inner skin at one pass. The added work of taking down the vacuum system, adding the rest of the layup, and replacing the perf, bleeder, and bag is the price of high performance.*

The inner skin of the layup can be added immediately after the core panels are placed as part of the ***continuous layup process, then vacuum-bagged in a single operation.*** The bag forces the entire layup to match the curve. The builder must take care that the beveled core plates do not creep out of position while applying and wetting the inside layer. (The mold marks will be visible).

> ***Note: Do not interrupt the vacuum on the vacuum-formed tool surface before the second (inside) skin in added and cured!*** *This means that there will be two separate layups made on the tool surface, one for each skin layer, and the pump may be left running for what seems like a very long time. This is an argument for a true industrial-grade vacuum pump!*

In cases where the curve may be too sharp to bend the flat foam directly, as near a leading edge, the builder can use the slotting techniques described in Chapter 5.

Very Sharply-curved Molds - No Plug

Aircraft wing leading edges are one application which requires ***long, single-curved, accurate female molds.*** The ideal composite leading edge would be laid up in a female mold with whatever stiffeners or varying thicknesses the designer might require. The ***inside finish is unimportant,*** but the outside should have the smooth finish of a high quality molding tool. The builder could, of course, follow the classical process of making an excellent stiff plug, then transferring the form of the plug to the actual mold. By the time all this work is completed, the builder may wish that he had done it another way.

You may not be able to use the "bent sheet" method described above with a form as sharply curved as a typical leading edge, because most forming sheets cannot be bent to a sufficiently sharp curve. The method can be adapted. The general idea is to form the female mold by bending a smooth sandwich sheet of fiberglass over an open form. The ***stiff sandwich is made bendable along the axis of the form by slotting***, as previously described for unskinned foam. The sheet of slotted sandwich material is similar to a child's toy snake, which bends around one axis only, and remains stiff on the other axis.

Leading Edge Bending Form

The builder should have an accurate reference for her finished wing section. This reference can be used to produce copies of the leading edge portion of the reference. Allow at least a 4-inch additional margin beyond the finished size of the part. Half-inch chipboard or particle board is satisfactory. See the sketch below.

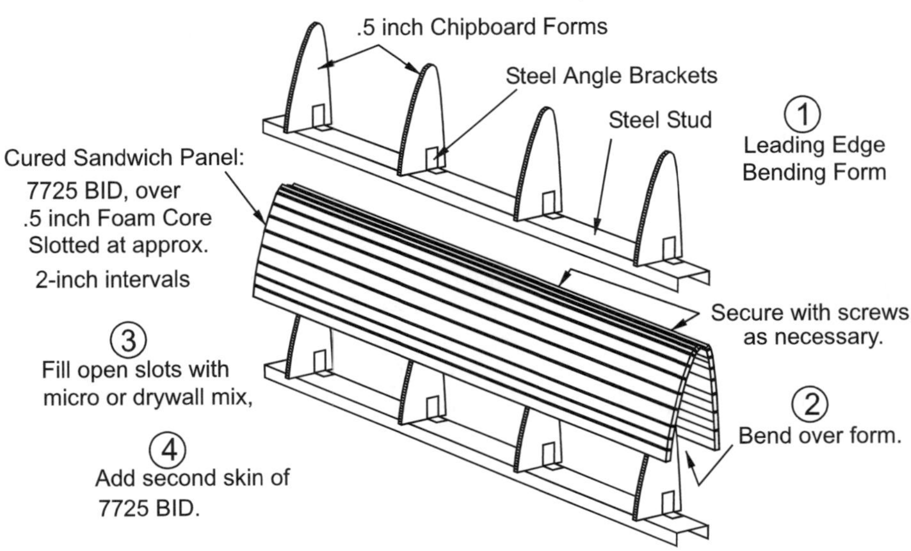

Sharply Curved Leading Edge Mold

Mount the leading edge sections on a mounting rail, such as a steel stud, as shown in the sketch, and align with a straight edge, a laser, or both. Use a generous number of sections, to assure accuracy. Spacing at 12-inch intervals will produce a good result.

The figure above shows a typical case for a wing of approximately the dimensions of the *Crosswind*. The sandwich sheet which will form the mold face should be made in the "usual way," vacuum-bagged on a flat tool to produce an excellent surface. A single layer of 7725 BID glass will be a reasonable compromise between skin stiffness and ability to make a smooth bend over the relatively sharp nose. Slotting is best done with a small battery-powered circular saw. Freehand accuracy following ruled lines is sufficient. The carbide blade is a bit wider than necessary, but the narrow steel blades are worn very rapidly when cutting glass.

The builder will not be able to see through the sandwich skin because some thin web of foam must be left in the slot to avoid cutting through the skin. Special care will be necessary to ***be sure that the sandwich fits the bending form perfectly***. The sketch shows the case in which the outer skin and most of the core is completely removed in the area of the bend around the nose. It will be possible to scrape away the core so that the bending form can be seen through the skin and the skin can be secured to the form with screws. (We accept the fact that the screws will make blemishes in the mold which will need to be repaired).

After fitting the slotted sandwich, it will be necessary to fill the open slots. Dry micro might be the ideal filler, but this mold is not going to fly, so expensive material can be saved by filling with ***drywall compound.*** An excellent smooth fill results, and the surface can be sanded easily because the first glass skin is not soft, like unskinned foam.

Wet out and apply a single layer of 7725 BID over the entire outside surface of the mold. (Leave provision for removing whatever screw were used to secure the skin -- these flaws can be repaired after the mold has cured. Cure thoroughly. The result will be a ***very stiff channel with a smooth interior surface that exactly fits the external shape of the leading edge***. BEFORE REMOVING THE MOLD from the form, make up appropriate supports, perhaps by bonding legs to the mold with scrap glass.

After a thorough cure, remove the mold from the form and repair the damaged areas where the screws were.

Using the Leading Edge Mold

The leading edge mold described above will be more difficult to use than the simple flat tools previously described. If it is sufficiently vacuum-tight, a vacuum-formed release film can be used in the same manner as with a flat tool. If not, standard mold release compounds can be used. Sections of glass can be wetted out and placed in the mold to get the required coverage. The builder will recognize this process as an open mold with definite possibilities of having uncured epoxy run down to the bottom of the mold. This "splash mold" technique, which is analogous to children's "papier-mache", but using using small sections of wet glass instead of glue and paper, is describd elsewhere as a technique for reproducing complex shapes like engine cowls.

The author has seen one case of a very complex shape,in which it was not practical to make the mold air-tight, was successfully vacuum-bagged by enclosing the entire mold in an envelope and sealing the envelope. For the very stiff mold described here, it may be possible to use an envelope bag if the support legs are omitted. The mold can be supported on sandbags after the envelope is put into place.

Chapter 13

Molding Techniques

Small-shop Commercial Practice

The mainline world of high production composite parts, typically automotive, marine, and aircraft, use molding techniques to make parts like automobile bumpers by the thousands. Tooling, materials, and development costs supporting these techniques are out of reach for the amateur. Readers interested in these methods should refer to standard engineering texts.

There is an intermediate position: Small shops serving custom race car builders, specialty boat builders, and ***parts suppliers for kit airplanes*** make use of ***simpler molding techniques*** which will be of interest to the home aircraft builder and should be understood, even if the builder does not chose to use them. Some of these techniques will be useful to the home aircraft builder for such applications as cowls, wheelpants, and wingtips. In almost every case, female molds for these applications will be made of glass composite, usually with much thicker skins than is typical of the parts to be produced, and reinforced with an external steel frame.

Composite parts made from commercial molds ***nearly always have double curves***, i.e. the surfaces are ***not defined by straight lines***, as described in previous chapters. Double-curved surfaces are stiffer and esthetically more attractive than corresponding single-curved surfaces for the same application. Even though the kit aircraft industry is typically composed of small firms, manufacturers of popular composite aircraft usually choose to invest in full-scale double-curved molds, partly to get that smooth "fast-glass" look with small labor cost and partly because they anticipate that enough parts will be sold to amortize the high cost of the molds.

General Principles

The Rough Plug

In most cases, the process starts by making a form called a ***plug***, which is the exact shape of the final molded product, and which may have a quite ***complex double-curved shape***. A good example is an aircraft fuselage which has the sleek and graceful tadpole shape so admired in many of the current popular kit aircraft designs. The plug might include all or part of the cowl form, wing roots, belly radiator shell, tail fairings, etc. The plug is essentially ***sculptured from a very stable material*** that lends itself to shaping. and can be given a ***very smooth surface finish***. Many materials can be used for the supporting structure of a plug. The author has seen at least two commercial plugs for current kit plane fuselages that used ***reinforced concrete,*** somewhat like for a swimming pool, as the supporting layer. These plugs required a very heavy steel frame and must have weighed well over a ton.

Occasionally, it is practical for the builder to ***reproduce an original part*** by using the part itself as a form. In these cases, the finished prototype can serve as its own plug and considerable work might be saved. One example might be copying an engine cowling or fairing from an existing aircraft. The discussion below will assume that it is necessary for the builder to work from scratch. In principle, it is practical for a builder to go to an aircraft salvage yard and obtain used wings or other parts and use them as plugs to make molds. The presence of rivet heads and holes will generally require quite a bit of repair, but the trade-off may be worth it.

The basic shape of the plug may be formed by using a series of wooden formers, or station templates (essentially cross sections of the desired shape) held in place on a firm supporting frame. Steel lath may then be bent over the formers. Chicken wire and/or metal lath may be added as needed to define a rough surface slightly smaller than the finished surface. Common plaster or a wide variety of commercial wall repair compounds is then used to cover the lath. The next challenge is to carve, surface, and finish the cured plaster, which is a very hard, slow, nasty, dusty job. (The author has experimented successfully with the use of ***"lite" sandable drywall finishing compound, which is much easier to***

shape than plaster, but not as strong.) However the plug is formed, a very sturdy means of support must be provided, usually some kind of steel frame, which may be attached to appropriate wheels to facilitate movement around the shop.

The "nearly final" surface is achieved with an easily-shaped material that lends itself to an accurate finish, like drywall compound or polyester auto body filler. The "final final" surface is auto body filler. The builder will find that there is a nearly exponential relationship between the cost of a filler material and the ease of use. Basic automotive "body filler," like Bondo, is sold everywhere, is fairly cheap, is a good workhorse filler, and sands out fairly easily. The material sold as "Non-Clog Lightweight Filler" costs somewhat more, but shapes quickly enough to easily justify the added cost. "Polyester Glazing Putty" costs enough to get the builder's attention, but is appropriately used in small quatities to adjust and clean up a surface with a minimum of work. ***All these styrene-based materials are compatible with each other;*** the differences are cost and workability.

Foam-based Plugs

Prototype and custom shops often base plugs on shapes carved from very cheap expanded styrene foam, the same material used for coffee cups and shipping wine bottles. The soft blue or pink styrene foam sold in slabs in home improvement stores for heat insulation is a little easier to work with and is cheap enough. These foams have the great advantage that ***they can be rough-shaped very quickly with a hot-wire cutter, then trimmed with a "cheese grater"*** tool to a form which is close to the final shape. Blocks of foam can be hotwired or cut with a bread knife and glued together as needed with urethane glue or contact cement. Very large forms, like sport car bodies, can be made by this method.

Polyester body fillers cannot be used directly on styrene foams. The foam is instantly dissolved. This is not really a problem, because additional rough shaping is almost always necessary and a thin coat of ***drywall finishing compound*** will protect the foam from the surfacing compound.

An alternative protection for the carved foam plug is to ***paint it with epoxy slurry containing a generous amount of flox.*** The resulting hard shell will facilitate handling while finishing with styrene-based body fillers. This alternative will be useful for small parts like air ducts.

Plug Surface Finish

After the builder has carved and filled to reach the perfect "rough" surface form, it is necessary to prime-seal the filler, then sand and polish, test paint, then polish and sand again until the smooth and accurate "final" surface is reached. The eventual ***final working surface will be a very smooth layer of shiny and fragile white lacquer.***

The following will describe only one of several different processes that are described in the literature. Given that the amateur home builder is probably wasting his time even considering this massive undertaking and can pursue the details elswhere, the description will be brief.

Surface finishing starts by sanding with ***#120 grit*** sandpaper and using the sense of touch to detect small differences in surface level. (It is quite amazing what a good differential micrometer the human hand is!) The worst low spots can be filled with a very thin coat of "glazing putty," applied with a squeegee, and sanded out again.

After the "touch" process (and the builder) is exhausted, the surface is sprayed with a ***thin coat of black "indicator paint."*** (Such paint is sold in the trade for the purpose, but cheap spray-can quick-drying ***lacquer*** is satisfactory.) Repeated sanding off of the paint will take off high areas and identify low areas. If the process is repeated three or four times, the desired surface can be arrived at fairly quickly. (We have assumed that the plug has no significant "straight line" areas, so the "smudge stick" technique for identifying high spots is not applicable.)

After the polyester surfacing material has been brought very close to the final form by the above process, the raw plug is washed with acetone and painted with multiple coats of ***lacquer-based primer***, wet-sanded, followed by a heavy coat of ***glossy black lacquer***. The following stages essentially consist of ***shaping and forming the black lacquer.*** Sanding proceeds from ***#240 grit, through #400 grit***, and requires at least one ***repeat painting with black lacquer*** and resanding. The process is finished with several final coats of ***white lacquer*** applied at the #400 grit stage and ***wet sanded with #600 grit*** to leave the final white surface.

At this point, the home builder may wish to think again about the rewards of doing all this for a one-off job.

Making the Mold

Surface Preparation of the Plug

The thoroughly dried surface of the plug must be completely coated with ***mold release wax*** and buffed and polished out, using at least three coats, preferably five. Usual conservative commercial practice is to then spray the plug with ***several VERY THIN coats of PVA***, a commercial water-based mold release compound. Upon drying, the PVA leaves a very good (but not quite perfect) shiny green suface which will ***almost*** certainly allow a clean release of the mold from the plug.

The Tooling Gel Coat - Polyester

*In commercial molding practice, **USING POLYESTER RESINS,** the plug is then **sprayed** with **tooling gel coat.** As discussed previously, any gel coat **does not** contain reinforcement fibers. Tooling gel coat is harder than the gel coat ordinarily used for finished parts. Black or orange color is usually included to allow good contrast between the mold and the parts to be taken from the mold. In general, aircraft builders will be using **EPOXY RESINS, NOT POLYESTER RESINS.** The mainline composite industry **does not** ordinarily supply epoxy gelcoats.*

Gel Coat Substitution - Epoxy

Ordinary ***EXTERIOR ALKYD latex house paint*** can be used as a "gel coat" ***over a waxed plug*** and will allow a safe separation of the mold. The paint ***must be sprayed*** and will require some thinning with water. The thoroughly dried latex paint layer will be ***very thin***, but will be rugged enough to ***protect the plug and the release wax*** during a direct layup. The thin paint layer will transfer when the mold is pulled, but will polish out reasonably easily.

Laying Up the Mold

The actual layup of the mold is basically the same as any other open layup; the builder wets and applies strips of glass to the plug so that the plug is completely covered, with particular attention to assure complete wetting and ***complete contact by the first layer*** (***NO BUBBLES).*** On the assumption that the builder is working with aircraft epoxy instead of the far more common polyester, standard colored tooling gel coat will not be available. ***Tinting with carpenter's chalk***, as previously described, will work for the purpose, even though the color density will not be as great. Additional glass is built up as needed to produce the required stiffness. A generous amount of fumed silica (1/3 volume) should be added to the epoxy mix to prevent running and add some hardness. Additional thickness (and therefore stiffness) can be provided by bonding scraps of foam to the mold and skinning the foam with scrap glass, forming the usual sandwich.

> *The standard literature for making polyester molds assumes the use of gel coat and recommends that the first layer of glass should be mat (which is essentially non-woven glass felt) to avoid print-through.* ***Print-through*** *is the appearance of the weave pattern of the first underlying layer of glass due to shrinkage of the matrix.* ***Epoxy has much less shrinkage*** *than polyester and usually is not subject to print-through. Where applicable, the builder who is using epoxy has the option of using* ***extra-wet very fine satin-weave cloth,*** *like Style 120, as the first layer.*

After a thorough cure, the mold should be bonded to a suitable supporting frame, such as welded steel tubing. The finished mold can then be removed from the plug.

Partitioning the Mold

The mold must be designed so that the finished part can be removed from it after cure. This is not a problem with open shapes like boat hulls, which can be lifted straight out of the mold, but a reentrant form like an egg or a fuselage tub will require that the mold can separate to allow removal of the part. Also, the molding might include such things as small fairings which, if they were not part of the main piece, would have to be

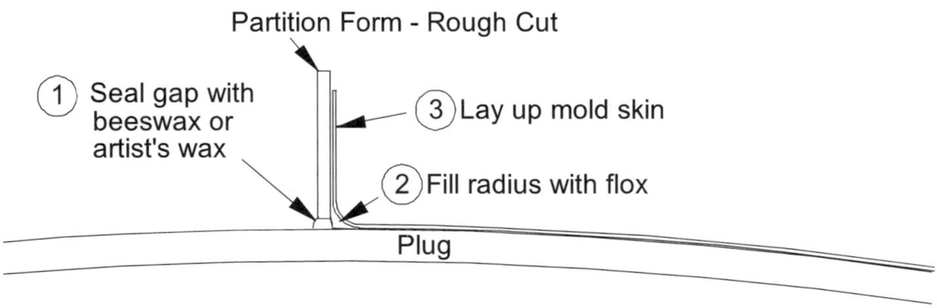

Partitioning a Mold - First Stage

attached separately, adding weight and complexity. Required partition lines will be fairly obvious, at changes of curvature and at high spots.

Partitioning is handled by adding ***matching flanges at the partition line.*** Making the flanges is a two-step process. First, a rough form is cut out from a sheet material, such as Masonite, which has been coated on one side with mold release, and attached to the plug at the partition line, using wax or some external support depending on the ingenuity of the builder. Second, the ***sharp right angle corner is filled with flox*** to provide a smooth fillet to support the layup. See the sketch. The rough saw

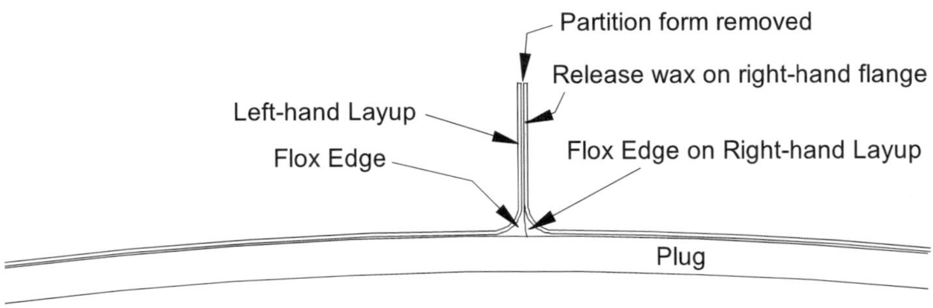

Partitioning a Mold - Second Stage

cut at the flange edge can be only an approximate fit, so the gap is filled with ***beeswax or artist's modeling wax.*** The mold section is then laid up against the plug and the flange and allowed to cure.

After the first section has cured, the Masonite flange and sealing wax is stripped off, leaving the first section in place on the plug.

The new flange is then coated with mold release compound. Again, a flox fillet is necessary to assure a sharp edge to the mold section. The second section is laid up and cured in the same way. After a thorough cure, the mold sections can be separated. Having been formed against each other, the ***two mold sections will be a nearly perfect match*** at the flange line. Even so, in real life, small differences in shrinkage will cause a slight visible line at the partition.

Mold Flanges

Whether or not it is necessary to partition the mold, it is essential to add ***mold flanges around the periphery of the mold to provide a place to attach a vacuum bag.*** The mold flange must be sealed to the mold and, typically, will be formed at the same time. There is no reason not to be generous with flange area, so a ***mold flange should be at least five inches wide.*** It is usually convenient to make a mold flange horizontal, but it is not strictly necessary.

Producing a Part from a Mold

Small-shop Commercial Practice

Production shops, even the small ones typical of the kit airplane industry (with at least one very notable exception), ***usually use "prepreg"*** materials, which are stored cold and which have a working time of many hours at room temperature. The prepregs are shipped between layers of separating plastic, like delicatessen cheese slices. The layers are marked as called out by the designer and cut very accurately through the separation plastic. After the mold is inspected and prepared, production workers peel off one layer of separation plastic and place each cut piece as required by the layup schedule. The uncured material is somewhat sticky and stays where placed, though limited repositioning is possible. The layers are built up fairly quickly and the process seems to be facilitated by C&W music in the workroom. Where called for, sandwich core material is added. After the layup schedule is completed, a vacuum bag system is rigged and the mold is placed in a ***non-pressurized oven.*** The vacuum bag is pumped down and the job is heated at atmospheric pressure to produce the cure. Very beautiful parts result.

Mainline aerospace companies may use a similar setup process, but the materials chosen may be different and curing takes place under pressure at much higher temperature. We are not going there.

Home Builder Molding Practice

This book has generally assumed that the home builder will make almost no use of standard molding techniques, but will use single-curved molds and release films almost exclusively. Still, the builder may prefer to make such double-curved parts as engine cowls and wheel pants from a complex mold, in situations where use of a release film is not practical.

In some situations, perhaps in the case of sections of an engine cowl, it may be practical to use ***stretchable bagging film as release film*** by the "double-bagging" method discussed in previous chapters. For moderate double curves, the bagging film can be ***drawn down into the cavity mold*** to form the tool surface and held by vacuum while the layup is made.

The author has seen one particular example where extraordinary light weight was required in a very complex form for a human-powered aircraft: an entire cockpit canopy and wing root fairing was made as a single molding by drawing down release film into the mold cavity. The workers were unusually skilled in this case.

For the home builder, the basic method of laying up in a mold is essentially the same as in the production shop. The amateur is not likely to reproduce the carefully planned cutting schedule found in commercial practice, but he can come fairly close by placing short pieces of wet glass tape to get the orientation and layer distribution he requires.

Exercise 1

Basic Open Flat Layup

Objective

This is a very important beginning exercise to demonstrate and gain practice with basic tools and materials that will be used many times in composite construction. In a single work session, we will lay up a four-layer flat panel section, of fairly typical size, using several typical composite materials. We will gain some experience with the different handling characteristics of common materials that we will use many times. This exercise will produce the first of a series of comparison test panels, all of 10 x 14-inch size.

Introduction

The process will be described in considerably greater detail than will be necessary for the more advanced exercises to follow, because the techniques demonstrated are so basic. The essential challenge is how to get just enough epoxy into the stack to fully wet it and get all of the air out, while keeping the several layers of the stack in the proper position. The procedure below might look a little frightening to the beginner who hasn't done it, but a beginner can do the layup in under an hour if the materials are ready. In fact, it is a very good idea to do this simple exercise at least twice - once to get the inevitable mistakes out of the way and again to get it right! Practice ***does*** make perfect, and rather quickly in the composites field.

General Description

The Workbench

All of the exercises to be described in this book can be carried out on a bench approximately 36 inches wide, 90 inches long, and 30 inches high. Smooth, unfinished particle board is convenient for the top because, later on, we will wish to bond temporary attachments to the table. (Particle board allows the attachments to be knocked off after use without splitting or raising the surface.)

The "Tool"

The composites industry refers to the molding surface that supports or forms a layup as the "tool." The tool may be a shape as simple as a smooth flat plate or as complex as a mold for half an engine cowl. For this first exercise, the tool will be any convenient stiff, smooth surface, such as a plate of sheet metal or plastic coated wallboard. We will make life easier for this first experience by avoiding the problems of "mold release" techniques and will make the layup on a piece of smooth, non-stick, plastic film.

Layer 1 is the main working layer of ***7725 bidirectional glass cloth***, (BID) 0.015 inch thick, with the weave pattern set at +/- 45 degrees to the edges of the structure ("on the bias") for optimum resistance to torsional (twisting) loads. This bias-cut material often surprises the beginner when he first tries to position it. This exercise is a good place for a first experience.

Layer 2 is ***7715 unidirectional glass*** (UNI), 0.010 inch thick, with the fibers running in the long dimension of the panel. We will presume that the designer of this panel requires the full strength of the available glass fibers in this layer in the "long" direction, and can accept the fact that the load-carrying ability of this layer in the "short" direction is zero. Note that the flat yarns of the UNI glass are held in position with light threads running crosswise at widely spaced intervals. We will ignore these crosswise threads in this layup.

Layer 3 is very light, tightly woven, ***Style 120 glass cloth***, 0.004 inch thick, cut at the 45 degree bias angle like Layer 1. It adds some longitudinal and torsional strength and produces a smooth surface to the final layup, covering the crosswise stitching of the UNI layer. It also serves to protect the layer of unidirectional glass from having the glass yarns moved or bunched up while "working" the epoxy.

Layer 4 is polyester ***peel ply***, which will be removed from the stack after it has cured. Peel ply is usually laid up in the same operation with the structural glass layers, but does NOT bond to the epoxy matrix and is NOT intended to remain as part of the finished structure. (We will use a slightly different procedure here.) In this case, we assume that we will wish to bond other structures to this cured panel in the future and need to make a very good bonding surface. Peel ply serves to protect the surface below and leaves a clean, slightly roughened surface, which makes an excellent bond. This layer ***must*** be removed (peeled) from the cured structure when the time comes to make the bond.

Detailed Procedure

1) Work Area: Prepare the general work area. Be sure that the temperature is above 70 degrees F (65 degrees F may be tolerable for some epoxies, but it considerably extends the working time).

2) Cut the layup layers: Cut the four layers of the different woven materials to be used to 12 by 16 inches, or a little larger. See that the 7725 BID and the Style 120 BID glass is cut "on the bias" at 45 degrees and that the 7715 UNI glass runs in the long direction. It is convenient to cut the bias-cut layers between two sheets of 4-mil polyethylene (PE) plastic. As each layer is cut, place it in a paper grocery bag for protection and label each bag with a marker pen. Be careful not to stretch the two bias-cut pieces. ***On the peel ply layer (4) use a red wax crayon to make four of five wiggly lines to clearly identify this layer and identify the side that carries the wax writing.*** Include some orienting symbol so that you can tell which side is written upon - like the letter "F."

SAVE ALL THE GLASS SCRAPS IN A CLEAN LABELED BAG. YOU WILL USE THEM EVENTUALLY .

3) Tool surface: Place a sheet of any smooth, flat material, approximately 3 feet by 4 feet, on the work table. (This is larger than necessary here, but we will reuse this tool surface many times.) The 1/8- inch plastic-coated particle board sold for kitchen and bathroom cabinets is cheap, light, and convenient. Sheet plastic or sheet metal would work as well. This flat surface will be the mold, or "tool."

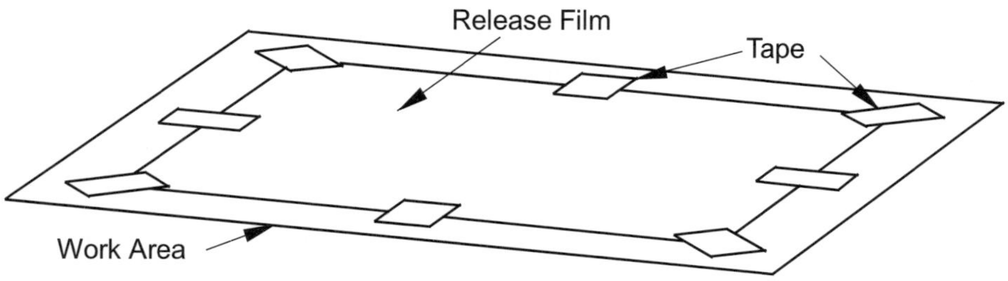

Release Film on Work Area (Tool)

4) Tool release film: Cut a piece of 4-mil PE plastic to approximately 18 x 24 inches. This is sufficient to cover the work area and serve as a non-stick release film.

Use a felt-tipped marker pan, such as the "Sharpie" permanent marker, to draw a 10 x 14-inch rectangle on the release film. This area will serve to locate the layers of the layup.

Place the smooth plastic in position on the tool with the ***marked side down*** (away from the layup). Secure in position with small pieces of brown mailing tape. Stretch the plastic and reposition the mailing tape as needed until the plastic film lies smoothly. The cheap PE plastic sold for painter's drop cloth material is perfectly satisfactory except that you may have difficulty in finding an adequate section that is not creased or wrinkled from the roll. For this exercise, a ***smooth work area at least 14 x 18 inches*** is sufficient. (If a small wrinkle remains, don't worry about it - we aren't going to fly this panel.) If the stretchy nylon bagging film used in professional composite shops is available, the wrinkle problem can be avoided. For this exercise, we will use this "release film" to avoid the problem of poor release of the cured layup from the tool, or avoid damage to the tool. (We will consider direct mold release in later Exercises and will save our mistakes in mold-handling until then.)

See that the surface of the marked work area is smooth and straight. With molds, ***"What You See"*** is exactly ***"What You Get"***, when the work is done.

5) Mix epoxy: Wear gloves! Measure out and mix thoroughly about 6.5 to 7 ounces of epoxy. Add just a little bit of red or blue carpenter's chalk to serve as an inert tinting agent. Mix again to disperse the chalk. (A very small amount of India ink can be used as an alternative tinting agent.)

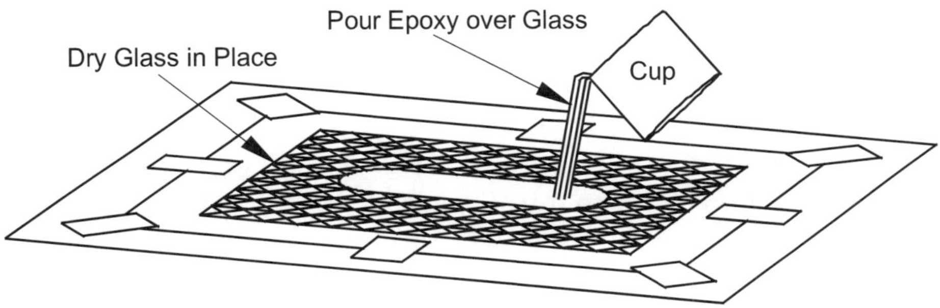

Applying Mixed Epoxy

6) Wet the layup stack - first layer: Pour a little mixed epoxy on the middle of the work area. Use a plastic squeegee to chase the epoxy around and wet the work area. Identify and place the 7725 BID (Layer 1). Be careful about stretching the bias-cut cloth. As a beginner, you will almost certainly pull one side of the BID after it has contacted the wet epoxy surface, distort the cloth, and suffer from frustration. The BID cloth may be repositioned by sliding it with the gloved hand, working from the middle of the cloth. This sounds harder than it is. With a little patience, you can slide the flattened BID into position and smooth it out.

Add a little more epoxy to the center. ***Cut a scrap of PE "spreading" plastic about ten inches square.*** Use the squeegee to work through the plastic to chase the epoxy around. Drive epoxy into any areas that contain trapped air, which will appear to be white. Drive the air out to the side and drive unused epoxy back into the layup. You can do a surprisingly complete job of driving out the air just by squeegeeing through the spreading plastic, but some air bubbles tend to remain. The roller brush and the grooved roller are very useful tools used by professional composite shops to drive out air. If you have these tools, use them, but you can do very well with just the squeegee by lifting up the spreading plastic and repeatedly pressing down on the layup surface with the edge. Use of an electric heat gun (hair dryer) is usually helpful to locally heat the work area to reduce the viscosity of the epoxy.

7) Wet the layup stack - second layer: Place the layer of 7715 UNI (Layer 2) on the work area. Use the edge of a squeegee to press the UNI down into the wet epoxy. Lift the spreading plastic and add epoxy as needed. Again, if you have a roller brush and/or a grooved roller, work out whatever air you can find. If some areas do not have enough epoxy, use the squeegee to "shovel up" excess from the sides of the work area and move it to where it is needed. If the stack is sufficiently wetted, the entire work area will appear to be evenly tinted.

8) Wet the layup stack - third layer: Place the layer of Style 120 0.004 inch BID on the stack and wet out as above. The best procedure is to unroll the layer onto the layup from the edge of the stack. There is no substitute for practice here. If the light cloth gets away from you and makes a mess, throw it out and use another one. At this stage, you should have used nearly all of the mixed epoxy. If necessary, mix and tint more.

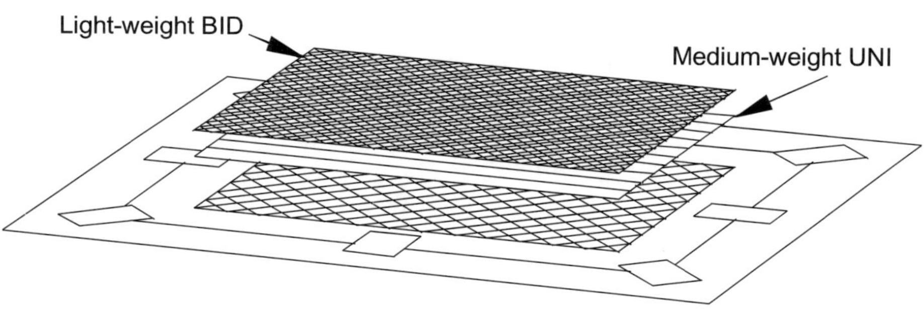

Place Following Layers for Wetting

9) Removal of excess epoxy: Use the squeegee to work through the scrap of spreading plastic to drive excess epoxy out to the sides. A little practice will soon develop the technique. The plastic is strong and you can be fairly aggressive in forcing the excess out. Any area where you have been a little too enthusiastic, allowing the stack to take up air, will appear cloudy. If that happens, chase a little more epoxy into the area until the tint is uniform. The goal is to fully wet the stack without having any more epoxy than necessary. If you need to recover, use the edge of the squeegee or the roller brush again. Ignore the ragged edges of the stack - these will be trimmed after cure.

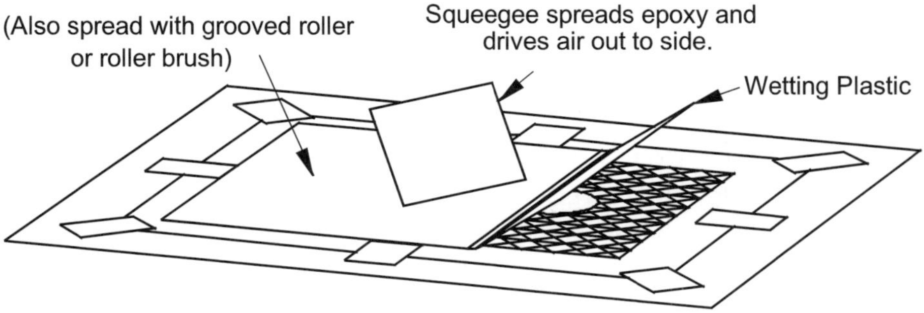

Spreading Wet Epoxy

10) Apply the peel ply: Place the peel ply layer (4) on the stack, being careful that the ***wax crayon orientation marks are right-reading*** and are away from the layup (on top). Use the squeegee to bring up epoxy from the wet stack into the peel ply. Unless you have done an extraordinary job of removing excess epoxy in the previous step, there will be sufficient epoxy from the lower layers to wet the peel ply. Be sure that the entire peel ply layer is wetted. If there is a spot with insufficient wetting, just add a little epoxy and squeegee it in.

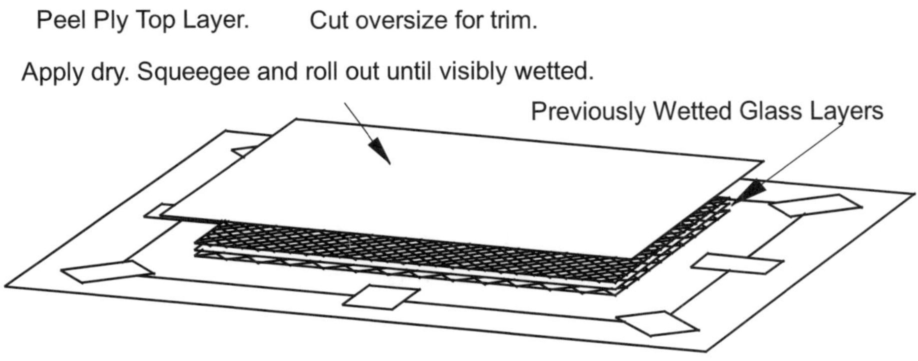

Application of Peel Ply

Please note that the procedure just given here differs from that described in most of the literature on composite techniques for the home aircraft builder. Most authorities say to include the peel ply layer in the main layup, exactly as with the structural layers. The procedure given here accomplishes the same purpose, but is a bit simpler and quicker and offers the advantage of saving some weight by removing a little more excess epoxy.

11) Cure, trim, weigh, and label: Allow to cure 24 hours. Remove from the tool. Trim to 10 x 14 inches, using a band saw or spiral saw (Appendix 1). A table saw with carbide teeth is convenient for rough trimming, but allow a full 1/2-inch margin from the final bandsaw cut, because the ***table saw tends to delaminate the layup*** near the cut. Start the edge of the peel ply with a fingernail or a razor blade and strip off the peel ply. Weigh. (A postal scale is convenient.) The finished panel, after trimming, should weigh something less than 150 grams. Use a felt-tipped pen, such as the "Sharpie," to label the panel with the date, exercise number (and trial number), and the weight.

Comments

If this is your first experience working with composites, you may feel disappointed at the results of this first effort. You will have made a mess, used up an unreasonable number of latex gloves, buried way too many bubbles, gone over the one hour time quoted, and ended up over weight. This puts all your mistakes in one place.

The cure for the disappointment is to do the exercise again, while congratulating yourself that you weren't working on an actual part called out in the plans of your composite aircraft of choice. The important thing on the second trial is to take the time to work out all the air, making more use of the roller brush and the grooved roller. The heat gun is helpful if the workroom is a little on the cool side. Your speed and efficiency will increase remarkably.

Don't throw this first effort away. In fact, you may want to hang it on the wall of your shop to remind yourself of how things were before you got good at this art.

Exercise 2

The Basic "Transfer" Layup

Objective

This exercise will extend the work done in Exercise 1 to demonstrate the important technique of assembling a layup on the worktable and transferring it as a unit to some other structure.

We will make exactly the same 4-layer layup as demonstrated previously. In this case, the transfer will be made to a tool surface, as in Exercise 1.

Introduction

The layup technique to be demonstrated will be the use of layers of plastic film to hold temporarily the four layers while the stack is wetted with epoxy. After the layup stack has been wetted out and the epoxy distributed reasonably evenly, the entire stack is moved to the tool. Again, this project should take the beginner about one hour. There is a little more work to be done than in the previous exercise, but we have had the benefit of experience!

Detailed Procedure

1) Work Area: Same as in Exercise 1. Prepare the general work area. Be sure that the temperature is above 70 degrees.

2) Cut the layup layers. Exactly as done in Exercise 1, cut the four layers of woven materials to be used to 12 x 16 inches, or a little larger. As each layer is cut, place it in a ***labeled*** paper grocery bag for protection. ***As before, mark the peel ply layer with a wax crayon.***

3) Cut the "wetting" plastic: Cut a piece of 4-mil PE plastic to about 15 x 36 inches. This "wetting" plastic will be used to confine the epoxy when wetting out. Fold the wetting plastic to make a double layer about 18 inches long. Use a marking pen to mark the 10 x 14 inch work area. Put aside.

4) Tool surface: Use the same 3 x 4 foot tool as in Exercise 1.

5) Tool release film: (same as Exercise 1). Use a marker pen to mark the 10 x 14 inch work area on a sheet of smooth PE plastic. Place the film on the surface of the "tool" so that the work area is generously covered, ***marked side down***, as before. Hold it in position with small pieces of brown mailing tape. Stretch and reposition the film as needed until it lies smoothly. See the figure.

6) Mix epoxy: Wear gloves! (Same as in Exercise 1.) Measure out and mix thoroughly about 6.5 to 7 ounces of epoxy. Add just a little bit of red or blue carpenter's chalk to serve as an inert tinting agent. Mix again to disperse the chalk. (A very small amount of India ink can be used as an alternative tinting agent.)

7) Wet the layup stack - first layer: Place the folded wetting plastic over a convenient work area away from the marked tool. Unfold and temporarily position the plastic over the workbench with small pieces of brown mailing tape. Pour a little mixed epoxy in the middle of the marked work area, just as you did directly on the tool in Exercise 1. Use a plastic squeegee to chase the epoxy around and approximately wet the work area.

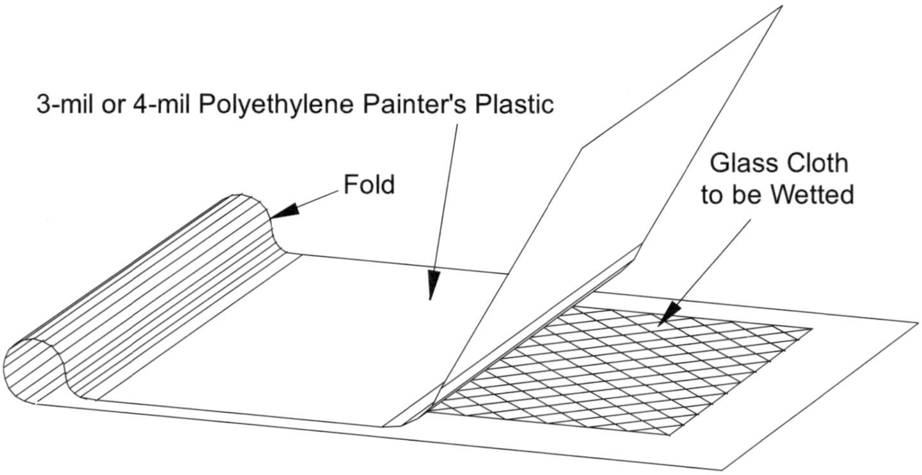

Wetting Out In Plastic Sandwich

Identify and place the 7725 BID (Layer 1). Be careful about stretching the bias-cut cloth. (You'll probably do a little better this time!) Add a little more epoxy to the center. Close the folded wetting plastic so that the layup is covered and work through the plastic sandwich to chase the epoxy around with the squeegee. Drive epoxy into any areas that contain trapped air, which will appear to be white. Drive the air out to the sides of the plastic envelope. As before, use the squeegee edge, roller brush, and grooved roller.

Lift the entire layup from the working surface, held by the wetting plastic, and inspect the layer to see if there has been full wetting. Turn the stack over and chase out some more air, if necessary. Try not to waste too much epoxy while getting the air out. (A little practice will teach the necessary technique.) If the stack is quite wet, the next layers will absorb the excess. Turn the stack over again and replace the stack on the work surface.

8) Wet the layup stack - second and third layers: Open the folded wetting plastic. Place the layer of 7715 UNI (Layer 2) on the work area. Use the squeegee edge to tap UNI down into the wet epoxy (Don't use the squeegee to stroke the surface directly because of the risk of pulling the unidirectional yarns out of place.) Add a little more epoxy if it doesn't wet readily. Add the Style 120 glass (Layer 3) and use the squeegee edge to stroke and tap the glass into place. Close the wetting plastic and work through the plastic to chase the epoxy around, driving out air. Lift the wetting plastic and add epoxy as needed. As before, use the roller brush and grooved roller to remove air. At this stage, you should have used nearly all the mixed epoxy. If some areas are short of epoxy, use the squeegee to "shovel up" excess from the sides of the work and move it to where it is needed. If the stack is properly wetted, the entire work area will appear to be of uniform tint. As before, turn the stack over and inspect for any voids.

9) Removal of excess epoxy: Use the squeegee to work through the wetting plastic, to drive excess epoxy out to the sides. A little practice will soon develop the technique. The wetting plastic is strong and you can get fairly aggressive in forcing out the excess. Any area where you have been a little too enthusiastic, allowing the stack to take up air, will appear cloudy. If that happens, chase a little more epoxy into the area until it is clear again. The goal is to fully wet the stack without having any more excess epoxy than necessary. Work from both sides of the layup.

You will find that lifting the wetting plastic tends to lift the peel ply layer and introduce air. Work slowly and smooth the layers back with the gloved hand.

10) Transfer the stack to the tool: Turn the stack over so that Layer 1 (the 7725 BID) is on top. Use a scissors, knife, or razor blade to cut the fold line of the wetting plastic. Carefully, and slowly, peel the wetting plastic from the 7725 side and discard it. The act of pulling the wetting plastic away will tend to lift the first layer and introduce small air bubbles. After removing the wetting plastic, smooth the exposed layer with a squeegee until the bubbles are driven out and the layup appears entirely clear.

NOW PAY ATTENTION! If you have not done this before, the following is a high-risk operation:

Lift the stack, now without the bottom layer of wetting plastic, and place it over the work area on the tool so that the edge of the stack overlaps the edge of the tool work area. The wet layers may have a tendency to slide relative to each other, depending on the viscosity of the epoxy. You will probably need to hold the bottom layer in place by pinching the stack with the gloved hand as you move the stack. Place the edge of the wet stack over the edge of the work area and carefully lay the stack down, working from the edge, trying not to trap air between the tool and the stack. Don't worry if you haven't hit the work area squarely - we have allowed sufficient margin. Use the squeegee edge to force the stack down against the tool, working out from the center. Try to remove a little more excess epoxy at this stage, but do not go so far as to allow air into the stack.

If the transfer has gone well, the wet stack will be correctly placed on the tool and the entire layup will be nice and clear, showing no trapped air.

If the transfer has ***not*** gone that well, and some air remains in the layup, slowly lift the wetting plastic and accept some possible disturbance of the top layer. Reposition the top layer as needed with a gloved hand or the squeegee. Use the squeegee directly on the top layer to remove a little more excess epoxy and get a smooth, CLEAR surface. Use the brush and roller, as needed.

11) Peel ply: Add the peel ply (Layer 4) to the wet stack. As before, use the squeegee to bring up epoxy to wet the peel ply.

Please note again that here we are not following the general industry practice of wetting out the peel ply with the structural layers.

12) Cure, trim, weigh, and label: Allow to cure 24 hours. Trim to 10 x 14 inches, using a band saw or a spiral saw. Strip off the peel ply. Weigh. The finished panel, after trimming, should weigh something less than 150 grams. (With the experience of Exercise 1 behind you, a weight of 140 grams would be reasonable.) Label the panel with the date, exercise number, and the weight.

Comments

The panel produced by this exercise should be identical to the panel produced in Exercise 1. In reality, it will probably be a little lighter because the builder will be a little more skilled in wetting out and won't have quite so much excess epoxy. In this case, we chose to transfer the wetted stack to a tool covered with release film. In practice, this technique is often used to ***place large areas of wetted glass onto large or complex structures*** like fuselage forms or wing skin molds. The method lends itself to ***two people working together*** to accurately place large wetted layers, particularly when access is awkward or the builder must work on a vertical surface. After some skill is developed, ***it is even feasible to work from underneath a horizontal surface, pressing the stack upward into position.***

In this exercise, we chose not to trim the wet layup to the final size ("green" trim), because a final, exact trim will be made after cure and we will tolerate the ragged edges until then. Also, it is fairly difficult to cut a 3-layer stack with scissors. When this technique is used to move and place large areas of wetted material, the stack is usually trimmed wet, under the wetting plastic, following marks previously made on the wetting plastic. We will do this in Exercise 3. Trimming under plastic produces a clean edge and greatly reduces the pain of trying to manipulate wet BID directly!

Exercise 3

Basic Vacuum Bagging

Introduction

This exercise will duplicate exactly the same 4-layer layup made in the previous two exercises, but will demonstrate basic vacuum bagging using very ordinary materials, typically from the home supply store, rather than the aircraft supply house.

This is a very convincing exercise. If the reader has completed Exercises 1 and 2, he will already have all of the consumable materials required to complete this project, except for the "bleeder" fabric, which will be described.

A very important new requirement is that a ***vacuum pump and some associated small hardware will be needed.*** The subject of appropriate vacuum pumps is discussed in Appendix 1. This demonstration will require the pump, a vacuum pass-through, a vacuum gauge, and some quarter-inch plastic automotive fuel line for plumbing. This description will assume that the reader will use a small commercial vacuum pump suitable for long-term continuous operation. ***Please read - or at least skim - Appendix 1 before starting this Exercise.***

General Description

As previously, we will protect the flat tool surface with a sheet of release film, which can witstand considerable harsh treatment without release problems.

The vacuum bagging operations demonstrated here are in addition to the basic layup operations of Exercise 1. There are two more layers to cut and slightly more assembly time. Experience makes a big difference, but adding the bagging layers and taping the whole thing down should add no more than 20 minutes to the working time, even for a beginner, and considerably less after the experience of a few trials.

Detailed Procedure

1) Work Area: Same as in the previous exercises. Prepare the general work area, at least 4 x 3 feet. Be sure that the temperature is above 70 degrees F. Cover the general work area with an unbroken layer of 3-mil PE plastic. Secure the cover to the worktable with small pieces of brown mailing tape. ***(This worktable cover will become part of the vacuum bag, so it cannot have any punctures.)***

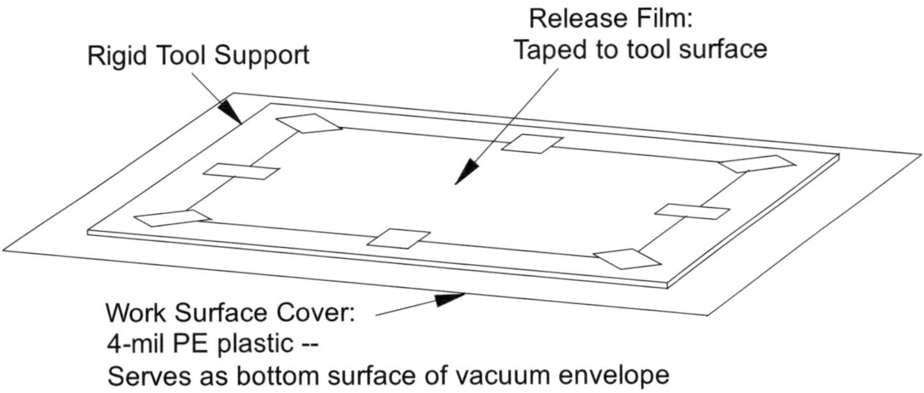

Preparation of Flat Tool and Vacuum Bag

2) Cut the layup layers: Exactly as done in the previous Exercises, cut the four layers of layup materials to be used to 12 x 16 inches. As each layer is cut, place it in a labeled paper grocery bag for protection. ***As before, mark the peel ply layer so you can tell which is the marked side.***

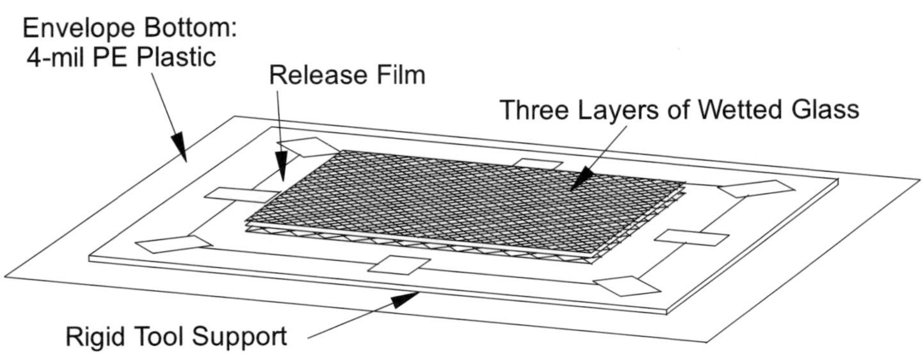

Layup Prepared for Vacuum Bagging

3) Tool and tool release film: Find a piece of smooth sheet material about 18 x 28 inches or larger, to serve as a tool support. Plastic-coated wallboard, sheet aluminum, glass, or plastic window stock will be satisfactory. It is only necessary that it be fairly smooth, larger than the layup area, and smaller than the area of the worktable that was covered with plastic sheet. Cut a wrinkle-free piece of 4-mil PE film about one or two inches smaller than the tool support to serve as the tool release film. Use a marker pen to outline an area 10 x 14 inches to serve as a "target" for the layup. Center approximately on the tool support and secure with small pieces of brown mailing tape, ***marked side down.*** Work from the corners and stretch the film slightly to make it lie as flat as possible on the tool

4) Cut the vacuum bagging materials: Cut a piece of bleeder material 18 x 22 inches and place in a labeled bag. Bleeder may be almost any absorbent fabric. Polyester "plush" or cotton sweatshirt material from a yardage store works well. Agricultural felt sold in garden stores for weed suppression is quite good. (Use three layers of weed felt.) Cut a piece of 4-mil PE plastic to serve as the vacuum bag, large enough to cover the entire layup, but at least two inches smaller than the sheet of plastic which covers the worktable. The idea is to make the bag large enough to cover the work while leaving a margin all around so that it may be tape-sealed to the plastic film covering the worktable. Place in a labeled grocery bag.

5) Mix epoxy: Wear gloves. As before, measure out and mix thoroughly about 6.5 to 7 ounces of epoxy. Tint with carpenter's chalk.

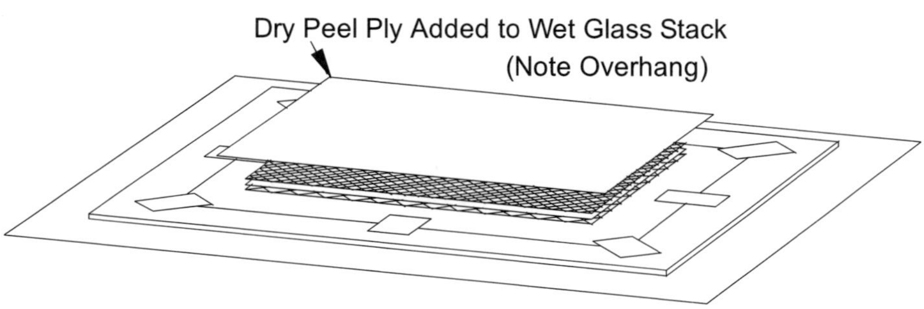

Peel Ply Added to Stack

6) Lay up the stack: Assemble and wet out the stack exactly as described in steps 6 through 10 of Exercise 1, including the peel ply layer. See that the peel ply layer overhangs the glass and is applied dry to the wet glass.

7) Place the bleeder and pass-through pad: Place the bleeder over the work area so that the excess hangs over the edge of the work equally. Find a scrap of bleeder and make a pad of two to four layers of bleeder, about 4 inches square or larger. Place this pad at the long edge of the large piece of bleeder so that it overlaps the main bleeder by one inch, or more. ***This pad is used to isolate the metal pass-through from the work and prevent the vacuum connection from making an imprint on the curing layup*** (see the sketch).

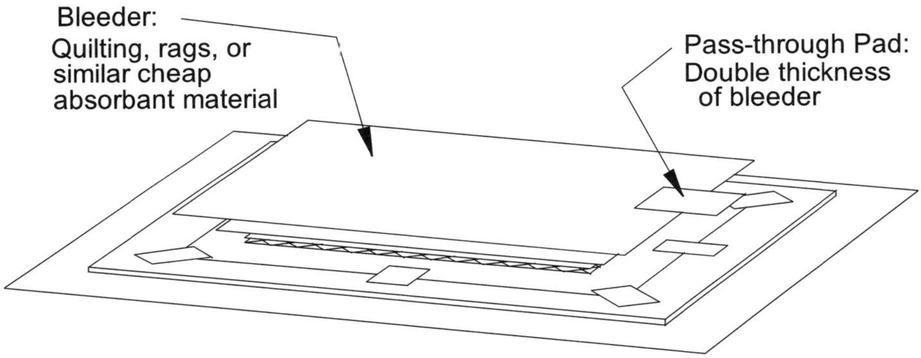

Bleeder and Pass-through Pad Added

8) Place the bag and vacuum pass-through: Place the sheet of bag plastic over the work area. Adjust the position of the pad so that the vacuum pass-through will not press directly on the layup and the pad will provide an air path to the rest of the bleeder. Use a marker pen to mark a quarter- inch "X" over the bleeder pad. Use a razor to cut through the bag film on the mark. Locate the vacuum pass-through, remove the top washer, push the pass-through in the hole, replace the top washer, and tighten the nut. Use brown mailing tape to seal the bag to the plastic sheet that covers the worktable, to make an air-tight chamber all the way around the tool.

> ***This arrangement is called an ENVELOPE BAG. It is a lot easier, in this case, to enclose the entire work, including the entire flat tool,***

in a soft bag than it is to get an airtight seal against the tool itself. Also, this system does not require that the tool surface itself be airtight. Note that it is perfectly satisfactory to tape the soft plastic bag to the soft plastic worktable cover. Under vacuum, the bag and worktable cover will just lie flat together. The only rigid surface within the vacuum is the tool.

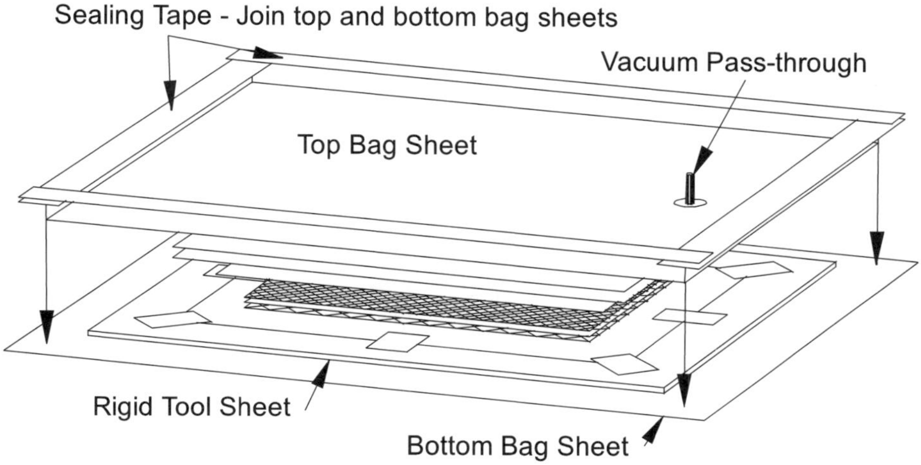

Closing and Sealing the Envelope Bag

9) Pump down: Connect a vacuum hose to the vacuum pass-through and start the pump. The envelope bag system used here is unlikely to leak because it is easy to get a good seal by flat taping over the edge of the bag. If there is a leak, it is probably from not forcing the tape down solidly or from not fully tightening the nut on the pass-through. Unless your pump is quite inefficient, you should be able to pull over 15 inches of mercury. If you show less than 7 inches, hunt for leaks. With vacuum applied, you will immediately start to see excess epoxy forced out into the bleeder. This is a very revealing display. Areas that have a substantial excess of epoxy may flood the absorbent bleeder. Areas that have a shortage of epoxy may show only a little flow into the bleeder, or even none. It is very unlikely that excess epoxy will be drawn into the vacuum pass-through, with the parameters given here. The thick pad under the pass-through provides a vacuum path to the rest of the bleeder and extra absorbent capacity to keep epoxy out of the plumbing. Run the pump until the epoxy has become fairly firm - typically about five hours.

10) Cure, trim, weigh, and label. Allow to cure overnight and remove the bag. Strip off the peel ply and bleeder together. Trim to 10 x 14 inches, using a band saw or a spiral saw. Weigh. The finished panel, after trimming, should weigh something less than 140 grams. Label the panel with the date, exercise number and the weight.

Comments

The panel produced by this exercise should be somewhat lighter than the panels produced previously, because the vacuum does a very complete job of forcing out excess epoxy. Assuming that the vacuum-bagged result can be taken as "perfect," note that ***an experienced builder, who has done Exercises 1 and 2 a few times, might get quite close to the vacuum-bagged result.*** Reducing weight is always welcome, of course, but the most important benefit of the vacuum-bagging technique is the prevention of voids and increased confidence that the structure is very well bonded. The excess epoxy will be drawn into the bleeder.

This exercise uses the common industry practice of routinely using peel ply on the top layer and placing the bleeder directly on the wet layup. The bleeder, of course, will bond to the peel ply and may be left for added protection until the work is to be assembled into some practical structure. Leaving bleeder in place delays inspection and requires removal if the part is to be trimmed, which exposes the part to shop contamination.

One important conclusion from this exercise is that the vacuum-bagging process is very little extra work over the work of a simple open layup. If the builder anticipates the need to bag an item, the only extra material required would be sufficient bleeder to cover the work and enough mailing tape to make the seal. Very often, the tool, tool release, and bag from one layup can be easily reused for another.

Exercise 4

Making Accurate "Prepregs"

Introduction

This Exercise will extend the work done in the previous Exercises to demonstrate the important technique of wetting out layers of a layup on the worktable, cutting them accurately to size, and assembling them to form another structure. This process is the amateur builder's low-cost answer to the professional's use of expensive pre-impregnated fabrics and is very important when assembling large structures like wing skins and fuselages.

We will make exactly the same 4-layer layup demonstrated previously, but in two separate parts, using "prepregs." The transfer will be made to a flat tool surface, as in the previous Exercises.

The layup technique to be demonstrated will be similar to that used in Exercise 2, except that we will wet out only two of the four layers at one time and will precut them close to final size. We will then assemble these two separate stacks into one structure, vacuum bag, and cure. Again, this project should take the beginner about one hour, but we should be getting rather good at the basics by now! There is only a little more work to be done than in Exercise 3.

We will also ***introduce the use of perforated release film***, which is release film that contains many small pierced holes, which allow the flow of air and liquid epoxy when the wet layup is placed under vacuum. This technique is not absolutely necessary for this exercise, but is easy to do and facilitates disassembling the layup.

Detailed Procedure

1) Work area: Same as in Exercise 1. Prepare the general work area. Be sure that the temperature is above 70 degrees F.

2) Cutting layup layers: Exactly as in Exercise 1, cut the four layers of different woven materials to be used to 12 x 16 inches, or a little larger. As each layer is cut, place it in a labeled paper grocery bag for protection. As before, mark the peel ply with the letter "F" with a wax crayon.

3) Cut the "wetting" plastic: Cut two pieces of 4-mil PE plastic to about 15 x 36 inches. This "wetting" plastic will be used to confine the epoxy when wetting out, as in Exercise 2. Fold the "wetting" plastic to make a double layer about 18 inches long. Use a marker pen to mark an 11 x 15 inch work area on each piece of the wetting plastic. (Note that this is a little larger work area than used in Exercise 3. This is the trimming margin.) Put aside.

4) Cut and perforate the release film: You will need to prepare a sheet of perforated release film, 14 x 18 inches. Use 2 or 3-mil PE film as before. Find a small steel pointed object like the point of a compass or the stylus on a carpenter's square. Use the pointed tool to poke holes in the PE sheet on roughly half-inch centers, working against a short-nap rug or a folded towel. It is convenient to fold the release film twice and punch four layers at a time. Note that these are not punched holes; they are ***pierced holes.*** Spacing is not critical and placement by eye is satisfactory. (Piercing doesn't take long for this small teaching project, but it is a bother when you do larger layups. ***You will want to build the perforating tool described in Appendix 1.***) Store in a labeled grocery bag.

5) Work surface and tool: Use the same work surface and tool as in Exercise 3.

6) Tool and tool release film: Use a marker pen to mark the 10 x 14 inch work area on a sheet of smooth PE plastic (or use a marked paper pattern). This mark will serve as a "target" when placing the layup. Place the film on the surface of the tool so that the work area is covered, marked side down, as before. Hold it in position with small pieces of brown mailing tape. Stretch and reposition the mailing tape until the plastic film lies smoothly.

7) Mix epoxy: Wear gloves! Measure out, mix thoroughly, and tint about 7 ounces of epoxy.

8) Wet the layup stack - first two layers: Place the folded wetting plastic over the work area, away from the tool area. Unfold and temporarily position the plastic over the workbench with small pieces of brown mailing tape. Pour a little mixed epoxy in the middle of the marked work area. Use a plastic squeegee to chase the epoxy around and approximately wet the work area. Identify and place the 7725 BID (Layer 1). Be careful about stretching the bias-cut cloth and assure that the marked work area is covered. Add a little more epoxy to the center. Close the folded wetting plastic so that the layup is covered and work through the plastic to chase the epoxy around with the squeegee. Drive epoxy into any areas that contain trapped air. Drive the air out to the sides of the plastic envelope. As before, use the squeegee edge and grooved roller.

Open the folded wetting plastic. Place the layer of 7715 UNI (Layer 2) on the work area. Use the edge of a squeegee to tap the UNI down into the wet epoxy. Try to place this layer fairly accurately, but don't worry if you are a bit off - this is why we left a margin. Add more epoxy if it doesn't wet readily. Lift the entire layup, held by the wetting plastic, from the working surface. Inspect the layup to see if there has been full wetting. Chase some more, if necessary, with the stack inverted. Turn the stack over.

9) Trim the first stack: Use scissors to trim the entire sandwich along the marked lines on the wetting plastic. Discard the trimmed margins. Put aside.

10) Wet the second stack - third and fourth layers: Similarly use the second "wetting plastic" to wet the light 0.004 inch Style 120 glass (Layer 3) and use the squeegee edge to stroke and tap it into place. Close the wetting sandwich and work through the plastic to chase the epoxy around, driving out air. Lift the wetting plastic and place the peel ply layer. (Here, we are following industry practice and including the peel ply as one of the structural layers.) As before, use the roller brush and grooved roller to remove air. At this stage, you should have used nearly all of the mixed epoxy. If some areas are short of epoxy, use the squeegee to "shovel up" excess from the sides of the work area and move it to where it is needed. As before, turn the stack over, inspect for any voids, and correct them.

11) Trim the second stack: As in Step 8, trim the second stack to the mark.

12) Transfer the first stack to the tool: As in Exercise 2, turn the first stack over so that Layer 1 (the 7725 BID) is on top. Carefully, and slowly, peel the wetting plastic from the 7725 side and discard. After removing the wetting plastic, smooth the exposed layer with a squeegee until the bubbles are driven out and the layup appears entirely clear. Place the two-layer stack on the tool and remove the plastic. (Get the help of a second pair of hands, if necessary.) Be sure that the marked work areas overlap. Pat down with the edge of the squeegee, if necessary.

13) Transfer the second stack to the tool: Similarly, transfer the second stack to the tool, adding the layers to the first stack. Be sure that the marked peel ply layer is on top.

14) Bag and cure: Place perforated release film, bleeder, and bag as described in Steps 7 through 9 of Exercise 3. Pump until firm (about five hours).

15) Finish cure, trim, weigh, and label: Allow to cure 24 hours. Remove the bag and bleeder. ***(Note that it is very easy to separate the bleeder from the perforated release film and that it is possible to salvage and reuse the perf film. Note also that the bleeder may be only partially filled with epoxy and can be used again over new bleeder.)*** Trim to 10 x 14 inches, using a band saw or spiral saw. Strip off the peel ply. Weigh. The finished panel, after trimming, should weigh something less than 140 grams, very nearly the same as in Exercise 2. Label the panel with the date, exercise number, and the weight.

Comments:

The panel produced by this exercise should be identical to the panel produced in Exercise 3.

The Exercise is useful because the method of assembling accurately-cut sections of wetted glass into larger structures will be important in assembling aircraft-sized layups.

The perforated release film allows convenient inspection of how the excess epoxy is distributed because you can clearly see it coming through the holes in the perf film and making a pattern in the bleeder. Most important, it provides an immediate visual check that there are no places where the layup is too dry.

Exercise 5

The Flat Sandwich Laminate

Introduction

This demonstration will depart somewhat from common aircraft kit practice, because the skin thickness will be less and the core thickness considerably more than we usually see in current composite designs. This example will emphasize the strong effect of the thickness of the structure to get the most stiffness from the least weight and the least number of components.

We will make two symmetrical foam-cored laminated sandwich skin structures at the same time, in the same panel size we made in the previous four exercises, ***in a single work session***, using vacuum bagging. We will demonstrate the use of comparable samples of the two foam families widely used in composite homebuilt aircraft. The result will typically weigh about 1.5 times as much as the 10 x 14-inch panel we made previously, but will be many times stiffer. Most people who are not familiar with composite techniques are quite surprised when they compare the flat panels made previously to the sandwich panel made here.

This exercise is also useful in familiarizing the builder with the handling of larger glass areas.

General Description

We will make a simple 10 x 14-inch symmetrical sandwich with one layer of 7725 BID glass, cut "on the bias" at 45 degrees to the weave, on each side of a half-inch thick core of 2 lb/ft^3 closed-cell foam.

As explained in Appendix 3, two major families of foams are used in most current composite designs: (1) polystyrene (Styrofoam) and (2) urethanes. The aircraft supply houses offer both of these families in a wide range of densities, foam cell sizes, and shapes. They have somewhat different characteristics and handling properties and it is imortant that active builders have some understanding and experience in the use of both types. We will make two otherwise identical panels, one with polystyrene foam and the other with urethane foam.

We may briefly summarize the handling characteristics of the two major foam families as follows: (See Appendix 3, Foam Materials, for a detailed description of core foams, their properties, and their handling methods.)

Polystyrene: Light blue, coarse surface feel from the relatively large cell size in the most-used 2 lb/ft^3 form, which is usually shipped as blocks. It is also available in a small cell form in sheets. It is very convenient to cut with a hot wire, but still easy to knife-cut, carve, and sand. Requires that the large open cells in the surface be filled in order to make a reliable bond, a process that takes a substantial amount of lightweight filler. It is reasonably tough and is fairly easy to handle in the shop without damage. Very widely used in the Rutan designs and their many derivatives. Dissolves in fuel and polyester resins, so must be kept away from these fluids. It deteriorates in ultraviolet light.

Urethane: Pink or tan, smaller cell size than the corresponding polystyrene, but still requires filling. Must not be cut with a hot-wire because high heat releases an extremely poisonous gas. Very easy to cut and carve. Gives the impression of being brittle and crumbly. It has good stiffness, but less strain to failure than polystyrene. It is easily damaged in the shop; edges and surface are very fragile. Urethane is completely inert to petroleum and styrene and is easily damaged by ultraviolet light. The stiffness for its weight is very good and it is an excellent core material when protected by a skin because the skin will limit the deflection under load.

This exercise will describe the process of filling the surface of the foam cores in detail.

Detailed Procedure

1) Work area: Same as in previous exercises. Prepare the general work area, at least 4 x 3 feet. Be sure that the temperature is above 70 degrees F. Cover the general work area with an unbroken layer of 4-mil PE plastic. Secure the worktable cover with small pieces of brown mailing tape.

2) Cut and perforate the core panels: Cut two 12 x 16-inch panels from half- inch 2-pound foam, one from polystyrene, one from urethane. Perforate the two panels on approximately 7/8 inch centers, either with a perforation tool (Appendix 1) or by hand, using a large needle or a piece of piano wire with the end ground sharp. (Consider building a perforation tool while this Exercise is curing!)

3) Cutting layup layers: Cut two pieces of 25 x 17-inch 7725 BID glass at 45 degrees, as follows: Place PE plastic on each side of the dry glass cloth, leaving a cutting margin around the selected area. Use a marker pen to mark the plastic for cutting. Use scissors to cut the entire sandwich consisting of the top layer of PE, the layer of glass cloth, and the bottom layer of PE, all at the same time. Roll up each sandwich and place it in a labeled grocery bag. (This technique will be extremely useful when it is necessary to handle and wet out large areas of bias-cut glass. Once you try it, you will be convinced!)

4) Cut the vacuum bagging materials: Cut a piece of bleeder material 18 x 26 inches and place in a labeled bag. Cut a piece of 4-mil PE plastic to serve as the vacuum bag, large enough to cover the entire layup, but at least two inches smaller than the sheet of plastic that covers the worktable. The idea is to make the bag large enough to cover the work while leaving a margin all around so that it may be easily sealed to the covered worktable. Store in a labeled grocery bag. (This is the same envelope bag used in previous Exercises.)

5) Cut and perforate the release film: Use the same 2- or 3-mil PE as before, cut to 28 x 18 inches. Perforate with the perforation tool or the spike. (Spike rollers are sold for the professional shop, but they are very expensive and they tend to tear the PE.) Store in a labeled grocery bag.

6) Tool and tool release film: Find a piece of smooth sheet material about 18 x 28 inches, or larger, to serve as a tool. Cut a wrinkle-free piece of 4-mil PE film about one inch smaller than the tool support to serve as the tool release film. Center approximately on the tool and secure with small pieces of brown mailing tape. Work from the corners and stretch the film slightly to make it lie as flat as possible on the tool. Use a marker pen to outline an area 24 x 16 inches to serve as a “target” for the layup.

7) Mix epoxy and micro: Mix and tint about 9 ounces of epoxy. Pour about half into a second cup and add approximately two volumes of micro balloons to the second cup to make "slurry." This 2:1 mixture should run off of the mixing stick. Set the slurry cup aside.

8) Wet the layup stack: Pour a little epoxy on the work area. Identify and place the peel ply over the work area, marked side down. Use a plastic squeegee to chase the epoxy around and wet the peel play. (Peel ply is thin, but its fine weave requires more attention to ensure thorough wetting.)

Identify the 7725 BID. (You will have discovered that handling and wetting bias-cut bidirectional cloth can be difficult.) Place the dry 7725 BID on a piece of scrap PE plastic and draw it into approximate proportions by stroking it with you hand. Pour some epoxy on the BID and smooth out with a squeegee. When the cloth is fairly well wetted, lift the support plastic and carry the wetted cloth to the work. Place and smooth into position with your gloved hand. Be careful about stretching the bias-cut cloth. (You'll probably do a little better this time!) Add more epoxy to finish wetting the layer. As before, use the squeegee edge, roller brush, and grooved roller. By now, you should have enough experience so that the glass is evenly wetted without a large amount of excess epoxy.

9) Fill the bottom face of the two foam core panels: Use a vacuum cleaner with a soft brush attachment, as used for cleaning furniture, to vacuum any dust from one face of the two core panels. (This is particularly important for the urethane panel, which is subject to minor scuffing and damage at the surface.) Paint the slurry on the two core faces. Observe that it tends to run into the perforation holes, which gives a rough indication of how evenly the slurry is spread. Use a heat gun at the low heat setting to warm the slurry and increase penetration. Set aside. (By the time the following steps are completed, the slurry will have thickened a little.)

10) Place the core panels: Place the two core panels side by side, approximately in contact, on the wet glass layer. Do not worry about any small gap that exists between the panels. This gap will be filled with epoxy and micro and will be trimmed away when the cure is complete.

11) Fill the top foam surfaces: Again, vacuum the surface of the cores with a soft brush vacuum attachment. Paint the surfaces with slurry. Warm with the heat gun. Add tinted micro as needed to get a thin, even coat. The texture of the foam surface should show through the coat of warmed slurry.

12) Place the top layer of glass: As before, wet the 7725 BID glass on a temporary piece of plastic. Carry the wetted glass to the work and transfer the glass to cover the foam cores.

13) Place the perforated release film: Center the layup so that the excess hangs over the edge about equally (not critical).

14) Place the bleeder and pass-through pad: Place the bleeder over the release film so that the excess hangs over the work about equally. Find a scrap of bleeder and make a pad of double or triple thickness about four inches square or larger. Place this pad at the long edge of the main bleeder so that it overlaps the bleeder by one inch or more.

15) Place the bag: Place the sheet of bag plastic over the work area. Adjust the position of the pad so that the vacuum pass-through will not press directly on the layup. Use a felt-tipped "Sharpie" pen to mark a quarter- inch "X" over the bleeder pad. Use a razor to cut through the bag film on the mark. Locate the vacuum pass-through, remove the top washer, push the pass-through in the bag, replace the top washer, and tighten the nut. Use brown mailing tape to seal the bag to the plastic sheet that covers the worktable to make an airtight chamber all the way around the tool.

16) Pump down: Connect the vacuum hose to the pass-through and start the pump. If you get less than about 8 inches of vacuum, hunt for leaks. Usually, pressing down on the line of tape around the bag and gentle heating with a heat gun will fix the problem. With vacuum applied, you will immediately see excess epoxy forced out into the bleeder through each of the perforations in the perf release film. Run the pump until the epoxy is firm - about five to eight hours.

17) Cure, trim, weigh, and label: Allow to cure overnight. Remove the bag. Rough cut the two panels apart at the line of meeting, using a fine-tooth jig saw or a spiral saw. Final trim to 10 x 14 inches, using a band saw or spiral saw. Strip off the peel ply. Weigh. The finished panel, after trimming, should weigh something less than 210 grams. Label the panel with the date, exercise number, foam type, and weight.

Comments

The superior stiffness of the two panels produced by this exercise is immediately apparent. As they stand, the panels are not completed structures because the exposed foam edge is vulnerable to damage and must be covered and sealed for use in a practical structure.

Exercise 6

The Cored-skin Sandwich Laminate

Introduction

This is the most important exercise in the entire book.

This exercise will produce a pair of very light panels that could actually be used in practical aircraft structures. The techniques used will be described in complete detail, because they will be required many times in the examples to follow.

We will make two ***non-symmetrical*** foam-cored laminated sandwich skin structures, in the same 10 x 14 inch panel size we made in the previous five exercises, using both polystyrene and urethane cores. The cores will be totally enclosed in a glass-to-glass joint and no edge filling will be required.

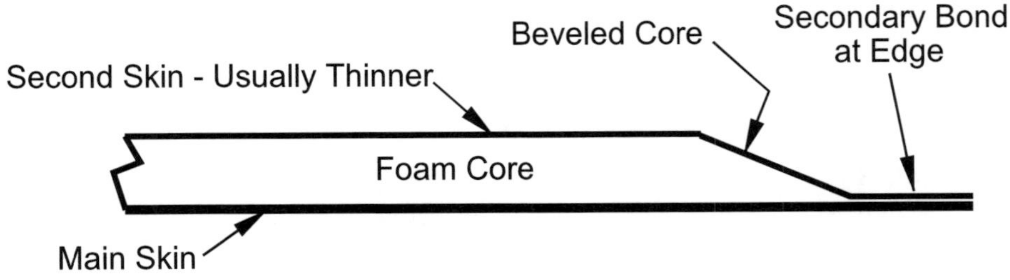

Glass Edge

This time, we will demonstrate how very light layers of glass, working with a much thicker, light-weight core, can produce a very stiff panel. The effect of the stabilizing layer is roughly similar to the way a tall sailboat mast uses a light spreader strut half way up to shorten the effective length of the mast and prevent buckling.)

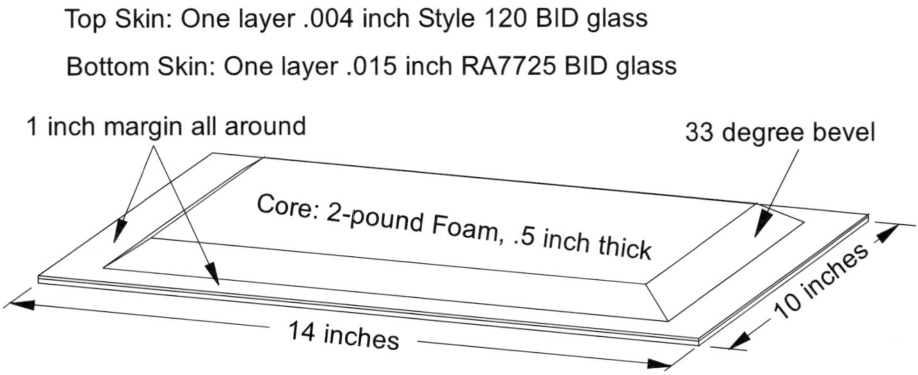

Basic Cored-skin Sandwich Panel

The author has demonstrated this Exercise at Workshop sessions of the Experimental Aircraft Association. The entire operation, not including curing time, takes about 45 minutes for a single panel. The two-panel case described here will take about one hour. This will take the beginner longer, but that is why we do this as a training exercise!

General Description

The thin composite sheet made with a single layer of the 7725 BID we have been using in the previous exercises will carry a ***very large load in tension*** if that load is distributed along the entire edge of the sheet. In fact, if we take a design load of 25,000 lb/sq in for E-glass cloth and use a thickness of 0.015 inch, the ten-inch edge of our test panel has an area of 0.15 sq in, so should carry 3,750 pounds in tension! Unfortunately, though this "tensile strength" is frequently quoted, it is an ***almost meaningless figure for practical structures***. Very ***thin plates*** of any practical material - composite, aluminum, or steel - all ***buckle freely under quite low compression*** loads. Designers of fabric covered aircraft simply assume zero strength in compression for the entire skin; necessary compression loads must then be carried some way other than in the skin. Many designs in aluminum also basically treat the thin skin as "aluminum fabric" and depend on separate structures to carry the inevitable compression loads.

In this Exercise, we will make a pair of test panels that can carry practical compression and twisting loads, using a single layer of the same 7725 BID we used previously, stabilized with very light Style 120 BID, at 3.16 ounces/sq yd. We will make the two panels in a single layup and separate the two panels after the layup has cured. It is important that the 7725 cloth is totally flat, with no wrinkles, so this ***layer will be cured first*** and the cores will be added in a second operation.

We will assume that we will wish to bond other structures to these test panels, so we will use peel ply on the bond areas. ***Peel ply will be used ONLY at the bond areas and NOT on the light Style 120 stabilizing area.*** This is because we wish to avoid concentrated local peeling stresses caused by removing the peel ply, which might break the bond to the core. The peel ply will also serve to remove thin ridges of cured epoxy that form in any wrinkles in the release film.

Detailed Procedure

1) WORK AREA: Same as in the previous exercises. Prepare the general work area, at least 4 x 3 feet. Cover the work area with an unbroken layer of 4-mil PE plastic. (This layer will serve later as part of a vacuum bag.)

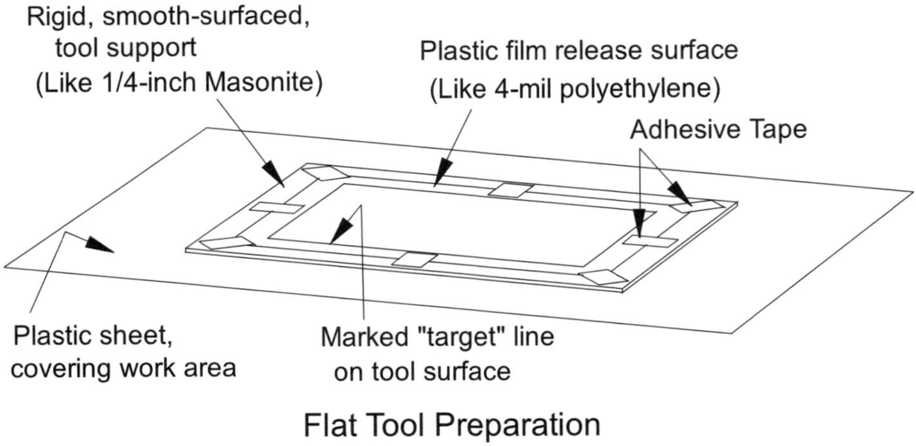

Flat Tool Preparation

2) TOOL AND TOOL RELEASE FILM: Same as Exercise 5. Mark a target area of 21 x 14 inches on the release film and secure it, marked side down, on the 28 x 23-inch tool, as before.

3) CUTTING LAYUP LAYERS: Cut one 22 x 18-inch piece of 7725 BID at 45 degrees. Cut a second layer of Style 120 crowfoot weave, 0.004 inch thickness, at 45 degrees. Place in marked grocery bags and set aside.

4) PERFORATE, CUT AND BEVEL THE CORE PANELS: Cut two core panels, slightly larger than 8 x 12 inches, from half- inch 2-pound foam, one from polystyrene and one from urethane. Perforate the two panels on approximately 7/8-inch centers, either using the perforating tool (Appendix 1) or by hand, using a large needle or a piece of piano wire with the point ground sharp. (Consider building a perforator while this exercise is curing!)

Cut a straight bevel edge on all four sides of the two core panels, as described previously. It is very convenient to use a ***beveling tool*** as described in Appendix 1, but if the reader hasn't gotten around to it yet, the same effect can be obtained by clamping the foam between two flat steel strips. Use a "cheese grater" forming tool like the Stanley "Surform" to cut the bevel. This is a very quick operation. (The author can bevel all four sides of a panel in under one minute.) The ***polystyrene can also be cut with a hot wire*** in seconds. DO NOT HOT-WIRE THE URETHANE. If you use a hot wire, be sure that the setup does not allow electrical continuity between the steel strips. Set aside.

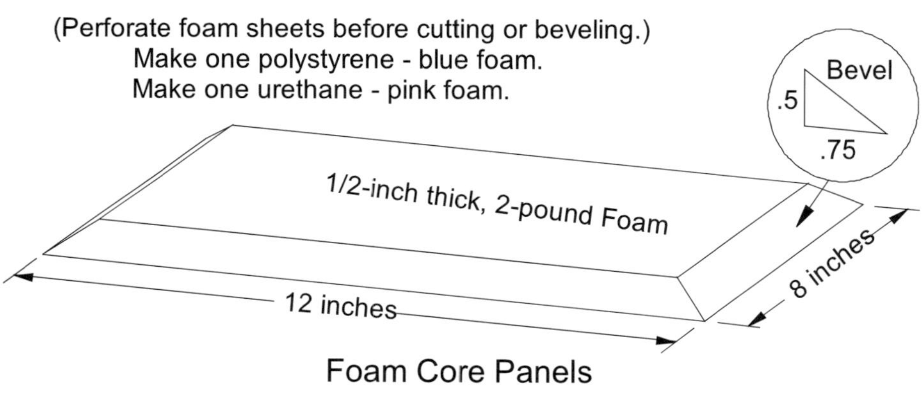

Foam Core Panels

5) CUT THE VACUUM BAGGING MATERIALS: (1) Cut a piece of bleeder material 18 x 24 inches and place in a labeled bag. (2) Cut a piece of 4-mil PE plastic to serve as the vacuum bag, large enough to cover the entire layup, but at least two inches smaller than the sheet of plastic which covers the worktable. Place in a labeled grocery bag**.**

6) CUT AND PERFORATE THE RELEASE FILM: As in Exercise 4, prepare a sheet of perforated release film, 24 x 18 inches. Re-use the same perforated PE film as before, if convenient. Place in a labeled grocery bag.

7) MIX EPOXY: Wear gloves! As before, measure out, mix, and tint about 6 ounces of epoxy.

8) PLACE AND WET OUT THE 7725 BID: Place the 7725 BID over the "target" area of the tool surface and wet it out as in Exercise 1. Work with a squeegee through a "spreading" plastic. (This should be quick and easy with just the single layer!)

> ***SPECIAL NOTE:*** If wetting dry glass in place, as described, be especially careful to wet out thoroughly. The following ***vacuum-bagging operation will remove excess resin, but will NOT transfer resin to dry areas.*** Inadequate wetting will appear in the cured layup as spots which show the weave pattern of the cloth.

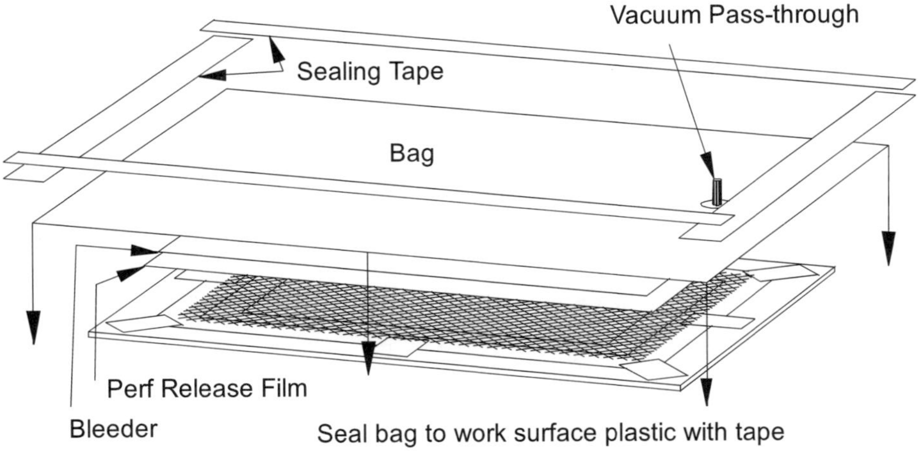

Bagging of Main Skin

9) BAG THE 7725 BID: Bagging may seem to be gross overkill for one thin layer, but we need to cure this layer absolutely flat and we have all the vacuum facilities available, so we might as well use them while we wait for the cure. Place the perforated release film, the bleeder, the bag, and the vacuum pass-through, and pump down. There is other work to do while waiting for the cure

> *Do not skip this intermediate curing step to go immediately to placing the cores on the wet glass. Under vacuum, the thin single layer of 7725 glass will **form a small kink at the edge of the foam core plate,** which will greatly reduce the tensile and compressive strength of the structure.*

10) CURE THE LOAD LAYER: Cure the 7725 BID until it is firm. This will be at least 5 hours in warm conditions, or 10 hours in a cooler room. Ideally, the epoxy will not be completely cured, so that the secondary bond to the next layer will be better. Remove the bag, bleeder, and perforated release film. If the bleeder is not too badly soaked, put it aside for another layup. After cure, strip off the vacuum rig. Everything but the sealing tape will be reusable, even the bleeder.

11) MIX EPOXY: Measure, mix thoroughly, and tint about 8 ounces of epoxy.

13) FILL THE BOTTOM FACE OF THE TWO FOAM CORE PANELS: Use a vacuum cleaner with a dust brush attachment to vacuum any dust from the bottom (wide) face of the two core panels. Use a little of the epoxy from the previous step to mix some ***flowable slurry.*** Paint and squeegee slurry on the two core faces. Use a heat gun at low setting to warm the resin and increase penetration. Use a squeegee to gently spread the warm resin and force it into the foam. Be careful of the fragile edges, especially with the urethane core. If you chip away the edge, don't worry abut it; it is easily repaired at the next stage. (By the time the following steps are completed, the resin will have thickened a little.) ***Use enough slurry to visibly cover the surface.*** Any excess slurry will be drawn through the perforation holes by vacuum in the next step.

14) PLACE THE CORE PANELS: Carefully place the two core panels on the marked area of the tool. ***The marks will show clearly through the glass already on the tool.*** Work from one edge to minimize trapped air and press into position. The light cores will stick quite well in the wet epoxy.

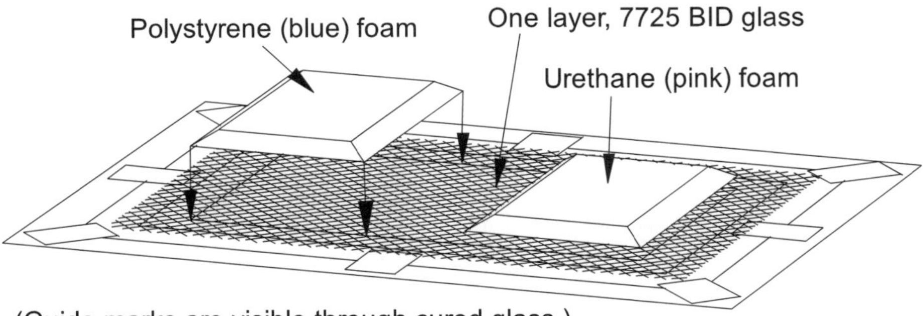

First Glass Skin and Core Placement

This is a tricky operation. The builder has a wet panel with very fragile edges and needs to turn it over and place it accurately without breaking the edges. The solution is (1) to turn the panel over (gloved hands!) and park it, wet side down, on a temporary support, like the rim of a 3-pound coffee can, (2) stick small nails into the dry face to serve as handles, (3) use the nail handles to pick up and position the panel. The nail holes will be filled with slurry as the work continues.

15) FILL THE TOP FOAM SURFACES: Vacuum, paint with epoxy, and squeegee out, as before. The cores will stay in position quite well.

16) FILL ANY NICKS IN THE CORE EDGES: If there has been damage to the beveled edges of the cores, mix up a little dry micro and fill the damage, being careful not to get micro on the bonding surface. Wipe up any loose micro. Smooth out with a chisel edge. *It is not necessary to wait for these small patches to cure.*

17) WET AND PLACE THE STABILIZING LAYER OF GLASS: Identify the 24 x 18-inch piece of Style 120 0.004 inch glass. This crowfoot weave cloth is very light and, when bias cut at 45 degrees, ***may be difficult to handle*** to get accurate and smooth placement. Even though wetting out such thin material is almost trivial, it will save work to use the "sandwich" method of wetting out. Use 2-mil PE for wetting. Cut the wetting sandwich to make two pieces 12 x 18 inches.

For each core panel, peel off one side of the wetting plastic, discard, and place the glass. Be sure that the glass completely covers the marked area. Discard the remaining wetting plastic. Smooth out and remove bubbles by prodding or "stippling" with a squeegee edge or a cheap paint brush. There will be a slight ridge at the corners, which can be ignored for this very thin glass. The important thing is to ***get the glass smooth over the entire surface***, including the bonding area around the cores.

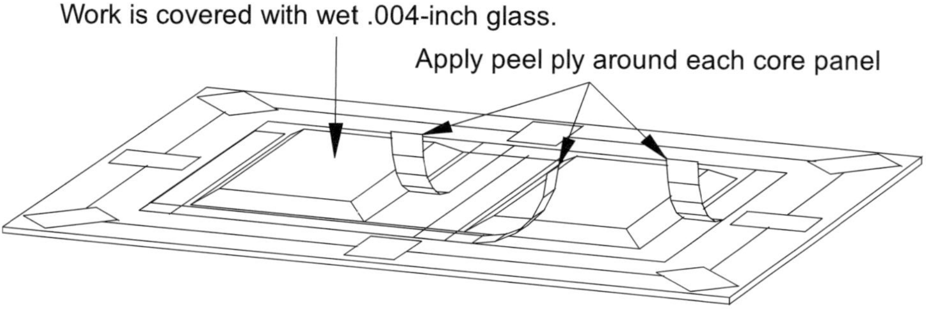

Preparation for Final Bagging

The fine-weave .004 inch BID glass used here has only a very limited range of adjustment by "ooching" the weave. The beginner will find it difficult to fit the wet glass over the two thick cores and form the glass to a smooth fit over the beveled corners, even with assisstance. It can be done, but it is ***much easier to cut the wet glass into two or three strips and place and shape each strip separately.*** *Allow an overlap of about 1/2 inch between strips.*

18) PLACE THE BOND AREA PEEL PLY: place a 1 1/4-inch strip of peel ply, previously marked with a wax crayon, all the way around both cores, ***covering the bonding area only.*** Do not place peel ply on the beveled area. Use the paintbrush and squeegee to assure full wetting.

19) PLACE THE PERFORATED RELEASE FILM: Place the perf film centered over the layup so that the excess film hangs over the edge of the layup about equally. Allow enough slack in the release film to ***avoid bridging*** between the two cores. Smooth out to remove creases. Where necessary to get a smooth fit, use a razor blade to cut the release film.

20) PLACE THE BLEEDER: Place the bleeder over the release film so that the excess hangs over the work about equally. Find a scrap of bleeder and make a pad of double or triple thickness about four inches square. Place this pad at the long edge of the layup so that it overlaps the main bleeder by at least one inch.

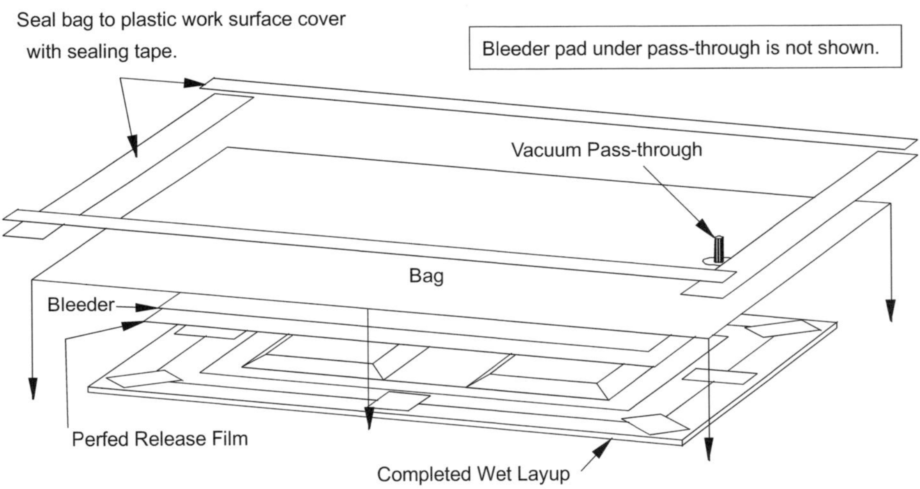

Final Bagging Setup

21) PLACE THE BAG: Place the sheet of bag plastic over the work area. Again, allow enough slack in the bag to avoid bridging between the two cores. Adjust the position of the pad so that the vacuum pass-through ***will not press directly on the layup*** and the pad will provide a vacuum path to the rest of the bleeder. Use a marker pen to make a 1/4 inch "X" over the bleeder pad. Use a razor to cut through the bag on the mark. Mount the vacuum pass-through in the hole. Use brown mailing tape to seal the bag to the plastic sheet that covers the worktable and make an airtight chamber all the way around the tool.

22) PUMP DOWN: Attach the hose and start the pump. With vacuum applied, you will immediately start to see excess epoxy forced out into the bleeder through the holes in the release film. Run the pump about five hours.

23) CURE, TRIM, WEIGH, AND LABEL: Allow to cure overnight. Unbag. Rough cut the two panels apart near the separation line, using a fine-tooth jig saw, spiral saw, or small battery-powered circular saw. Use a band saw to make a final trim at the marked line. Strip off the peel ply. Weigh. The finished panel should weigh something less than 140 grams. Label the panels with the date, exercise number, and the weight.

Comments

The two test panels produced here weigh about the same as an equivalent area of 1/16-inch aircraft plywood, but are quite significantly stiffer. At EAA demonstrations, it is entertaining to pass around the room comparable samples of 1/16-inch plywood, the samples from Exercises 1 through 5, samples of cured 7725 BID *without* the stabilizing core, and the stabilized cored panels produced here. Experienced builders are always impressed by the obvious stiffness and the fact that the glass edge is ready to be bonded without further operations. (In future Exercises, we will make full-scale assemblies of bonded panels.)

It may seem to the reader who has not done the work described that there is a lot of work involved just to produce two small panels. This is quite true. These methods become efficient when applied to large areas, such as wing skins and fuselage panels, where a single operation may cover most of an entire wing. This Exercise has described very basic and simple operations in complete detail. Once done, they become obvious and the builder can proceed without further instruction. In addition, many operations lend themselves efficiently to two people working together.

The operations described in this Exercise are so basic that the beginner should consider repeating the work. This is an excellent place to get mistakes out of the way quickly and cheaply before real damage and loss of time and materials occur.

Exercise 7

Bonding Panels

Introduction

This is a very important exercise, which will demonstrate several essential techniques, including the use of structural adhesive.

We will practice wetting out and placing glass bonding tape to carry loads between panels. This will demonstrate the great load-carrying strength that can be achieved by composite structures by distributing the load over a large area, avoiding the use of discrete fasteners, such as rivets or bolts.

The techniques used will find application in making fuel tanks or bonding wing ribs to load-carrying skins. Again, this is a good place to get your mistakes out of the way before undertaking a flyable structure! The techniques demonstrated in this exercise are so important that it is well worthwhile to go through it at least twice.

Total working time for this exercise will be about 1 1/2 hours, not including three epoxy curing times and one structural adhesive curing time. With curing times, the project will take as long as four days. To paraphrase *Vision* designer Steve Rahm, "there is always something to do - never be idle during a cure." You will develop an appreciation for the convenience of fast-hardening auto body filler, used as a temporary adhesive.

General Description

We will make a very light, strong, airtight, box, using panels made in previous exercises. We will demonstrate: (1) use of "secondary" bonds and "tape joints," (2) use of adhesives to bond previously cured surfaces, and (3) closure of a previously bonded structure.

Detailed Procedure

1) Prepare the box surfaces: The 10 x 14-inch cored panels produced in Exercise 6 will be the top and bottom of the box. If necessary, use a belt sander or sanding block to true up and match the edges of the panels.

Select two non-cored panels from Exercises 1 through 4. Use a fine-toothed jigsaw or a spiral saw to split one of the panels to produce two 5 x 14-inch panels, which will be the sides of the box. For the ends, cut the remaining panel to produce two 5 x 10 inch panels. Make this cut across the panel to be trimmed (along the 10 inch dimension). Discard the remainder. True up and match the edges with a sander or sanding block. (The final dimensions are not critical as long as the sides match.)

2) Prepare the work areas: We will need two work areas about three feet square, covered with protective plastic, for wetting out glass.

3) Temporarily assemble the box: Use brown mailing tape to temporarily assemble the bottom and the four sides of the box, with the foam core facing inside, as shown below. Place the slightly textured surface, the one that had peel ply covering it, toward the inside. Fold the sides up to the vertical and secure the corners with tape. Use brown mailing tape to seal the joints all the way around the open box.

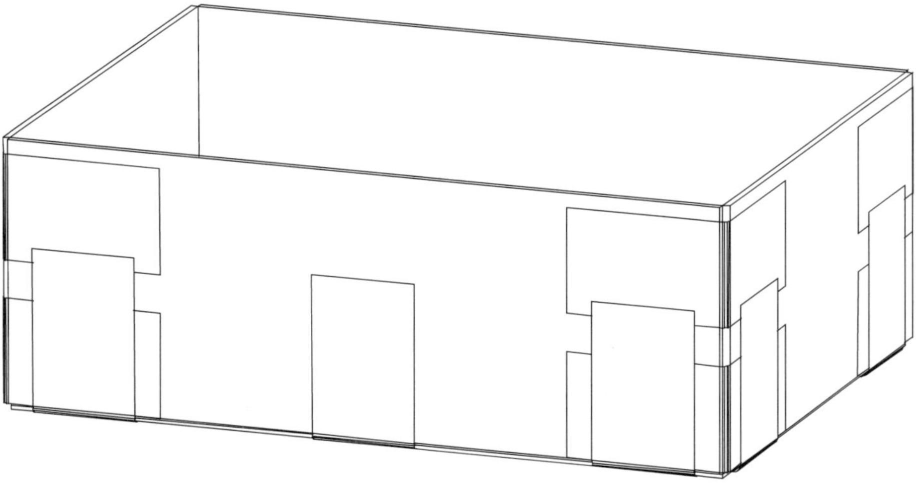

Temporary Assembly With Mailing Tape

4) Fillet the joints: We need to provide a gradual transition to carry loads between two perpendicular surfaces and we need to provide a radius for the corners that the bias-cut glass tape can follow (see below). Luckily, the human fingertip is just about the right radius. The problem is that the human finger tends to get excess micro on the bonding area. Use high quality clean-release masking tape (such as 3M Long Mask tape) to mask off the edges. Leave a clear strip about 3/8-inch from the edge.

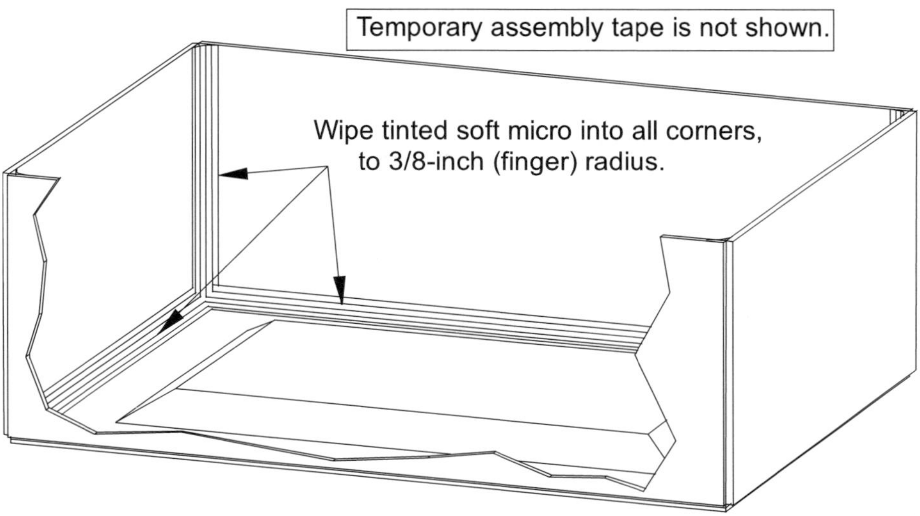

Filleting the Inside Corners

Mix a minimum amount of epoxy. Use a cheap paintbrush to paint the bonding areas with a thin coat of clear epoxy. Use the remaining epoxy to make dry micro, about the consistency of peanut butter. Use a clean mixing stick or your fingertip to place dry micro along the entire joint area. Smooth out with the gloved finger to make a smooth radius all along the joint. Remove the masking tape and touch up any disturbance in the micro. This is a very quick operation. *Note: This is a convenient stopping place. The smooth micro just placed will support the glass tape to be placed in the next step. You can allow the fillet to harden, if convenient, or "work it wet" and continue to the next step.*

5) Prepare and wet the bonding tape: Cut a piece of 2-mil PE plastic about 20 x 18 inches to use as a wetting sandwich. ***(Note that we are using the thinner 2-mil PE, not the 4-mil used in the previous exercises.)*** Fold the plastic double to make a sandwich 10 x 18 inches. Use a marker pen to mark six parallel lines spaced 2 1/4 inches apart and 17 inches long. Place the marked side of the plastic down on the layup area and secure temporarily with bits of brown tape. Mix and tint about four ounces of epoxy. Add a volume of Cab-O-Sil equal to about half the volume of the mixed epoxy and mix thoroughly again. Removing lumps will take a bit of patience. ***The Cab-O-Sil will prevent the epoxy from running down a vertical surface before it has had time to cure. Cab-O-Sil will allow wetting, but will make the mixture "sticky."***

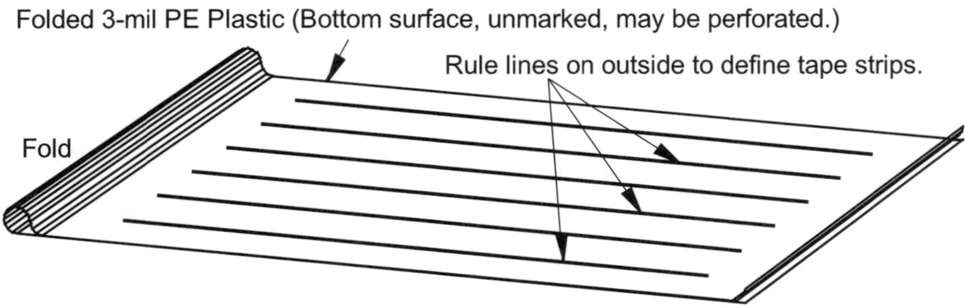

Marking Plastic for Wetting and Cutting Tape Strips

Cut a piece of 7725 BID glass about 10 x 15 inches, at 45 degrees. As before, place it on the marked wetting plastic and wet it out. Be sure that the cloth is completely wetted.

6) Place the bonding tape: Use scissors to cut the marked layup plastic sandwich into 2 1/2-inch strips. Cut four pieces 5 inches long. Remove one layer of plastic from a 5-inch piece.

Place this strip inside one of the end joints and smooth into place (see the sketch on the following page). Remove the remaining plastic and smooth again with the gloved hand. Do the same with the other three corners.

There is an excellent chance that you will spoil the joint on the first attempt by pulling the BID weave hopelessly out of alignment. Pull off the short piece of misaligned tape and try again. This is a ***very cheap lesson*** and the third try should get it right.

Assure that the edges of the tapes lie smoothly with the surface by placing short pieces of either peel ply or scraps of PE plastic over the joint.

Cure and trim. Smooth off the outside edges with a belt sander.

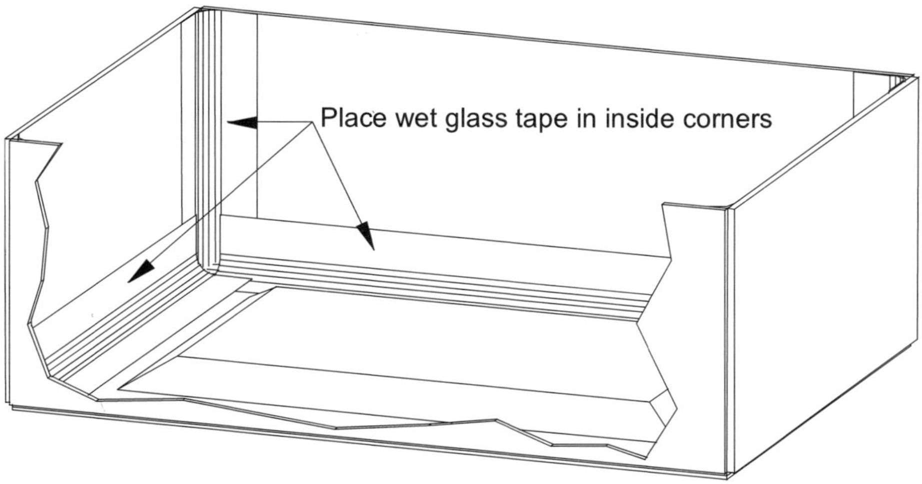

Bonding Tape over the Corner Fillets

8) Place a vacuum fitting for testing: Drill a 3/8 inch hole near the center of one side for a vacuum pass-through fitting. Attach the fitting and tighten the nut. Later, we will pull a vacuum on our new box for a load test. We will close the box in the following steps.

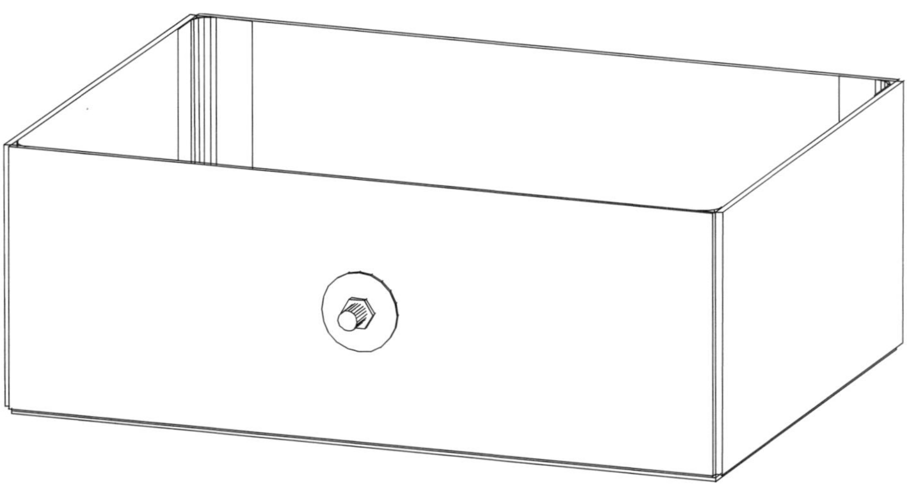

Vacuum Pass-through Inserted in Side of Box

9) Prepare the closure flange: Trim off any excess glass. Mask the top edge with clean-release masking tape, leaving a 3/8-inch clear area, as with the other edges. We now need a removable "flanging tool" which will form a border like a picture frame all the way around the box and which projects inside the box approximately one inch (see the figure). The quickest way to do this is to make such a "picture frame" from scrap wood. Use Bondo to attach little bits of scrap wood to serve as aligners. Make a release face of brown tape as shown in the sketch. Mask the release face. Place the "flanging tool" over the box. Use weights as needed. See the following sketch.

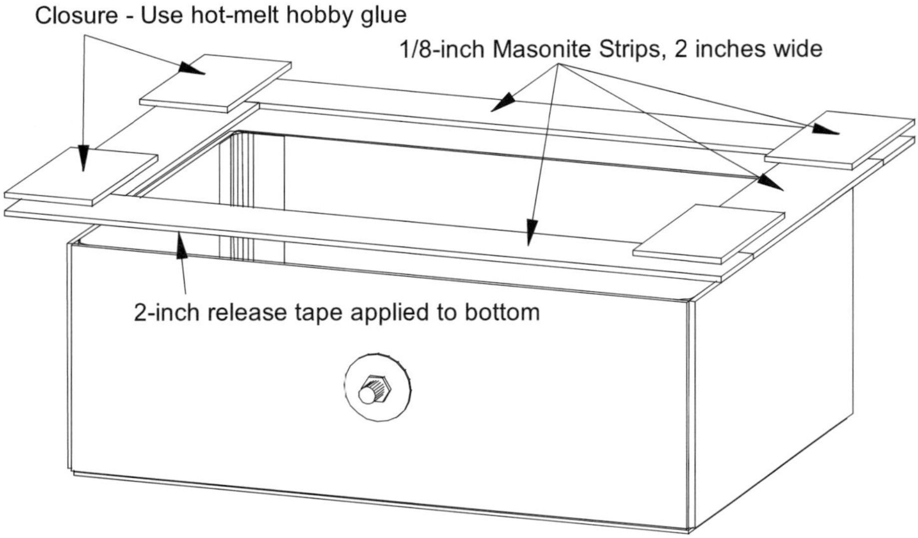

Setting up the Flanging Tool

Strips of clear cured fiberglass can be used instead of opaque Masonite. With clear release tape, it is much easier to inspect the flange and correct errors.

10) Make a fillet for the closure flange: As in Step 5, mix a small amount of dry micro and make a finger fillet all the way around the top edge of the box. This will be harder because you will be working blind and will ***not be able to see behind the flanging tool*** to verify placement of the fillet. Now you can be glad that you used masking tape to protect the bond area. Remove the masking tape. Check your work with a small, cheap inspection mirror, like a dental mirror, as sold in auto supply stores. (Consider that dentists do this all the time.)

11) Place the closure flange: Mix epoxy, tint, add 1/2 volume of Cab-O-Sil, and prepare 50 inches of 7725 bias-cut BID, 2 1/4-inches wide, as in Step 6. As before, place this wet tape all around the top of the box and smooth into place. If some tape goes beyond the flanging tool, don't worry about it. Inspect the work with a mirror. If this is the first attempt to work with a blind, inverted corner, it will be best to use shorter lengths of wet tape and allow 1/2-inch overlaps between sections. The corners will probably be messy: If so, cover them with short pieces of tape. Allow to cure 24 hours.

12) Close: Remove the flanging tool. Inspect the exposed flange. Trim the ragged edge with a spiral saw or Dremel tool. Note that the open surface will be glossy smooth because of the use of smooth mailing tape as a mold release. ***Rough up the surface with 150 grit emery paper and wipe with alcohol.***

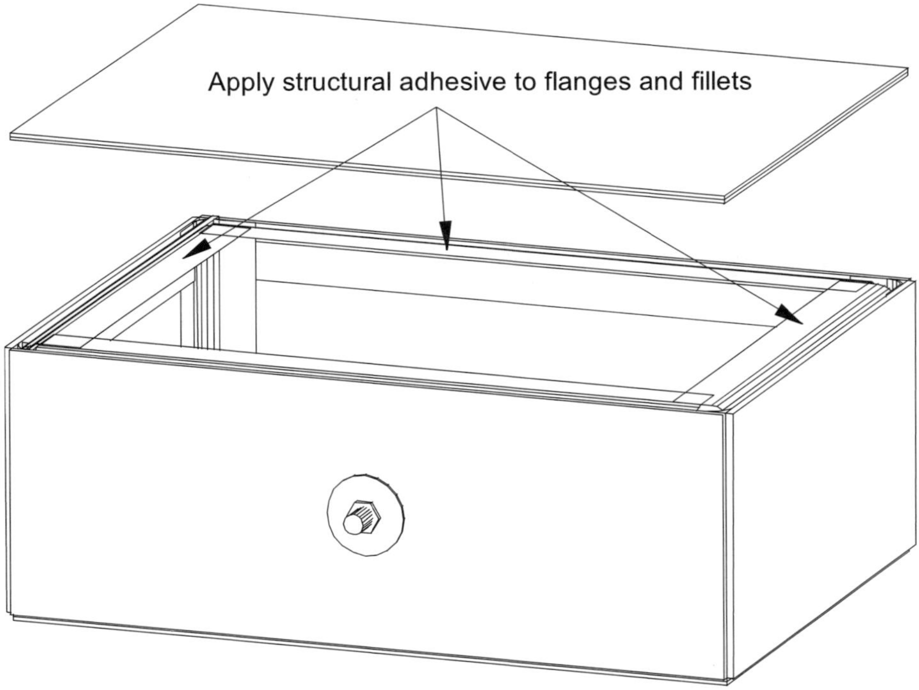

Closing the Box

Note that all joints made so far have involved wet epoxy bonding directly to previously cured epoxy - a ***secondary bond.*** *The conformation of the wet epoxy to the dry bonding surface is perfect, but there are no reinforcing fibers passing through the joint. We are now about to make a connection between two* ***previously cured surfaces*** *where the fit cannot be perfect. To do this, we will use a structural adhesive, which is designed to be significantly less stiff than the cured composite surfaces to be bonded.*

Mix about two tablespoons of a ***Structural Adhesive,*** such as AeroPoxy ES6279, A and B, and spread a bead of adhesive on the flanges and on the bonding area of the remaining cored panel. Place the panel over the flanges and slide it around a little to get good seating. The idea is to drive the air out of the joint as the bead of adhesive is squeezed down. ***The transparent glass allows you to inspect the distribution of adhesive in the joint.*** Clamp or hold in position with any convenient weights, such as sand-filled plastic bags. Let the structural adhesive cure about five hours.

13) Outside Finish: (This step may be skipped for the purposes of this Exercise, unless the builder is a genuine "neatnick" who insists on tidiness.) Remove the completed box from the worktable surface. Use a belt sander to grind off the sharp corner on all six edges of the box to about a 3/16-inch radius. As in Steps 6 and 7, cover the ground-off edges with 7725 bias-cut BID tape 2 inches wide. Mix epoxy, tint, add Cab-O-Sil, and wet out as before. ***(Note that this radius is about the smallest practical for 7725 BID tape. The wet tape will tend to lift off the surface and should be smoothed down after the epoxy starts to thicken.)*** This outside taping is very quick and easy. All the edges can be done at one time because the epoxy will not run down if Cab-O-Sil is used. Cure for 24 hours. Weigh.

14) Test: Connect a vacuum hose to the pass-through. Use a tee fitting and an auxiliary bypass valve. Open the bypass valve and start the pump. Gradually close the bypass valve so that the vacuum gauge reads approximately ***7 psi, corresponding to a distributed load of nearly 1,000 pounds on the top of the box.*** As described, the box will sustain this load. (The author stopped at this point. If the box has been well-cured by several days in a warm place, the reader might find it interesting to discover what the actual failure load is. The only practical value of this vacuum "test" is to easily generate an accurately-known large test load using equipment we already have in place. If this box had been intended to support pressure loads, we would have used a much larger radius on the edges of the box.)

Comments

The main purpose of this exercise was to demonstrate bonding inside corners and surfaces. We ended up with an efficient hollow box. The method lends itself well to building irregular hollow shapes like fuel tanks. Note that we compromised the ideal design somewhat in Step 1, because the UNI layer in the sides runs the wrong way, rather than across the preferred short dimension of the box. If this had been a serious application of the materials, rather than an exercise, we would have made the sides similar to the ends, but we took the easy way. We would have made an even better box if we had used cored skins for the sides, instead of the simple flat laminates we made in the earlier exercises.

Note that the layers of glass tape around all edges of the box, inside and out, make a strong frame and considerably increase the effective stiffness of the sides.

Exercise 8

Vacuum Release Film Experiment

Introduction

This experiment takes only a few minutes to do, but is well worth doing to convince the builder that he need not always suffer from the occasional problems of releasing a layup from a mold. This will ***introduce the toolside bleeder***. The toolside bleeder is a layer of permeable material which is placed between the tool and a non-permeable release film so that a vacuum can be pulled between the tool surface and the release film. The layup is then made on the relatively rugged release film instead of directly on the tool.

Mold release can be a problem in the commercial composites industry. The general practice is to use very expensive, very highly finished and polished molds, kept very clean, usually waxed, and wiped with special mold release compounds. In the boat and bathtub industries, the general practice is to use a sprayed gelcoat in the prepared mold, before doing the layup. Any bugs and dust that come along are simply included in the gelcoat. The gelcoat acts to protect the fragile release surface as much as it serves to provide the bright colors favored by boaters. We aircraft builders choose not to do that; as amateurs in a hurry, we do not want to spend too much time on building and finishing a mold.

General Description

We will demonstrate making a smooth flat mold surface using ordinary woven peel ply and bagging film, both of which usually have creases and folds, so are normally unsuitable for use as a layup tool. We will then simulate a vacuum-bagged layup which uses that mold surface. All this will take only a few minutes, using the setup we have in place for the previous exercises.

Detailed Procedure

1) Prepare the tool: Use the same 3 x 4-foot tool surface used in Exercise 1 and following. Brush it off. Don't worry if the tool has nicks, dents, and rough spots because we are going to cheat our way around this.

2) Prepare the toolside bleeder: Cut a piece of peel ply about 31 x 41 inches. Include any creases and folds that may occur.

3) Prepare the release film: Cut a piece of bagging film about 31 x 44 inches. Nylon bagging film is preferred for this purpose, but 4-mil PE film can be used.

4) Place the toolside bleeder: Lay the peel ply out on the tool and tape into place with small pieces of brown mailing tape. Start with the corners and pull gently to remove creases. Use more brown tape to make the peel ply reasonably smooth. This will take about 16 pieces of tape. Removal and replacement of the pieces of tape is made a bit easier by folding one corner to make a handle. Don't worry if small creases are still visible.

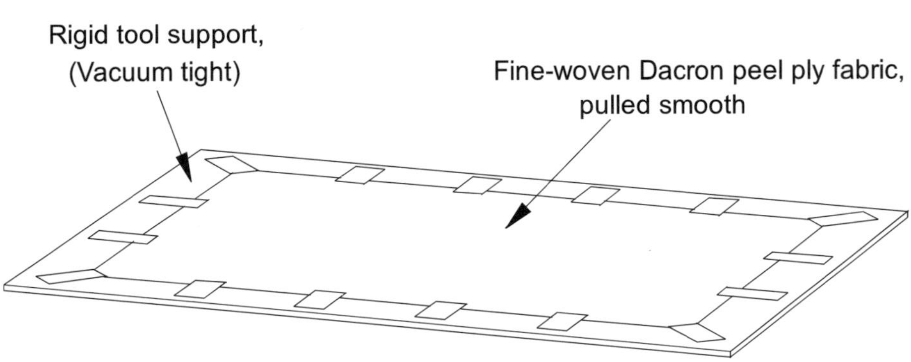

Preparing the Toolside Bleeder

5) Place the release film: Lay out the release film to cover the bleeder. Tape into place with small pieces of brown tape, starting with the corners. Mark a 1/4-inch "X" about 4 inches in from one corner and cut out with a razor blade. Lift one corner, add a 4 inch square pad of bleeder, and insert the vacuum pass through. Replace the tape and smooth out. Tape to the tool all around to make an airtight seal.

6) Pump down the release film: Connect the vacuum line, including a by-pass valve open to atmosphere. Check that the valve is open and start the pump. Close the by-pass valve ***very slowly***, so that the vacuum gauge ***barely reads.*** Use your hand and a squeegee on the surface of the release film to chase any creases and folds off to the side of the tool. (The squeegee is very effective.) Close the by-pass valve a little more and finish the smoothing operation. When the release film surface is very smooth, close the by-pass valve completely to clamp the surface securely. You should get a vacuum of about 24 inches or better with a typical mechanical pump. ***You have created a professional quality tool surface in a few minutes, which is good as long as the pump is running.*** Leave the pump running.

7) Place the layup bleeder: Find or cut some scraps of bleeder so that you can cover an area of about 2 x 3 feet and spread the bleeder on the release film. A bath towel will work well as a test bleeder.

8) Place and seal the layup bag: Find or cut a piece of 4-mil PE film sufficient to cover the bleeder above. As in Step 5, insert a vacuum pass-through near one corner of the bag. Be sure that you have a pad of bleeder under the pass-through. Seal the bag to the tool with brown tape.

9) Pump the simulated layup down: Connect a piece of vacuum hose between the by-pass valve and the bag pass-through. Watch the vacuum gauge and very slowly crack the by-pass valve so that the vacuum always reads 10 inches or above. Let the air in the "layup" bag evacuate. You will eventually see the vacuum increase. Open the valve some more until it is wide open and full vacuum exists in the entire setup.

Comments

This technique allows the builder to make a ***good mold from a bad surface.*** Exercise 3, Basic Vacuum Bagging, could have been done with this setup. Note that when the entire setup is pumped down in Step 9, there is no pressure difference between the tool surface and the work. The release film is held very firmly to the tool during the entire operation and the layup is held to the release film.

This exercise is a simulation. In actual practice, the finished work will tend to pick up the texture of the toolside bleeder. With 4-mil PE film, as described, and standard peel ply, the texture is very slight. If peel ply is to be used on the tool surface of the actual work, the peel ply will further isolate the tool texture and must have its own texture in any case. The builder must do his own trials, here.

The "toolside" bleeder is required because some air path is required to remove air trapped between the release film and the tool. If there were no such path, some areas of the release film would clamp down on the tool, trapping pockets of air, which could not reach the pump. The fine weave of the peel ply is not a very good air passage, but it is sufficient. ***Agricultural weed felt***, used for suppressing weeds in gardens, is a ***good, cheap bleeder*** material. It leaves a slightly textured surface on the finished work, which may be acceptable. It is interesting to see how effective the squeegee is in Step 5; it works for air bubbles under the release film much as it works for globs of resin or bubbles of air in a layup.

Note that we are not restricted to the use of a flat tool surface. Peel ply, as a toolside bleeder, does not lend itself to very much of a double curve, but any cylindrical curve would work. Where two cylindrical surfaces meet at a slight angle, as where a straight wing skin meets a gently-sloping wingtip, peel ply could be fitted to both sides of the intersection. For the larger layups, multiple vacuum pass-throughs will be required to get evacuation over the entire tool surface.

There is a ***significant risk*** in the use of this scheme: ***Very good leak control will be necessary.*** The builder should not proceed unless the vacuum that establishes the tool is excellent, very close to the best vacuum obtained by the builder's pump. If the tool vacuum were truly perfect, it would not matter how large it was, because none of the airflow capacity of the pump would be consumed. It is a very good precaution for the builder to use a stethoscope/microphone to sniff for leaks before going ahead with a full-scale layup.

The author has visited one very advanced application in which the requirement was to make a very deep three-dimensional female mold for an extremely fragile cockpit/wing root cover for a human-powered aircraft. Gentle mold release was crucial. The solution was to stretch bagging film into the mold cavity by a technique similar to the above. Leak control was very difficult and the mold was evacuated with its own bag. In this case, two pumps were used, one for the tool bag, the other for the layup bag.

Exercise 9

Flat Cored Panels

Introduction

This exercise and the three following exercises will produce a ***"high performance" mailbox***. The result is quite a good mailbox with the same dimensions as standard U.S. Postal Service specifications. One example now stands at the author's house. Some people may consider that such extraordinary light weight, corrosion resistance, and stiffness is somewhat unnecessary in a mere mailbox, but it does serve to show the neighbors that you are a builder of high technology structures! More important, this series of exercises will extend the methods used in earlier exercises to make a practical box, which is similar to an aircraft fuel tank. It will also introduce new layup techniques and edge joining methods.

This Exercise will also demonstrate the use of a *large area flat tool*, formed by vacuum.

General Description

The surfaces of the mailbox are essentially the same sandwich construction as demonstrated in Exercise 6, the Cored Skin Sandwich Laminate. This method produces extremely stiff and light panels. The fact that the ***cores do not extend all the way to the edges of the panels considerably simplifies joining the panels***. The bonding areas of the panels merge to form short, therefore stiff, glass columns.

The sides, bottom, and end panels will be made on a flat tool in a single layup, essentially the method used in Exercise 6. Easy tool release is guaranteed by using release film over the entire tool surface. This is a much easier and quicker way to get a safe and accurate release from a flat tool surface than the classical method of forming and polishing a perfect mold and applying mold release compounds.

The panels are made in ***three separate steps:*** (1) forming the outer skin of 7725 BID glass, (2) placement and bonding of the beveled cores of

2-pound blue polystyrene foam, and (3) flat-forming and bonding the inner skin of 7725 BID glass. The outer skin must be cured before the core is bonded with vacuum to avoid the possibility of forming a slight kink in the glass at the edge of the core, which would greatly weaken the panel. Also, it is "aircraft-quality" practice to use the vacuum to remove excess epoxy and any trapped air, before bonding the cores.

> *NOTE:* This three-step process breaks the total job into ***separate working sessions*** that can be ***completed by the beginning builder without help***, if necessary. It also allows time for "aircraft quality" workmanship and recovering from inevitable mistakes. The author has successfully done the entire cored sandwich skin layup - both sides - in one working session, without help, but regrets it.

After curing, the panels are cut apart to be assembled later, using the "lost Bondo" method of forming an edge (Chapter 7).

Detailed Procedure

1) Prepare the work area and tool support: Obtain a 4 x 8-foot flat panel of any convenient material, such as 5/8-inch chipboard, to support the tool. Place on the workbench.

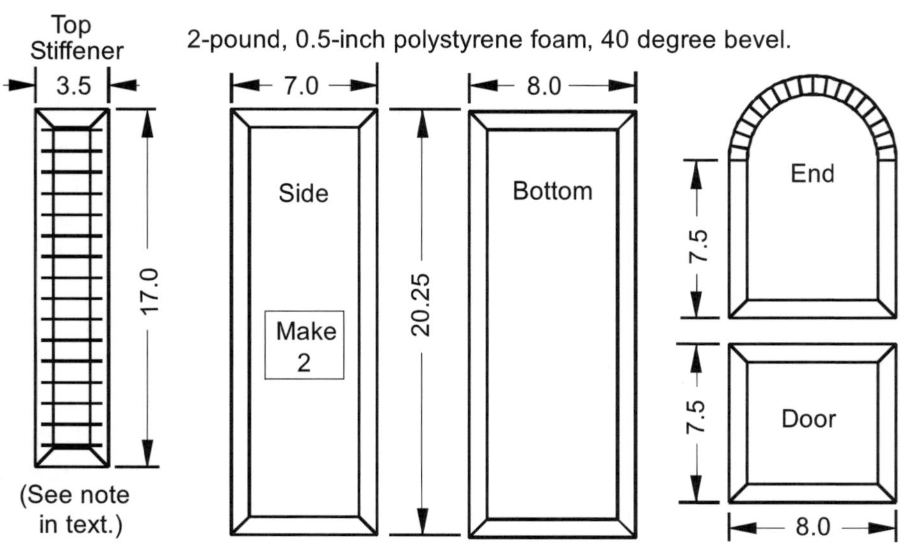

Mailbox Cores

2) Prepare the core panels: Perforate 1/2-inch 2-pound foam, then cut and bevel the six core panels as shown above. These operations go very quickly if the builder has constructed a hot wire setup, a perforating tool, and a beveling tool, as shown in Appendix 1. Set aside.

> *Special note:* The core panel labeled "top stiffener" will be used in Exercise 10 to make a stiffener for the top of the mailbox. The slots are ***saw cuts*** made freehand on a table saw so that an uncut web of about 1/8-inch is left at the bottom of each slot. Slotting the core allows it to be bent to a curve quite easily.

3) Prepare a layout pattern: Use flipchart paper or wrapping paper to make a pattern for the skin panels, as dimensioned below and show on the following page . Use felt-tipped marking pens to mark the glass boundaries with **black ink** and the core locations with **red ink.**

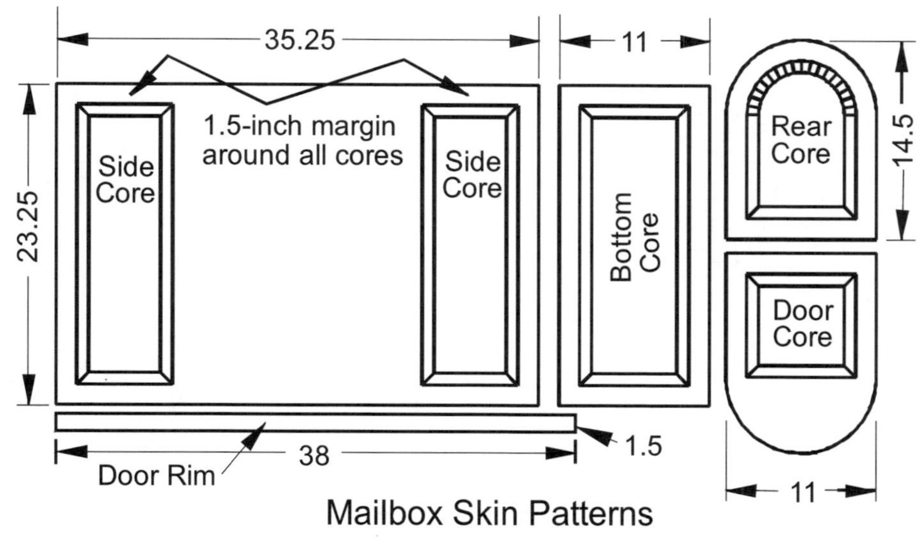

Mailbox Skin Patterns

4) Obtain the tool: Obtain a 4 x 8-foot panel of ***0.090 inch FRP plastic shower liner material.*** This material has a pebble-grain surface on the intended front. We will use the ***smooth BACK surface.*** This material is particularly suitable for our purposes because it is vacuum-tight, lies flat, is relatively cheap, can be re-used many times, and brown mailing tape sticks to it very well. The builder will find that it is a ***general purpose shop tool*** that will be used in many applications where a flat tool is called for. The only disadvantage is that it is awkward to store. Place on the tool support, smooth side up.

Alternative: If the vacuum-tight FRP sheet is not available, the ***envelope bag technique can be substituted.*** A sheet of 1/8-inch plain Masonite can be covered with good quality bagging film and sealed with mailing tape. (The bagging film chosen must also be a release film. Verify that the film used does, indeed, release.) The resulting bag can be pumped down to produce a wrinkle-free work area. The slightly porous Masonite acts as a toolside bleeder to permit full evacuation. The shower liner FRP tool is more convenient, so will be assumed here.

5) Place the pattern: Put the layout pattern on the tool support as shown. Hold in place with small pieces of brown tape.

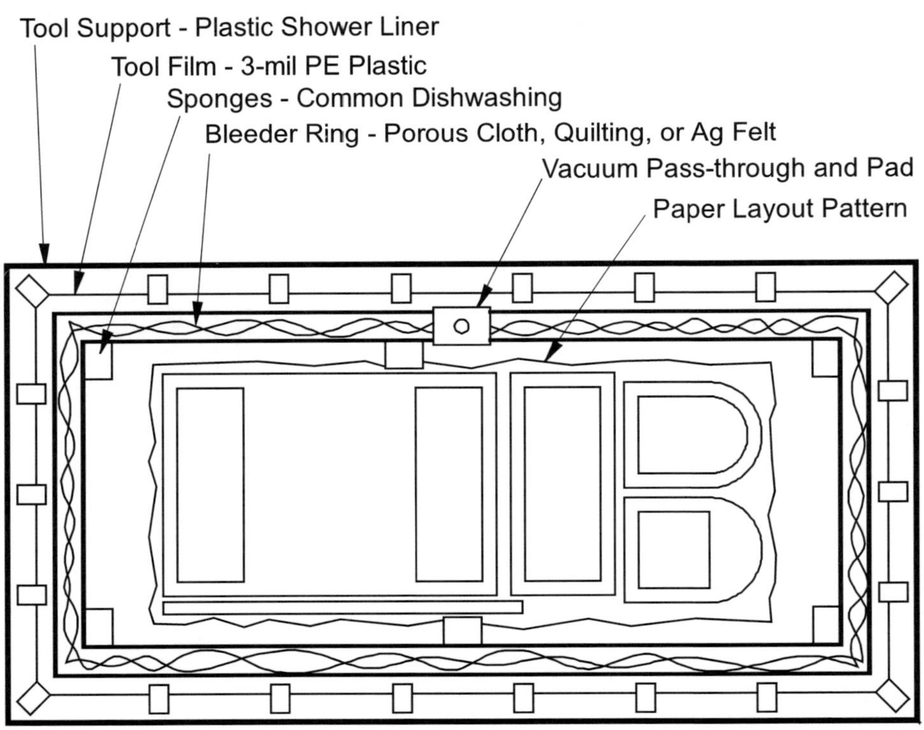

Vacuum-formed Tool Surface

6) Form the tool surface:

Bleeder Ring

See the sketch. Place the bleeder ring as shown, approximately 3 inches wide. A single layer of ***agricultural felt*** as sold in garden supply stores for weed suppression is sufficient, but two layers will produce a faster pump-down. Brown tape does NOT stick to weed felt, so it will be necessary to straddle the felt with tape to hold it in place. Cheap polyester quilt batting filler from a fabric store works well.

Alternative Bleeder Ring

Another good way to make a useful bleeder ring is to split sections of 1/2-inch PVC water pipe on a table saw and use the resulting half sections as stand-offs to lift the tool film off of the tool support. The tool film will lie flat without "pinchoff" when the vacuum is applied. It is not necessary to make a neat connection between sections of the split pipe. The unsupported tool film will bridge any gaps without breaking

Tool Film Placement

Cut a piece of ***good quality bagging/release film*** as shown and place it over the work area. Pull it snugly and hold in place with about six pieces of tape. Lift the edge of the film and place six ***common 3x4x1-inch dishwashing sponges,*** as shown. It will not be necessary to tape the sponges in place. Readjust the tapes as needed to pull the release film tight. The film should stand above the working tool surface like a trampoline, without touching. (The sponge system will not be necessary if the split PVC pipe was used.)

Vacuum feed:

Use bleeder material or knitted cotton cloth, such as an old tee shirt, for the pad. Place the pad over the bleeder ring. so that the side bleeder overlaps the bleeder line. Cut a 1/4-inch "X" in the film and mount the vacuum pass-through.

Complete and Test the Vacuum Seal

Seal all of the periphery of the tool area with brown tape. Connect a vacuum line and pump down. ***A perfectly smooth, wrinkle-free, tool surface will form.*** The sponges will be squeezed flat, but not before they will have served their function as standoffs. Note that the region far from the vacuum feed will form slowly, because of limited bleeder capacity, but the system will work as described and a very good tool surface will result. Verify that the system can hold an excellent vacuum. Do not proceed if the vacuum is only "pretty good." Turn off the vacuum, if desired.

7) Wet out the outer skin: Set up the vacuum system as shown. Turn on the tool vacuum and form the tool surface over the pattern. Sweep out any wrinkles and verify that the vacuum is excellent.

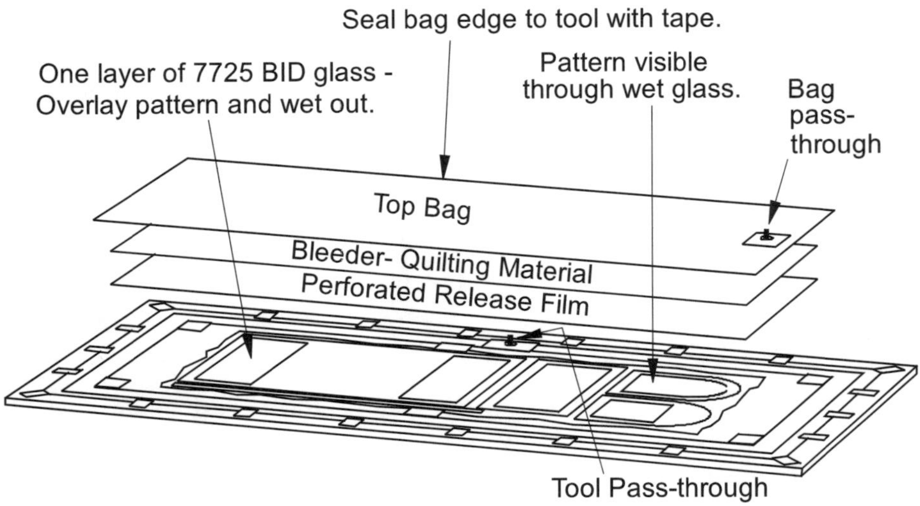

Layup and Bagging of Outside Skin

Cover the pattern area with a ***single layer of dry 7725 BID***. This is a good place to use up the glass scrap bag. Overlap pieces by a half-inch or more. A 45 degree orientation would be ideal, but is not important for this application. Mix about 6 ounces of epoxy, ***tint with BLUE chalk,*** and stir well. Wet out the glass over the pattern, as in previous Exercises. Use the grooved roller, the squeegee, and the paint brush to be sure that the single glass layer is thoroughly wetted and is in contact with the tool film. The tinting will allow easy inspection to find and eliminate bubbles. Use the heat gun to assist in bubble-killing.

8) Apply perforated release film: Cover the work area, plus a 3-inch margin, with perforated 4-mil PE film. Use the roller, squeegee, and heat gun to drive out all bubbles. See that the entire surface is smooth, with no wrinkles at all. Any wrinkles will be reproduced in a line of cured epoxy. (WYSIWYG!)

9) Apply bleeder and bag: Cover the work area with bleeder material. Cover the bleeder with bagging plastic. 4-mil PE can be used as bagging material, but there is some risk of leaks, especially at folds. Usually, the builder will "get away with it" using 4-mil PE and there will be time to recover if luck fails. To avoid doubt, use commercial bagging film. Install a vacuum pass through, as shown. Seal around the edge of the bag with brown tape.

10) Pump down and cure: Connect the vacuum line to the bag. Slowly open the valve in the bag line while monitoring the tool vacuum. Do not allow the tool vacuum to fall below 5 inches. ***Wait while the bag chamber evacuates, which may take a while.*** Shortly after the bag vacuum appears, spots of excess epoxy will appear in the bleeder, following the pattern of perf holes. Open the bag valve more, being careful not to allow the tool vacuum to fall. When vacuum in the two chambers is at equilibrium, open the bag valve all the way. Check for leaks as needed. Run over the tape bond line with a heat gun while pressing the tape into place. Cure for at least 12 hours, or at least until the epoxy is firm and smooth. It is not necessary to wait for a full 24-hour cure.

11) Mount the core panels: Open the bag on three sides or remove the bag and bleeder entirely. Mix epoxy for wet slurry and paint the bottoms of the five skin core panels with slurry. Squeegee in, wait about ten minutes, and verify that the entire surface is covered with a thin coat of slurry. Add more as needed. Put the core panels in place, as determined by the red marks in the pattern. Slurry should rise through the perf holes and be visible on the top. Remove any excess slurry that has escaped out the side. (***Cotton tipped medical swabs*** are useful for this operation.) The wet cores will stick quite well. Proceed immediately to the next step - adding the inside skin.

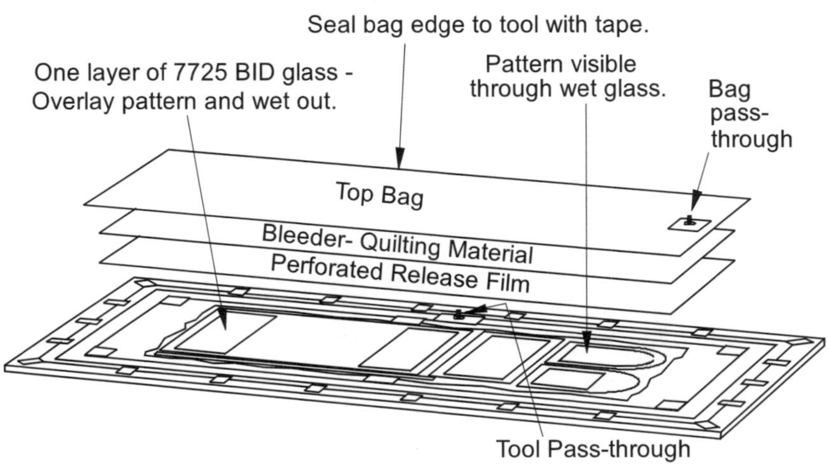

Layup and Bagging of Outside Skin

*NOTE: If it is **not** convenient to proceed with the layup of the inside skin, it is important that the core panels be very well bonded to the cured outside skin. **Skip directly to vacuum-bagging the cores,** without the skin. Do not be concerned with an accurate fit of the release film or bag. The intention is to secure the cores so that the inside skin can be added at a more convenient time. Slurrying and placing the cores is a quick operation and the beginner will usually be able to do the cores and the skins in a single working session.*

12) Wet out and place the inner skin: Remove the bag, bleeder, and perf film. Cut sufficient 7725 BID glass to cover the work and allow about 50% extra for trimming. Make up three folded "wetting sandwiches" from 4-mil PE, about 30 inches square. Mix epoxy and tint with RED chalk. Note the weave orientation of the glass. Place the dry glass in the wetting sandwiches so that the weave is at 45 degrees to the edge and wet out. Squeegee around and add epoxy until the glass is wetted completely, with only a little excess. Turn the sandwich over and verify that wetting is even and complete.

Use scissors to cut wide "tapes" of wetted glass about 8 x 16 inches. Remove one plastic surface and discard in a grocery bag. Carefully place the section of wet tape so that the edge slightly overlaps the work area. Remove and discard the second plastic layer. Wherever the "wide tape" overlaps a core plate, carefully smooth the wet cloth over the bevel and assure that the cloth is firmly positioned at the bottom of the bevel. Slight wrinkling at the corner of the bevel is acceptable, but the builder will find that the satin-weave cloth will conform quite well. A squeegee and the gloved hand will fit the cloth to the cores.

Continue placing "wide tapes" until the entire work area is covered. Overlap tapes by at least 1/2-inch.

13) Add peel ply: Use 2-inch strips of dry peel ply to cover the margin areas. Squeegee out smooth. Verify that the peel ply wets through with excess epoxy brought up from the wet layer below. If there are dry areas, add a little epoxy and heat with the heat gun until wetted.

14) Bag and cure the inner skin: As before, position and reuse the perf film.

NOTE:: *The perf film will not conform well to the bevel angles. A small ridge of epoxy will be left wherever the perf film fails to con-*

form, but can usually be ignored. (The underlying glass skin is unaffected.) Where desired, the builder can use a razor blade to slit the perf film to better fit the cores.

Add bleeder generously and close the bag. The builder will find that once-used bleeder typically has only a fraction of its volume occupied by waste epoxy and can be used a second time if fresh bleeder is placed next to the work. It is difficult to have "too much bleeder" and even partly used bleeder helps the air flow.

Connect the second vacuum line and pump down as before. Cure for 24 hours.

15) Unbag, mark, and finish: Remove the bag, bleeder, and perf film, and expose the finished layup. ***DO NOT REMOVE THE WORK FROM THE PATTERN!***

Leave the peel ply in place. Clean the surface with a rag wetted with alcohol. ***Trace the visible marks on the pattern*** to the finished layup with a black marking pen. NOW remove the layup from the tool surface.

Rough cut the separate panels with a spiral saw. Finish trim with a fine-toothed bandsaw. Touch up the exact line with a belt sander.

16) Prepare the post mounting attachment on the bottom: Assuming that the mailbox will be attached by four 1/4 inch bolts, it will be necessary to provide four hard points. One very easy way to do this is as follows:

Locate the four bolt holes and cut out four 1-inch or 1 1/4-inch holes to correspond to the final bolt locations. A hole saw or a fly-cutter can be used with a very cautious feed. (The single layer of glass and the 2-pound foam core will be very fragile with this transverse load.) Freehand cutting with a spiral saw is safe and quick, but makes an untidy hole. Close one end of each of the four holes with brown mailing tape. Mix a little epoxy and make a barely flowable paste of ***epoxy and coarse sawdust. Fill the four holes completely with the sawdust mix.*** Cover with a scrap of PE plastic and allow to cure. Remove the brown tape and the PE cover. If desired, drill countersinks at each of the four hole locations.

Mix and tint epoxy with about 1/6 volume of flox. Wet and place two patches of 7725 BID glass to entirely cover the hole area. Press the wet cloth into the countersink depressions with the gloved fingertip. Smooth out and peel ply the edges. Cure.

Drill out the required bolt holes.

17) Mask the inside joint surfaces: While the skin is flat, it is very easy to put masking tape or brown mailing tape along the inside edge of all joint surfaces. Place the masking tape ***about 3/8-inch in from the trimmed edge.*** The purpose of this tape is to protect the joint area from the rather sloppy process of applying Bondo to fill the area for the "Lost Bondo" joint in the following Exercise.

Exercise 10

Forming and Joining

Introduction

In this exercise, we will form and complete the shape of the "high performance" mailbox which we laid up in Exercise 9. While it is true that the requirements of a mailbox are not very demanding, the techniques demonstrated here could be used to form a wing leading edge or a fuel tank.

The essential point is that the actual layup is done on a ***flat tool***, which is very easy, then the actual forming of the finished part is made by means of a very simple bending jig. This process is ***much easier than making a full-scale precision mold*** capable of supporting all the operations necessary to make a closed part directly.

General Description

We will make a very simple bending form, or jig, bend the skin from Exercise 9 into the desired shape, form the top brace, attach the end plate and bottom plate, and complete the "Lost Bondo" corner joint.

Detailed Procedure

Making the Bending Form

1) Make the bending jig: Find some scrap plywood or chipboard of about 1/2- to 5/8- inch thickness, and build a wooden jig as shown in the sketch following. Dimensions are not critical as long as the ***curved sections contain semicircular cutouts of 5 1/2-inch radius*** to match the required curve of the top of the mailbox and the four 9-inch "legs" should project above the bottom edge of the mailbox. The two forming curves should be separated by about 16 inches. Diagonal bracing is not necessary.

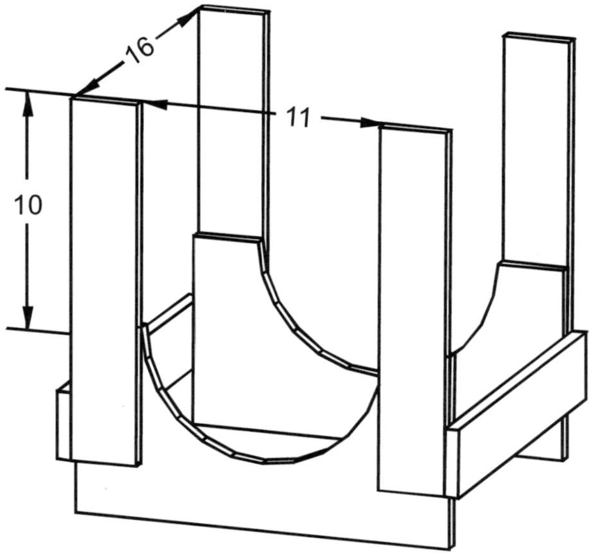

Mailbox Skin Bending Jig

Use small steel corner ties, like the Simpson Type 421, and self-drilling washer-head screws rather than bother with complicated wood joints. With the steel corner ties and self-drilling screws, this will be the work of only a few minutes.

2) Mount the top brace core: See the following sketch. Locate the slotted core constructed in Exercise 9. Prepare a small work area by covering a table with PE plastic. Mix medium slurry and coat the bottom of the core. Mount the slotted core on the flat skin, set back two inches from the edge of the skin. Use the edge of a hacksaw blade to ***clear the slots of excess micro***, to make things easier when the skin is bent. Cover with scraps of PE film and secure with sandbags. (We will not bother with vacuum bonding here - we will get an excellent bond with the sandbags.) Check to see that the wet slurry has not oozed out to form ridges. Smooth as necessary. Cure the micro 24 hours.

3) Bend the skin: Inspect the top brace core and see that the saw slots are open. If not, ***clear any hardened micro with a utility saw.*** This is a non-critical freehand operation and it does not matter if some foam is torn out. Mix up a little medium wet micro and fill the slots completely.

Place the skin in the bending jig, as shown, and secure in approximate position with sandbags or other convenient weights. See that ***the slotted core is easy to bend*** and ***excess wet micro will be squeezed out of the slots*** in the core. True up the bent skin with a carpenter's level and sandbags, as required. Clean up excess micro. The builder may find that it is necessary to secure the skin to the jig with self-drilling screws. The small screw holes in the edge will be covered when the "Lost Bondo" joint is made.

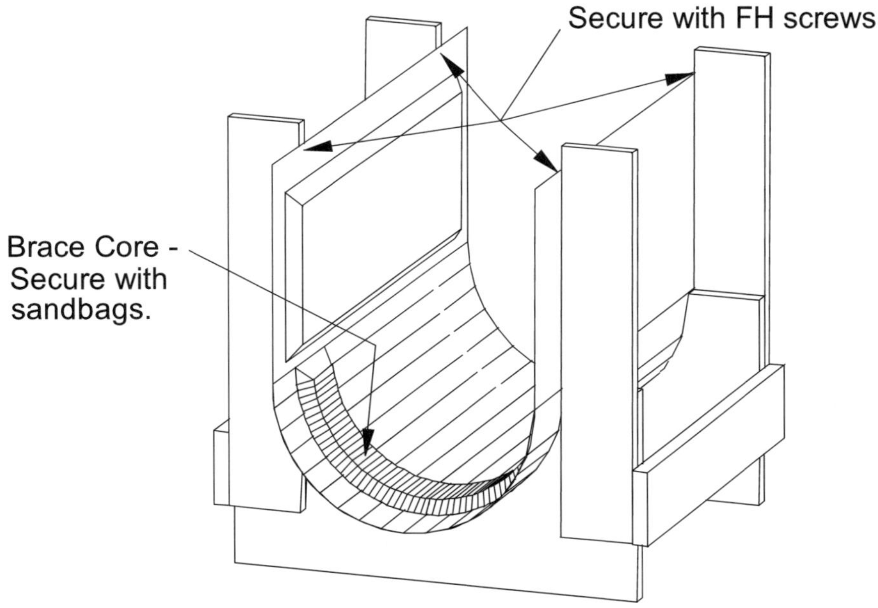

Formed Skin with Curved Brace Core

4) Complete the top brace skin: Find some glass scraps, mix a small amount of tinted epoxy and cover the top brace with ***one layer of 7725 BID.*** (It will be easier to use two overlapping sections of wet glass, rather than to try to fit a single piece.) Smooth out with the gloved hand. Cover the wet edge with scraps of ***peel ply to make a nice blended edge.***

5) Fix ("tack") the position of the back: Place the back into approximate position and secure with small pieces of brown tape. Ignore irregularities and find a good compromise fit. Mix up a small amount of 5-minute epoxy and add an equal volume of flox. Mix the flox quickly. Secure the back into position with ***small dabs of 5-minute flox carried on small pieces of cotton rag,*** about 1 1/2-inches square. Allow to cure.

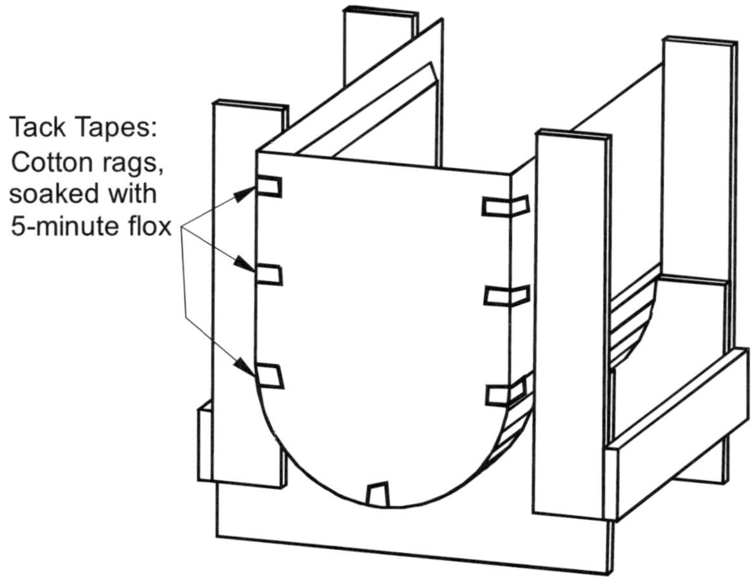

Tack-attaching the Back Panel

Note: If the builder does not mind waiting 24 hours for an epoxy cure cycle, it will be less rushed to attach the back using regular epoxy and either the "flox dab" method above or by using small patches of epoxy-soaked rag. The beginner may find the additional working time to be welcome.

As a second alternative, dabs of ***methacrylate adhesive*** can be used as mixed, without any flox. This alternative is the best compromise between ease of use, working time, and stability of the tacked joint.

6) Apply filler to the joint between the back and the main skin: Mix about four tablespoons of styrene-based body filler, such as Bondo, or similar material sold as "Light-weight Non-clog Filler," and ***tint the filler with chalk to enhance the red or blue color of the catalyst.*** (Use of too much catalyst to produce a deeper color will reduce the working time too much, so we will use chalk.) Place the wet filler in the joint and smooth out with a wooden tongue depressor and the gloved hand. Smooth the filler until the ***edge is neatly blended.*** Immediately remove the masking tape and blend again. Repeat with fresh filler until the entire joint area is filled.

Note: It is very handy to use a disposable ***pastry decorating bag,*** as is used for writing "Happy Birthday" on cakes, with a #3 or #4 nozzle, to place the mixed filler. The method is fast and accurate, but the builder will be limited by the very short working time of the styrene filler and must allow for quite a bit of waste of filler in the bag. Fortunately, the cheapest non-clog filler is quite satisfactory. Again, if the builder can accept a 24-hour epoxy cure cycle, use of "regular" dry micro, with about 1/3 volume of fused silica, will allow plenty of working time.

7) Fix ("tack") the position of the bottom plate: As in Step 4, above, attach the bottom plate. Note carefully that the ***cored side of the plate faces to the OUTSIDE***, so that the mail box will have a flat interior floor. Do not attach the floor (bottom) upside down!

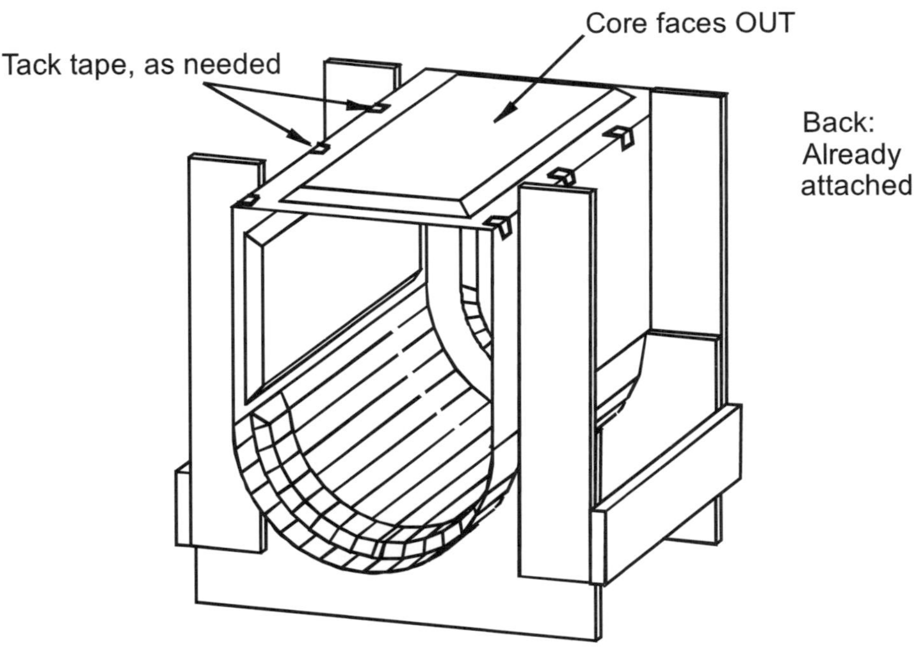

Tack-attaching the Bottom

8) Apply filler to the joint between the bottom and the main skin: As above, apply filler to make a radiused inside joint. Use the tongue depressor, the pastry bag, or the gloved finger to apply and distribute the filler. Immediately remove the masking tape and clean up. It will be a little harder to work at the back end (bottom) of the mailbox, so do not attempt a batch of filler that is too large.

9) Tape the entire inside joint: Use a folded sandwich of 2-mil PE plastic to prepare eight running feet of single-layer 7725 BID glass, 2 1/2 inches wide. Wet a panel of glass about 16 x 15 inches, plus a margin for trimming, which will produce 6 strips 2 1/2 inches wide. Tint the epoxy and use about 1/3 volume of fused silica and 1/6 volume of flox to make the tape "sticky" and avoid running out by the epoxy. Squeegee out well and assist wetting with a heat gun.

Work with short lengths of 2 1/2-inch wet tape - no more than 4 to 6 inches in the curved areas. Remove one layer of wetting plastic, and place the tape in position. Peel off the remaining layer of wetting plastic and smooth and relocate the tape as needed. Repeat along the entire joint, allowing about a 1/2-inch overlap between tape segments.

Place short pieces of ***dry peel ply over the edge*** of the wet tape and smooth down to avoid snags and rough edges after cure. Excess epoxy will soak into the peel ply and allow easy inspection.

Cure. Strip off the peel ply.

10) Grind off the outside corners: Notice that the structure is ***now quite stiff,*** even though there is only one layer of glass at the joint. Use a hand-held belt sander with a coarse-grit belt (about 50 grit) to grind off the outside corner to a smooth curve, so that the colored Bondo is just removed. Remove any eccentricities and overlaps. Grind to expose the underlying filler, removing most of it. There is no significant disadvantage to leaving small areas of filler in place; a thin layer of hardened filler will serve as a "core," in this case.

> ***Note:*** Grinding off the corners is a quick and easy job, but nasty, and is ***best done outdoors*** with a face mask and covering clothing.

11) Tape the outside corners: As in Step 9, make up about eight running feet of 7725 BID glass tape, 2 1/2-inches wide. Apply all around the outside corners. ***Use peel ply*** to assure excellent blending with the bonded surfaces. There is no problem of access here and an excellent joint will be the result. Cure. The mailbox will be ***very stiff in torsion.***

Exercise 11

Mailbox Door and Finishing

Introduction

Most readers of this book will probably not wish to actually own a mailbox as constructed in the two previous Exercises. The purpose of the mailbox is to serve as a useful practice project which offers considerable opportunity to make non-critical mistakes and profit from them. If the reader is seriously contemplating building an airplane, it may be a very good return on investment of time and effort to repeat Exercises 9 and 10 and build a second box, to polish basic skills.

The skills to make a door that will complete a practical mailbox are less important than the skills used to make and assemble the main skin, but the additional practice in precision fitting and assembly will be worth the small additional time spent.

General Description

We will form the door from the ***door panel*** and ***door rim strip*** made in Exercise 9, using the "Lost Bondo" joint described in Exercise 10. We will form the door to a ***quite accurate fit*** to the main mailbox structure and arrange a simple latch to hold the door closed.

Detailed Procedure

1) Shim the door clearance space: We will need a small but stable clearance between the door and the main skin. Note that the open-ended mailbox is quite rigid and it is practical to make a close-fitting door. Cut strips of newspaper or poster paper about 2 1/2-inches wide and make a shim over the end of the mailbox so that the shim is about 1/16- to 3/32-inch all around. Secure the shim with brown mailing tape.

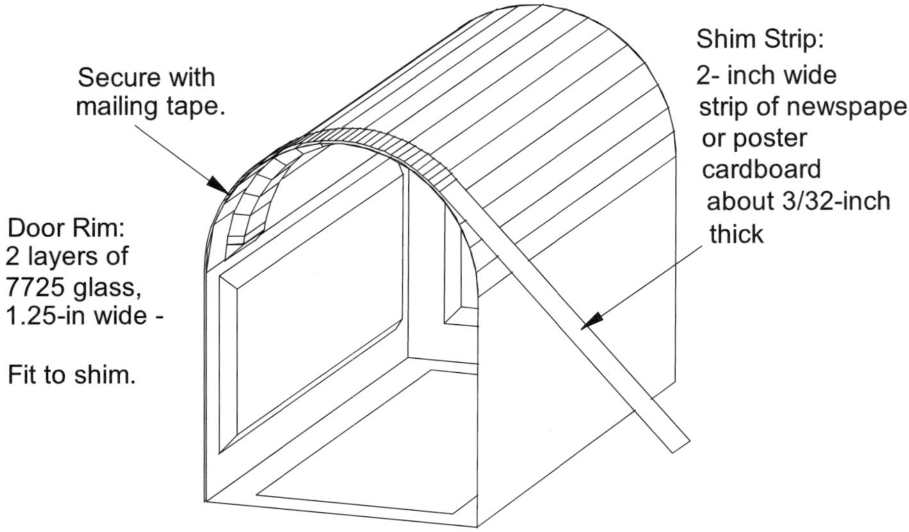

Shimming and Fitting Door Rim

Note on shimming technique: The builder may have occasion to require a ***very accurate shim,*** not important in the case of the mailbox, but possibly important in fitting a cowling or making a very accurate joggle. Layers of poster paper and/or newspaper may be ***coated with brown tape on BOTH SIDES*** and glued to the reference surface with ***rubber cement***, which has been ***dried in the air until non-sticky*** to the touch. After use, the shim will strip off cleanly.

2) "Tack" the front door panel: Find the 1 1/2-inch door rim of 2-layer 7725 BID that was laid up with the main skin in Exercise 9. Trim to exact size as needed with a table belt sander. (This will make glass dust! Work outside, if feasible.) Bend the door rim so that it fits tightly over the shim material on top. The springy glass strip will tend to bulge out at the sides, so ***secure the sides with a self-drilling #6 screw*** on each side. (The screw hole will be covered later in the taping operation.)

As in Exercise 10, "tack" the door to the rim strip with small 5-minute "tack strips," "flox dabs," or dabs of methacrylate adhesive. After the tack has cured, remove the door from the mailbox.

3) Fill and complete the corner joint: As in Exercise 10, apply filler to the joint between the door and the door rim and blend the curve smoothly. Wet out one layer of 7725 BID and tape and peel ply the inside corner, as before. After the inside tape has cured for 24 hours, place the door on the mailbox to serve as a support and grind off the edge and the "lost Bondo." Tape the outside, as done with the main box, and smooth out with peel ply.

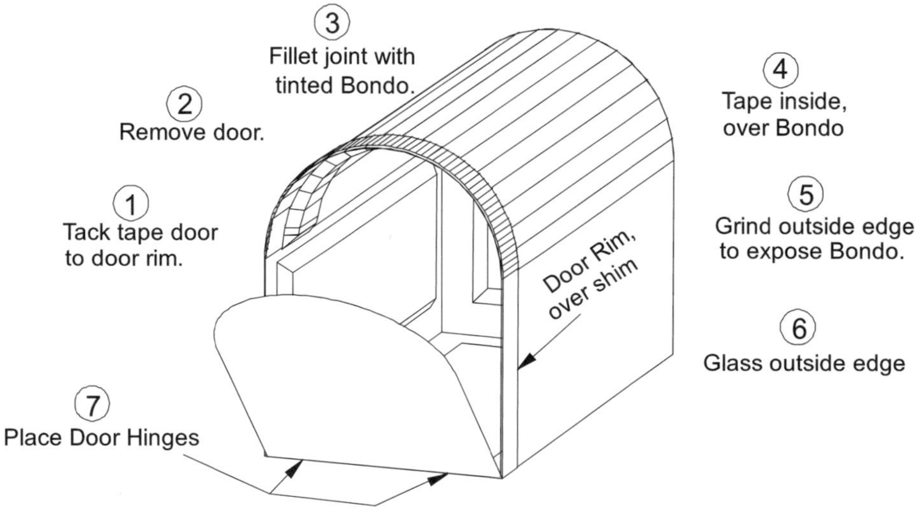

Mounting the Door to the Door Rim

4) Make the latch arm: See the sketch following. The less than expert welder will probably find it easier to weld the cross piece first and bend and trim later. A propane torch can be used for heating the steel rod at the bending points. The 1/8-inch steel rod is then easily bent hot with pliers.

5) Mount the latch arm: Drill a 5/16 inch hole in the door 1 1/8-inch below the top edge. Cut and deburr two pieces of 3/16-inch brass tubing, as shown, as sold in hobby shops for model airplanes. Use a cotton-tipped medical swab, with most of the cotton removed, to place a ***very small quantity of automobile bearing grease*** inside each tube, being very careful not to get grease on the outside of the tubes. Place the tubing over the pivot of the latch arm and slide the latch arm into position on the inside of the door. Mix up a small amount of flox (using 1/3 volume of fused silica) and ***secure the tubes into position with flox***. Cure.

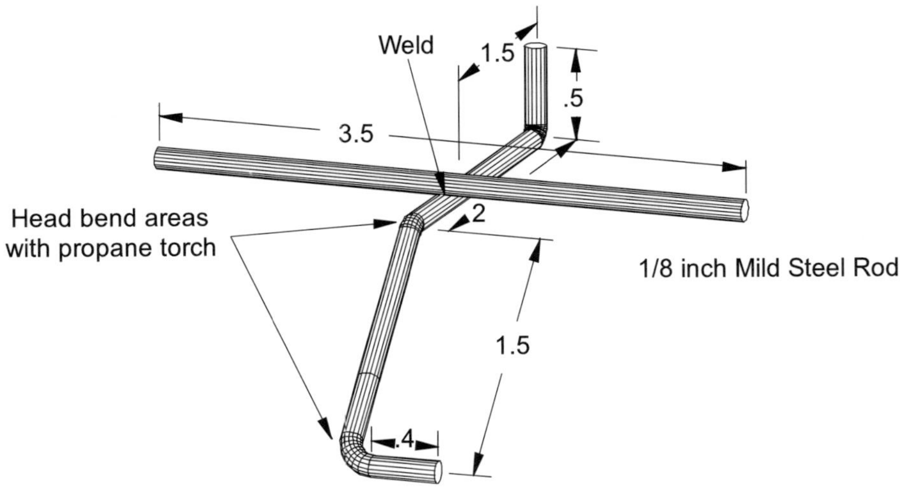

Latch Arm

6) Make a latch spring: Use 0.090 inch steel music wire as sold in hobby shops. Bend to a shape approximately as shown. Clamp in position, adjust for a "feels right" restoring force, and secure with globs of flox as shown. ***Cure thoroughly***, at least 24 hours, in a warm place.

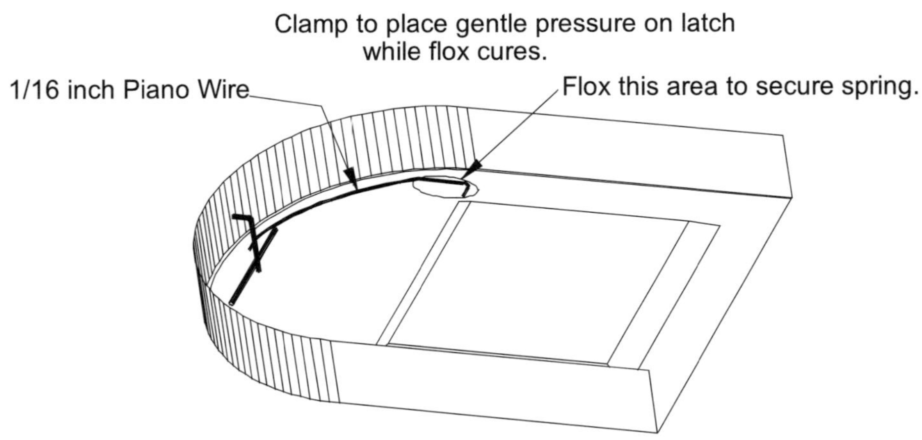

Door Latch Spring Adjustment

7) Make a latch stop: It will be necessary to provide a stop to limit the vertical motion of the latch arm so that the hook is just below the line of the edge of the door. (If the latch hook goes above the edge of the door, the door cannot be closed.) Shim the latch hook as needed and make a stop to define that position. It is convenient to use a ***very small scrap of cured glass in contact with the latch arm and to pack wet flox*** behind the scrap. Cure. (If methacrylate adhesive is available, use methacrylate flox and save a long cure cycle.)

8) Mount the hinges: Use standard hardware store brass tight-pin hinges, about 2 x 2 inches. Drill and mount with finger-tight nuts. Note that the ***hinge pin should be on the OUTSIDE*** of the bend.

9) Make the catch: Test fit the door to the mailbox and mark the exact position of the latch hook on the outside of the transparent mailbox. Find a fine tooth ***hacksaw blade*** and break off a length about 1/8-inch shorter than the distance from the edge of the mailbox and the latch hook - about 1 3/4-inches. Sand the piece of hacksaw blade to make it clean and bright.

Cut a small piece of metal about 1/4-inch long, such as a section of a #8 machine screw and use it as a spacer to set the angle of the catch. Use a very small quantity of 5-minute epoxy on the spacer and, ***before the 5-minute epoxy sets***, position the piece of hacksaw blade so that the catch has zero spacing at the edge of the mailbox skin and about 3/8-inch spacing inside the box. Adjust for an exact setting. Remove the piece of hacksaw blade. Let the 5-minute epoxy harden.

Mix flox and butter the bottom of the hacksaw blade. Return to position and fill the gap between the skin and the blade completely with flox. Smooth off the edges. Verify that the ***back end of the catch is smooth and flush with the steel edge.*** Cure.

10) Make the flag support adapter: The flag support is an adapter that matches the curved side of the mailbox to the vertical plane of the raised flag. An easy way to make such an adapter is as follows.

Tape a piece of PE plastic, about 8 x 10 inches, to the mailbox so that it fits the mailbox snugly.

Mix up a small batch of epoxy with about 1/4-volume of fused silica and about ***1/4-volume of flox***. Mix well. Add ***coarse sawdust*** so that the mixture is well wetted and stirred, but does not flow. Place the entire "glob" in position on the PE plastic. Tape another 8 x 10-inch piece of PE plastic on a scrap of plywood or other flat surface and force it down over the wet epoxy/sawdust so that the flat surface is in contact with the side of the mailbox. Move it around to drive out trapped air. Secure with a weight so that the weighted side is parallel with the side of the mailbox. Let the ugly mess cure.

Identify the part of the formed epoxy sawdust that matches the desired position of the flag. This will be a wedge-shaped form which is flat on one side, curved on the other, and fits the mailbox perfectly. Use a marker pen to mark a 2 1/4-inch square section. Use a bandsaw to cut out the finished adapter. Smooth the edge with a bench belt sander.

> ***Construction alternative*** This is a very general way to make adapters between different surfaces. A more conservative design, for applications more demanding than a mailbox, would be to use a ***glass skin on both surfaces of the adapter.*** To do this, wet out a layer of 7725 BID and place it on the curved surface over the release film. Mix up and place the wet sawdust "glob," as above, then ***tape PE plastic to a second flat surface and wet a second piece of 7725, which will stick to the PE plastic.*** As above, force the second flat surface against the wet sawdust and cure.

11) Make the flag: Use the glass scrap bag to locate glass scraps sufficient to wet out a three or four layer strip of finished layup 2 inches wide and 13 inches long. Wet out in the usual plastic sandwich, but use ***four times the normal amount of red tinting chalk.*** Smooth out, but do not trim while wet. Cure, with the plastic in place. Cut out the final form with a bandsaw and finish trim with a bench belt sander.

12) Drill and place the adapter: Drill two clearance holes for #10 machine screws perpendicular to the flat face. The center hole will be the pivot point for the flag. The second hole will be the limit stop for the flag. Make a trial assembly to locate the exact position of the adapter, tape in place, and drill the third #10 hole in the adapter and the mailbox so that it is perpendicular to the curve of the mailbox. Countersink this hole. Assemble the adapter finger tight, use the adapter as a drill guide, and drill the first two holes into the mailbox. Attach the adapter, using the countersunk hole, and tighten the nut to secure the adapter to the mailbox.

13) Complete the flag assembly: Use a spiral saw or Dremel tool to cut a slot in the flag. Cut a square piece of cured glass to serve as a cover and drill as shown. Make a trial assembly and adjust the end points of the slot to bring the flag to the desired limit position. When the limits are correct, tighten the bolts to produce the desired degree of friction to hold the flag.

14) Final finish: This will depend on how far the reader wishes to go with a project not directly related to aircraft building. Assuming that the reader will be satisfied with simple painting, it should be sufficient to smooth out the corner tape joints with a little polyester glazing putty from an auto supply store. Paint as desired and assemble.

15) Assemble the base and handle: Assuming that the mailbox will be mounted on a piece of 4-inch ABS pipe, redrill the mounting holes in the mailbox bottom and attach a standard 4-inch pipe cap, using countersunk screws. The finished mailbox will attach to the post with a gentle force fit.

A 2-inch piece of 1/2-inch PVC pipe can be used for a handle. Drill a hole in the middle, insert the latch arm, and fill the pipe with epoxy sawdust.

Exercise 12

Flat-sided Fuselage Simulation

Introduction

The author has assumed that the reader has a serious interest in applying the special techniques discussed in this book to build some kind of high performance structure, if not an aircraft. ***These techniques require practice to develop skills and the Exercises described previously are intended to provide that practice. The inexperienced builder is strongly advised not to attempt this Exercise "cold."*** It is not absolutely necessary to complete every preceding Exercise, but if the reader is disposed to "cut corners," he should, at least, have completed the basic layup work to the point where he is comfortable and reasonably quick at making flat layups. This Exercise requires full scale "taped" joints and will certainly save frustration if the builder has had some previous practice "taping."

The rewards of completing this Exercise are great because it will provide familiarity with working with full-scale parts, using a wide range of techniques, ***including the practical alignment and assembly of these large parts.*** The structure is extremely simple and allows easy access to all areas. Mistakes are inevitable, and this is an excellent place to make them, without a large expenditure of time or materials. (The author knows of builders who have ***twice*** made nearly-complete composite fuselages for their intended projects, only to scrap them and finally succeed on the third try. Better to nail the techniques first.)

General Description

In this Exercise, we will make a full-size, very simple, approximation to the tail cone of a small aircraft very similar in dimensions to the *Tailwind.* We will ***use only two nearly identical cored-skin plates, layed up on a flat tool.*** We will make the necessary corners by ***FOLDING the flat layups into half-sections, without a jig,*** and will fix the shape of the bent forms by placing bulkheads. We will then fit and assemble the two halves. This Exercise is a larger-scale extension of Exercise 9, the Mailbox, demonstrating folded corners.

The sketch following shows the concept.

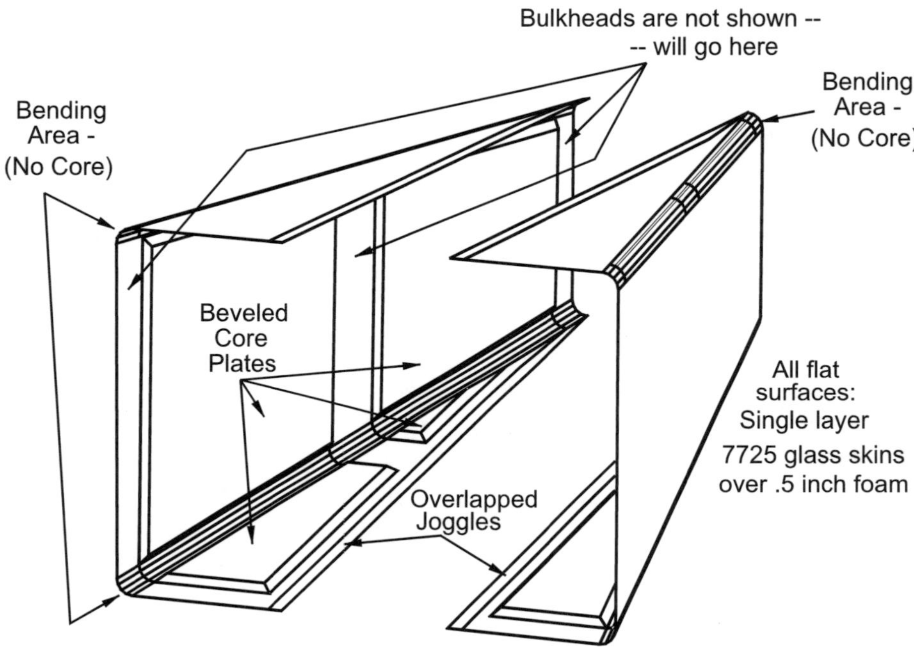

Fold-formed Fuselage Halves

Detailed Procedure

1) Prepare the layup tool: The sketch on the following page shows a dimensioned flat projection of one of the symmetrical skins. Note that the ***widest dimension is very slightly larger than 48 inches***, the width of the flat tool support we have used in previous exercises. We will need a maximum width of very close to 50 inches, plus a small working margin, so we will ***require a 6-inch splice in the tool surface***, as shown.

We will use ***plain, non-surfaced, 1/8-inch Masonite as the tool surface*** and form a working surface with release film, using vacuum, as in previous Exercises. Masonite is not vacuum-tight, so it will be necessary to use the ***envelope bagging*** technique to ensure an excellent vacuum when forming the working surface.

Exercise 12 - Flat-sided Fuselage Simulation

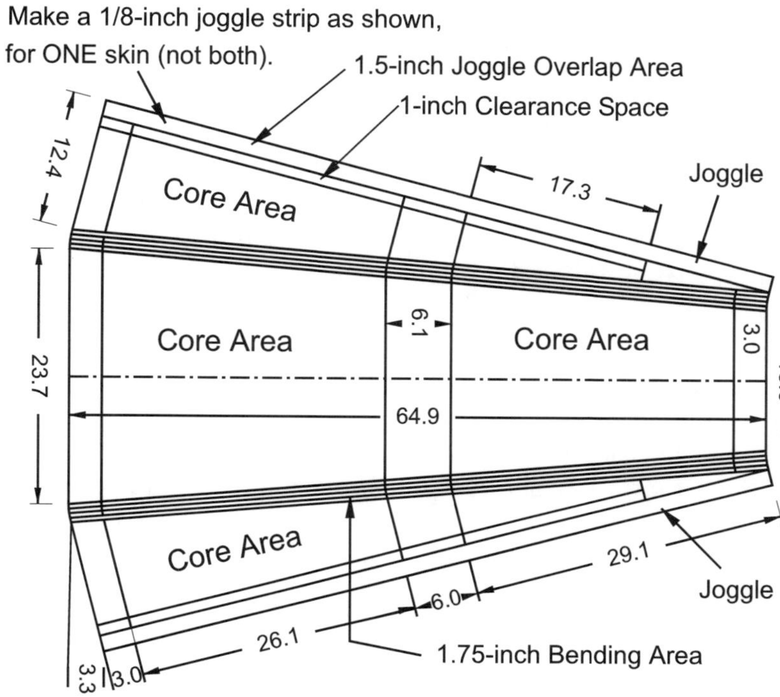

Fuselage Skins - Flat Projection

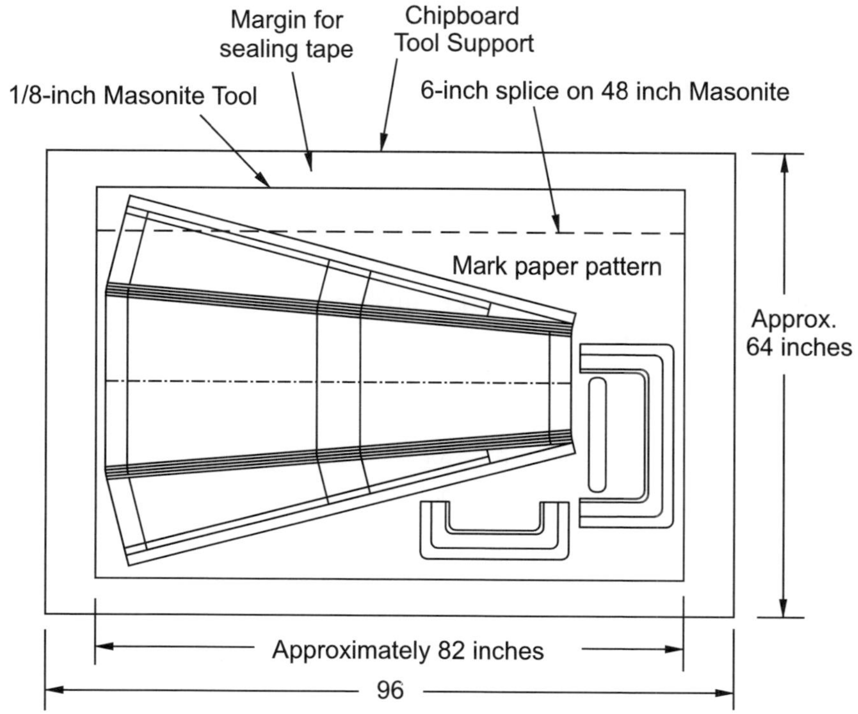

Tool Layout

Use a table saw to rip a 6-inch extension strip from a standard 4x8 sheet of 1/8-inch Masonite. Rip a second strip about 6 inches wide to support the splice. Cross cut the two strips to a length of 82 inches. Cut a second 4x8 sheet of Masonite to 48 x 82 inches. Lap splice the 6-inch extension strip to the second sheet of Masonite using a quick-setting adhesive such as contact cement or 3M Super 77. Sand off any small irregularities and place a strip of mailing tape along the entire length of the splice, to form a smooth surface. (We will ignore the small discontinuity caused by the thickness of the tape.) ***This widened sheet of Masonite will be placed in a vacuum envelope to form the working tool surface.***

It will then be necessary to ***support the Masonite tool*** with a flat surface wide enough to provide about a 5-inch margin for sealing the bag. The easiest way to do this is to use one piece of uncut 4x8-foot half-inch chipboard on one side of the splice and to rip a 20-inch strip from another piece of chipboad. The two pieces will then straddle the splice in the Masonite and provide full level support. See the previous sketch.

Place the chipboard strips on the work table and cover them with a sheet of ***industrial-grade bagging film or with 6-mil PE plastic.*** This plastic sheet will be the bottom sheet of the envelope bag that forms the working tool. Use short pieces of mailing tape to pull the plastic film into a reasonably flat position and secure the film to the tool support. Ignore small creases. Place the spliced Masonite tool over the plastic envelope.

2) Make the layup pattern: The previous figure shows the dimensioned flat layout for one fuselage skin. The following figure shows the three half-bulkheads. (Note that the dimensioning shows only single-digit precision. These parts are not critical and all will be trimmed to finish size.) Use a bold marker pen to make a paper pattern for the three bulkheads and the fuselage skin. Use mailing tape to secure the pattern to the Masonite tool approximately as shown in the figure. Use mailing tape to ***place 1/8-inch joggle strips***, where shown.

3) Form the tool surface: Cut a piece of industrial quality bagging film to cover the tool. Use mailing tape to seal the entire edge of the tool surface to the envelope bottom. Use a vacuum pass-through over a pad of bleeder in a blank space near one corner, as in previous Exercises. Seal and inspect.

Exercise 12 - Flat-sided Fuselage Simulation

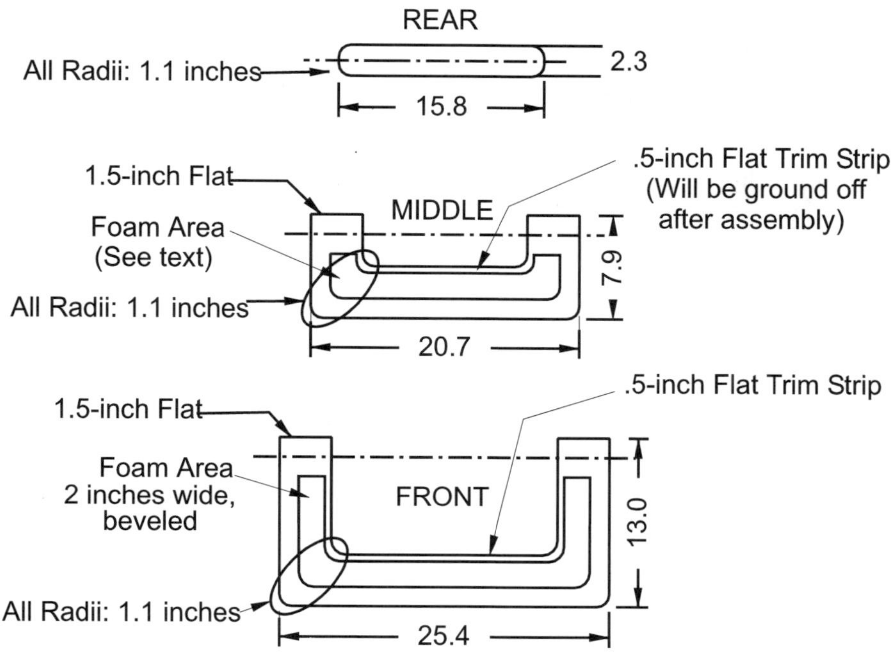

Bulkhead Layout

Connect the vacuum pump and apply vacuum. Sweep trapped air toward the pass-through with a squeegee. (A rubber squeegee works well here.) ***The bag will crumple and seal against itself.*** Continue to sweep air out of the working tool area. This is a considerably larger vacuum-formed tool than was used in earlier exercises and ***some patience will be required*** if the builder is using a fairly small pump. If the bag is adequately sealed, it will eventually pump down hard. If the bag is not adequately sealed, this is the time to find out and fix the bag. The final vacuum should be very near the best vacuum which the pump can produce.

> *If the builder cannot get a tool surface without wrinkles with reasonable attention with the squeegee, it will be helpful to use several dishwashing sponges and/or sections of split PVC pipe to raise the tool film above the working area. The sponges will squeeze nearly flat under vacuum and will not interfere with later sealing operations. If the tool surface and vacuum are satisfactory, turn off the pump.*

4) Prepare the glass: Cut 7725 BID glass sufficient to cover the tool and make two single-layer skins. Use scrap glass and overlap as necessary.

(The first skin will be easy, working with a dry tool.) We will make one skin at a time, with the corresponding bulkheads, in each of two separate layups.

5) Lay up the first layer: Start the vacuum pump, re-form the working tool surface and verify that the vacuum is very good. As in previous Exercises, place the cut glass on the tool and wet out directly on the tool. Tint as usual, even for this single layer. Cover with perf film and bleeder. Cover with bagging film, seal and pump down, as previously described. Cure.

> *Again, it may seem like overkill to vacuum bag a single flat layer of glass, but we have all the equipment in place and we will need it in following operations. We might as well get the benefit of good wetting control and prevention of amine blush. We can also re-use the perf film and bleeder for the second skin.*

6) Prepare the cores: While the first skin is curing, obtain half-inch 2-pound foam sufficient to cover the core areas shown in the sketch. ***Perforate before cutting.*** It will be necessary to use a number of separate pieces of foam to cover the large areas shown. It is unnecessary to fit the pieces accurately - reasonably square cuts and gaps of up to 1/8-inch are perfectly satisfactory. The beveling tool described in Appendix 1 will not be long enough to bevel the longest edges shown, so use sections to suit. The gaps and irregularities will be filled in with dry micro when the foam is placed.

> *See the preceding sketch of the bulkhead layout. The easiest way to handle the bulkhead cores is as follows: Assuming that hot wire cutting will be used, prepare several strips of beveled foam, 2 inches wide, by cutting a bevel on a convenient length of foam, then* ***inverting the piece of foam and making a second bevel cut.*** *The curved areas will be approximated with foam scraps and micro later, when the cores are placed.*

7) Place the cores: Keep the pump running to form the tool, but remove the bagging setup from the first skin layup. If the layup is fresh, no further preparation is necessary.

Mix medium slurry and paint the foam sections fairly generously. Squeegee out smoothly. (Excess micro will be drawn by vacuum through the perf holes and will spread out over the top surface of the foam.)

Place foam sections to cover the pattern. Mix dry micro and squeegee to fill the gaps.

> *In the case of the bulkheads, it is too much trouble to make a perfect blending of the radius shown. Use foam scraps and dry micro to fill out the radiused corner. The trim margin shown is very approximate. This margin will be ground off after cure, as will be described below.*

8) Bond the cores: Rig the perf film, bleeder, and bag again. Reseal, and pump down. This will ensure an excellent bonding to the skin and provide a means of inspection by seeing the perf hole pattern in the bleeder: we need the bagging hardware in any case. (If the bleeder has been reused, the new micro will be recognized by being white - not tinted as with the glass skin.) Cure. Keep the tool pump running.

9) Lay up the inner skin and bulkheads: Inspect the bonded cores. Clean up any sharp edges and cured micro that may be on the skin surface. As before, cover the area with one layer of 7725 BID glass, overlapping pieces as necessary. See that the glass follows the core bevels smoothly by adjusting with the edge of a squeegee and the gloved hand.

With this large area, it will be easier to use the sandwich method of wetting out the glass and applying the wet glass in sections, essentially as very wide "tape." It will also be necessary to apply the perf film in sections so that the perf film will fit the core bevels smoothly.

Replace the bleeder and bag and pump down as before. (The bleeder will probably be usable one more time, unless grossly excessive epoxy was used in the previous operations.) Cure.

Unbag and remove the cured skin. Use a spiral saw to separate and rough trim the molded parts. Clean up with a belt sander.

10) Lay up the second skin: Remove the sealing tape from the envelope bag sufficiently so that you can ***remove the joggle spacer.*** Reseal the tool bag. Turn on the vacuum pump and re-form the working tool surface, as with the first skin. Verify that the tool surface is smooth and that the tool vacuum is very good.

Make a second cored skin, ***without the joggle,*** as before.

11) Form one fuselage side. (This is the payoff step.) In this step, we will quickly and accurately "fold" the first fuselage skin, ***without a jig.***

Place the fuselage skin on the work table and add sandbags or other weights to hold it firmly in place. Bend the sides toward the vertical, folding the layup along the bending areas. Temporarily secure the position with aircraft safety wire strung between the sides through small holes drilled as needed in the margins.

12) Add the first bulkheads: As dimensioned in the figure, the bulkheads are cut to be ***sufficiently undersize to fit easily*** inside the folded

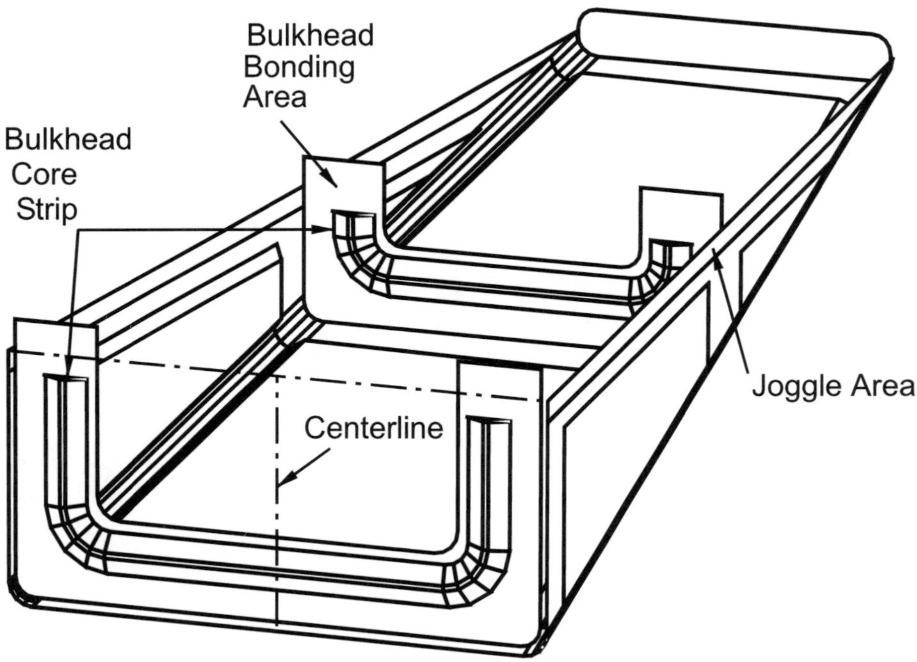

Bulkheads in Place after Folding

fuselage half. Place each bulkhead in position, secure temporarily with brown tape, and "tack" in position with small globs of 5-minute flox. Square up as needed. Verify accurate square angles with a carpenter's square.

After the 5-minute flox tacking has cured, remove any bracing and sandbags and verify that the fuselage half is correctly shaped. Using the now-familiar techniques, wet out and place 2-inch 7725 BID tape along the interface between the bulkheads and the skin. Cure. The half-section will now be surprisingly stiff and will ***serve as the assembly jig for the second half.***

13) Temporarily place the second set of bulkheads: Use #8 Washer Head Self-tapping screws to ***temporarily*** attach the second set of bulkhead halves as shown in the next sketch. (The third bulkhead may be trimmed to fit with a spiral saw.) Be sure that the screw heads are on the front side. Align with a straight edge.

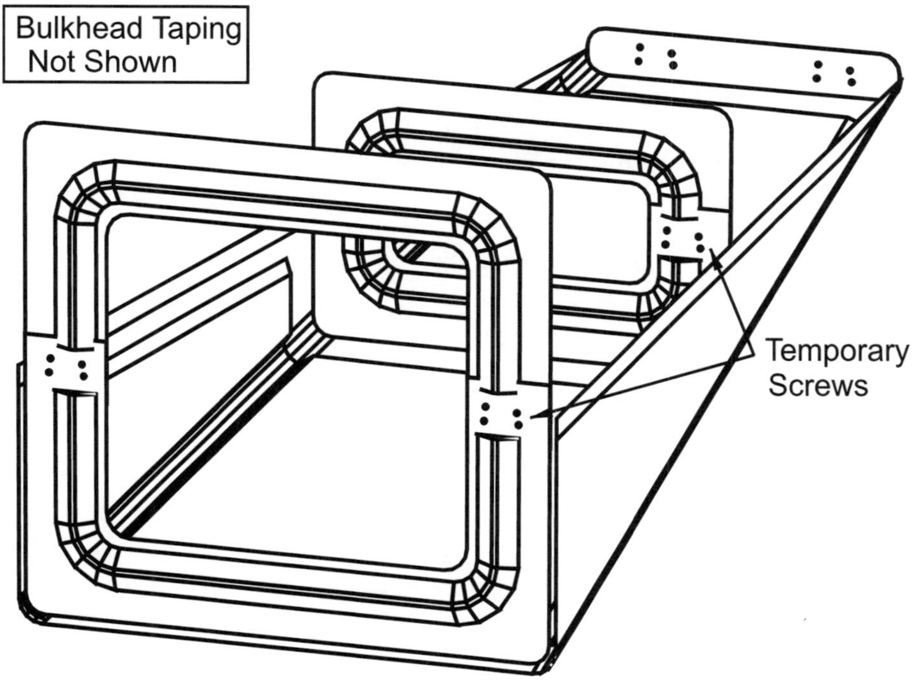

Second Bulkhead Halves in Place

14) Fold the second skin: Use any convenient support, such as cardboard cartons, to bring the first fuselage half to approximately horizontal. Weight with sandbags. Provide a ***spacing shim*** of newspaper strips or poster cardboard ***over the joggle area*** so that a ***space of about 1/16-inch (.060-inch)*** is created. The final assembly will be quite accurate and the parts will not bend significantly to allow for the adhesive layer. Secure with brown tape.

Place the second fuselage half approximately in position and weight with sandbags. As before, bend the skin to overlap the first fuselage half. Use #8 Washer Head Self-tapping screws to ***temporarily*** secure the overlapping edge to the joggle area of the first fuselage half. A temporary screw will be required at intervals of about six inches. The bond line will be straight and no further constraint should be required.

15) Tack the second fuselage skin: Mix 5-minute flox and use your gloved hand and mixing sticks to place globs of 5-minute flox at the corners and at 6-inch intervals along straight lines of the bulkheads. You will be able to reach the second bulkhead from the front. The third bulkhead - obviously - must be reached from the back. Cure.

16) Tape the second fuselage skin: Remove the temporary screws to separate the two halves. As before, tape both sides of the interface between the bulkheads and the skin with a single layer of 7725 BID. Cure. Trim and clean up as necessary.

17) Trial assemble the fuselage: Place both halves together and identify four or six of the screw holes as alignment holes. Use holes near the front and rear ends and a pair near the middle. A tapered steel tool such as an awl will be useful for bringing the holes into alignment. Mark the holes with a distinctive mark, such as an "x".

18) Bond the fuselage halves: In this case,we have the problem that we must join two surface that are essentially parallel, with very little allowance for closure. You will probably want an assisstant. (The author did it working alone, but does not recommend it.) A Structural Adhesive such as Aeropoxy ES6209 is satisfactory and the pot life of about one hour will be sufficient.

Mix well batches of structural adhesive in small cups and spread the adhesive on both sides of the bonding area. Use a notched spreader to assure full coverage. With help, lift the top fuselage half into position and place screws in the marked alignment holes. The transparent glass will allow inspection of the joint to assure that the bond area is fully covered. Place pull-up screws as needed in the other existing screw holes. Runout of excess adhesive can be scraped off with a chisel blade as the adhesive cures. Allow to cure.

Comments

The end result of the Exercise will be two full-sized, accurately formed, half-sections which are bonded together to represent a practical basic fuselage structure. (The Exercise does not make provision for specific attachments, like control rod pivots or elevator mountings, which would be required for an actual fuselage.)

If you have stuck with this long Exercise and completed it, you are well on the way to full competence in working with advanced composites. You have probably made many mistakes and the red and blue tinted epoxy is almost too effective in making minor irregularities look like an abstract painting. You will probably not wish to show this first full-scale work to your friends, but you can take pride in having produced a very light, very stiff, practical full-size structure.

Comparison with Classic Construction

The classic Wittman *Tailwind* steel tube fuselage is what structural engineers refer to as a ***truss***. (See Appendix 2 for a more detailed explanation of the principles of trusses.) A truss is a structure composed of short segments which, in principle, could be connected by ***pivoted joints*** without loss of stability. In the case of the *Tailwind*, and the general practice with aircraft, the short segments are actually steel tubing joined at the nodes by welds. The tube diameter is quite small compared to the segment length, so the welds are conservatively assumed to carry ***zero bending loads*** and to serve essentially as pivoted joints. All loads in the structure are passed only through the "pivoted" joints. A truss is ***"determinate,"*** meaning that the forces in each segment can be calculated easily (see Appendix 2). If tested to failure, the most probable failure mode would be buckling in some single segment, rather than failure of a weld, which can be identified and strengthened, if necessary.

The composite structure just completed is highly ***indeterminate***, with load paths spread in very complex ways throughout the structure. The only practical way to compare the performance of this structure and its classical equivalent would be to test both to design loads and/or failure. This training Exercise was not intended to be a suggested construction for an actual aircraft of similar dimensions, though it is the author's opinion that it does not miss a practical design by very far. As built by the author, the structure weighed approximately 16 pounds. In a practical application, the useful strength could be increased substantially at a small increase in weight by adding one more bulkhead, and by running a single additional strip of 7715 UNI glass in the inside corners. However modified, the result ***must be tested.***

Exercise 13

Single-curved Wing Skin Mold

Introduction

This Exercise, and the following Exercises 14, 15, and 16, will simulate the construction of ***practical cored-skin wing skins from accurate molds*** and the assembly of these skins to ***complete and close a wing.***

The method has the significant advantage that the shape of the wing is determined by an accurate external reference -- all ***variations in skin thickness and component alignment are kept inside the wing.***

The reader will apply most of the basic skills learned in previous Exercises, including production of molds from sheet material, wetting out and placement of multiple layers of cloth, and "double vacuum bagging." The experience gained from completing this final series of Exercises will enable the serious builder to ***proceed with confidence to a full-scale project.*** Obviously, this is more work than in the beginning Exercises, but the reader who has done the previous Exercises will have the skills to complete the necessary tasks quickly and efficiently.

The dimensions chosen for these Exercises produce parts which are ***small enough and light enough to be handled by a builder working alone.*** All layups and assemblies are smaller than a 4 x 8-foot sheet of chipboard, which may be used as a general workbench and wetting-out board.

The wing section to be demonstrated in the following Exercises will use the ***"torque box"*** approach, widely used in commercial jet aircraft, where ***very strong skins are used with two shear-carrying members to form a long "box" which will carry most of the required flight loads.*** The design takes advantage of the fact that it is very easy to use composite technology to make strong multiple-layer skins, with the material distributed for maximum efficiency.

The mold produced in this Exercise will serve as an ***assembly jig*** for closing the wing after it has been used to make a wing skin.

WARNING: Exercises 13 through 16 are intended solely for teaching techniques. The example chosen produces a very light and strong structure which is SIMILAR to the structure that could be used near the strut attach point of a small high-wing aircraft. In general, the skin thicknesses shown may NOT BE ADEQUATE for actual flight loads, but construction as described will require the reader to understand and perform the basic operations. The methods demonstrated can be used to produce a practical full scale wing which will require a full scale proof test before flight.

This Exercise will demonstrate several working shortcuts which will quickly produce a nearly full-scale mold for a curved wing skin with only a small fraction of the work necessary to make and polish a conventional plug and mold. The example chosen will be the *Crosswind* wing, partly because the materials used are conveniently within the standard dimensions as obtained at a home supply store.

The techniques demonstrated here are NOT typical of the mold-making methods used by the commercial composites industry or by aircraft kit builders.

Accurately Bending a Sheet Mold

We will make a ***simple mold from a sheet*** of smooth, semi-flexible material which will accurately define the main upper skin of the *Crosswind* wing, ***without the leading or trailing edges.*** The lightly-loaded leading and trailing edges can be formed and added later.

An important innovation demonstrated in this exercise is the use of large area release film, held in position by vacuum, as a substitute for the work involved in finishing the ***curved mold surface*** to commercial standards.

A ***convex bending form*** will be used to define the mold surface, then a sheet of semi-flexible, air-tight material will be bent over the form and bonded to supports which hold the sheet in place. A simple frame will be built over the bent sheet to lock the sheet in position. ***The bent sheet will be removed from the bending form***, then inverted on the work table. We now have an accurate concave form which can be used with vacuum to support a release film to serve as a molding surface.

Detailed Procedure

Making the Bending Form

Reference Wing Section

1) MAKE A MASTER REFERENCE WING SECTION: We will need a sturdy, accurate, stable, easily-reproduced reference for the wing section. This master will be used many times, so should allow repeated handling. We should also keep an identical backup. This step describes one practical way to make such a reference.

Obtain four pieces of ***Formica*** (or equivalent) kitchen counter material and two pieces of 1/8-inch ***Masonite*** about 45 x 6 inches. Use mixed epoxy, water-catalyzed wood glue, or contact cement to attach two pieces of Formica to each core piece of Masonite to make two ***laminated sandwiches*** about 6 mm thick. (Urethane glue can be used, but must be well weighted during hardening.) Spray one side of both laminations with flat white quick-drying enamel.

Make a very careful layout of the wing section, including the baseline, as shown on the following page. Use a bandsaw, spiral saw, or table saw to cut out the wing form about 1/16-inch outside of the outline. Use a ***belt sander to very carefully grind away material*** to just split the outline. If you make a mistake and go too far, repair the damage with 5-minute epoxy mixed with flox and grind again to the corrected line. THIS IS THE ***FIRST MASTER.*** Trace the first master to make an accurate copy, the ***BACKUP MASTER.*** Alternatively, bolt the two rib masters together and grind them to matching finished size with a belt sander.

Alignment Holes

Find two pieces of scrap welding steel about 3/4 x 3 x 1/8-inch (see the following sketch). Mount them on the first master about 10 inches back from the leading and trailing edges, with #6 flathead countersunk screws. Carefully clamp the two masters together. Drill out a ***1/4-inch***

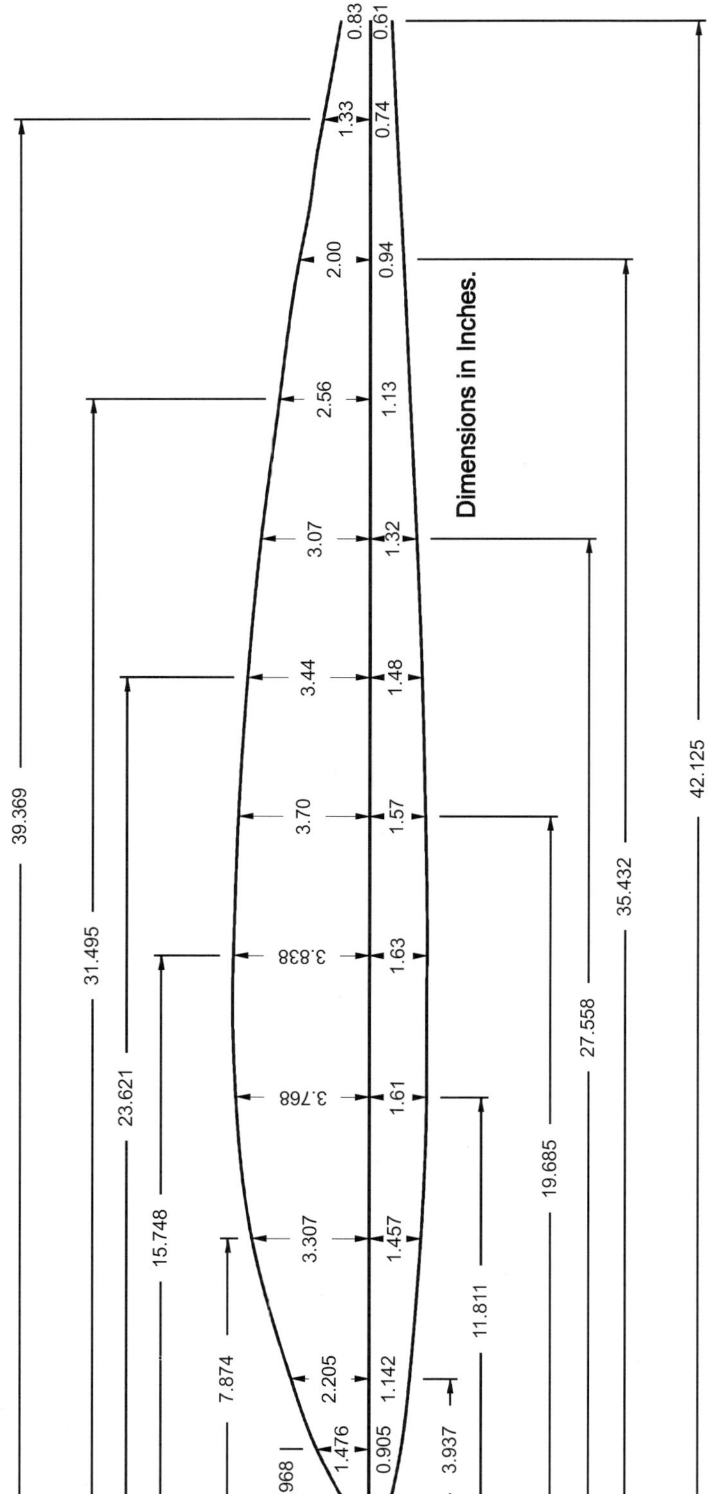

Crosswind Wing Section Coordinates

alignment hole through the steel, the first master, and the second master. (The steel will protect the alignment hole from wear as the master is used to make duplicates.) Very carefully drill out 3/32-inch marker holes on the baseline about one inch back from the leading and trailing edges and 3/32-inch leveling holes above the base line, as shown in the sketch below.

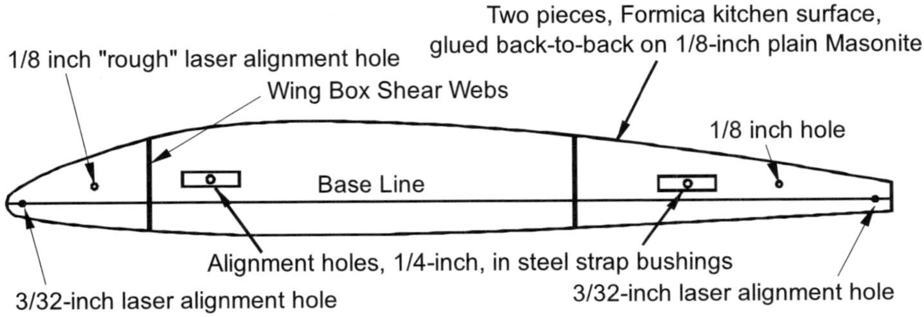

Master Wing Section Reference

2) MAKE WORKING WING FORMS: We will need ***four working wing forms that accurately define only the top surface of the wing.*** Use the master form to carefully trace the outline of the upper wing surface on four pieces of half-inch chip board, as shown below.

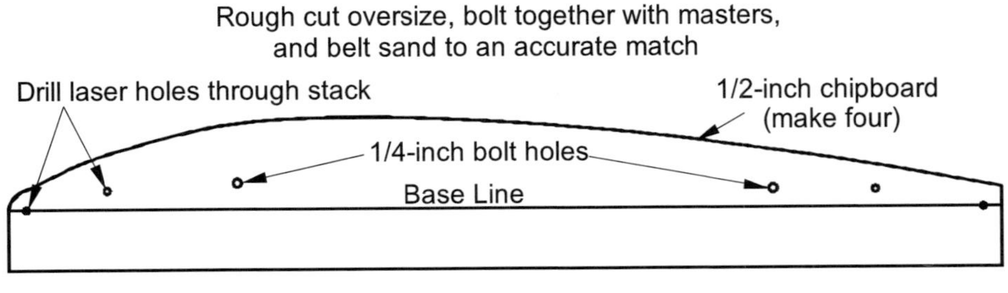

Bending Formers for Top Wingskin Mold

Locate Reference Holes

Stack all four wing formers together with the master form and average out the small misalignments for a best overall match. Clamp the stack securely. Use the master as a drill guide and drill ***two 1/4 inch alignment holes through the stack***, as shown. Bolt the stack together with two 1/4 inch bolts. Drill ***two leveling holes***, as shown, all the way through the stack, to snugly fit a 2-inch finishing nail. Using the same drill, make two or three ***laser alignment holes***, as shown. Countersink the laser holes in one of the end forms. See the previous sketch.

Use a bench belt sander to accurately form the stack of four rib forms to the scribed line.

3) MAKE A SUPPORT FRAME: We will use lightweight ***steel framing/ studding*** stock to make a frame that is essentially an accurate, stable pair of rails that will support the forms on the worktable.

Steel House-framing Material

Steel framing stock has many important advantages over wood: Most important, it is ***dimensionally stable*** and can be very quickly positioned and held with the ***special metal screws*** sold for the purpose. It is easy to work accurately and is much lighter than wood of the same dimensions, permitting easy adjustment. The material is available in 20-gauge (heavy duty) and 25-gauge (regular), in lengths of 8, 10, and 12 feet and in widths of 3 9/16 and 3 5/8 inches. ***Track*** is plain folded sheet, without any embossing or holes and with a sharp edge. ***Studding*** is material of the ***same dimensions***, but with embossing for stiffness and with smooth rolled edges. Steel framing is ***very easily cut*** with a 7-inch electric circular saw, using a standard ***abrasive disk.*** (In fact, cutting this material with the abrasive wheel takes about the same time as an equivalent saw cut in wooden studding. After some experience with this excellent material, you may wonder why it is not used more widely in residential construction. *(Only the larger home supply stores carry this material, so you may need to hunt a little or ask your local home improvement store to make a special order.)*

Assemble the Frame

Obtain one piece of 20-gauge "***track***" and one piece of 20-gauge ***studding***, each 10 feet long. Cut the ***track*** in half, to make two rails 60 inches long.

Cut two pieces of steel ***studding*** 28 inches long. Assemble a rectangular frame 28 x 60 inches, as shown on the following page. Use the remaining piece of studding to make a diagonal brace, as shown. You will find that the ***studding slips perfectly into the track*** and the overlapping corners can be easily fastened with the ***special self-drilling screws.*** No additional hardware is required.

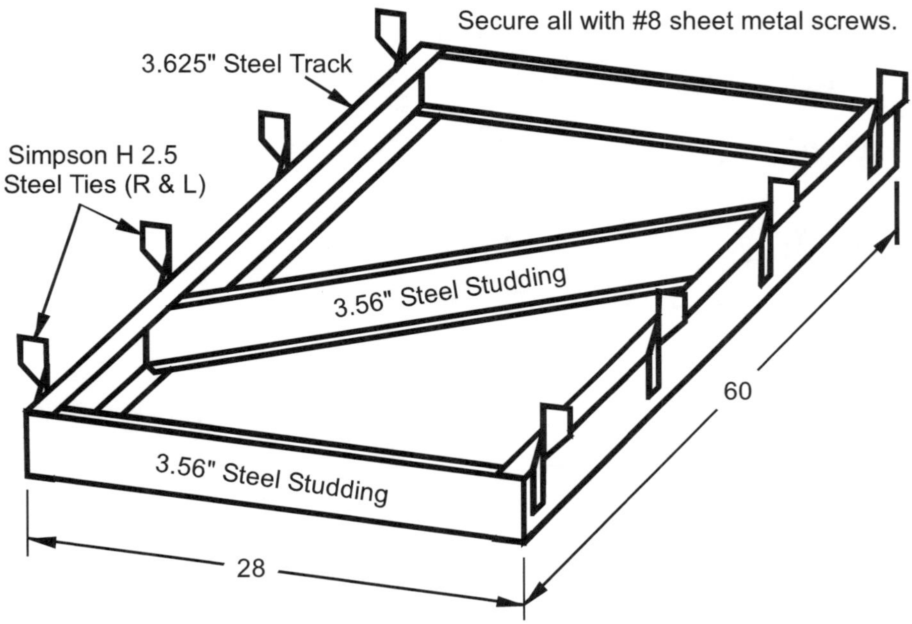

Wingskin Mold Bending Frame

Level the worktable with a carpenter's bubble level. Shim the table legs to level as needed. A level table is helpful, but is not critical, because final aircraft alignment will be a separate operation.

4) MOUNT THE WING FORM HOLDERS: We will use "hurricane ties", sold for reinforcing wooden house framing, to attach the wing forms, such as the Simpson "Strongtie" H2.5, Left and Right. Obtain four each of the Left and Right versions. Use two screws in the pre-drilled holes to mount them on the frame, as shown in the sketch. Space the wing form holders approximately equally.

5) MOUNT AND ALIGN THE WING FORMS: The forms at each end of the frame will be the ***reference wing forms.*** Temporarily mount the leading edge side of the one end wing form to the frame with a single screw, as shown in the sketch. Allow about 1/2-inch clearance above the frame. Insert snug-fitting nails in the "leveling holes." Place a carpenter's level across the leveling nails and adjust the trailing edge to level. (A ***soft-faced clamp*** will be useful, allowing the form to be gently tapped into position.) Fix in final position with a screw. Add a second screw to each holder to complete the mounting of the end wing form (see the sketch below).

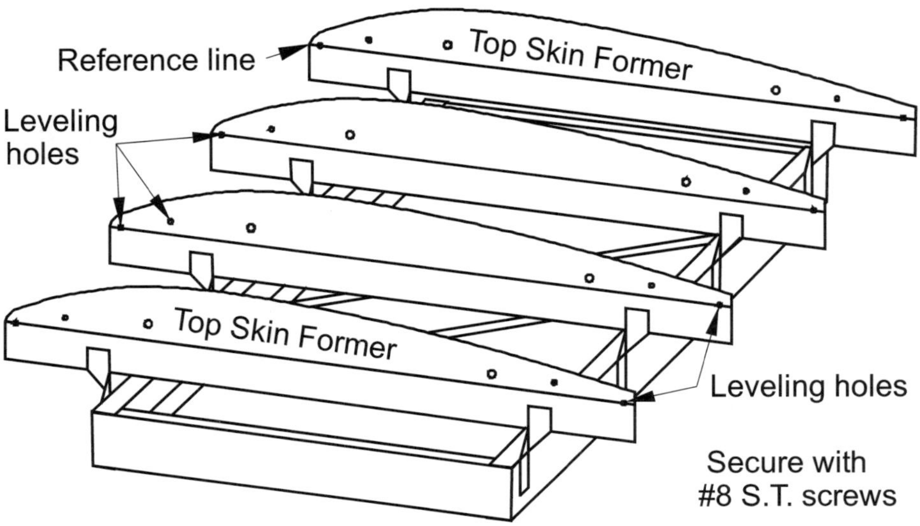

Wingskin Mold Formers Mounted on Frame

Similarly, mount a wing form at the other end of the frame. Use a long straight edge (span-wise) and a level to establish the vertical position of this wing form. (A piece of ***steel studding or track is a surprisingly good straight edge*** and will be good enough for this application.) Level chord-wise with leveling nails, as above. Ordinary measuring accuracy is sufficient; precision alignment will be done in the following steps.

Measure the diagonals between the four leveling holes and adjust the position of the end wing form so that the diagonals are equal.

Temporarily mount one of the intermediate wing forms using clamps with soft pads. This wing form can be aligned with the end forms by either of two methods:

Straightedge Method: Use two good straightedges, such as steel studs, laid across the forms at the ends of the frames. Carefully adjust the wing forms to just touch the straightedges, clamp, and secure with screws.

Laser Method (more accurate): Use a cheap pointing laser, as sold for classroom presentations. Point the laser through one of the 1/4-inch

countersunk alignment holes in the wing form at one end of the frame and find the corresponding hole in the intermediate wing form. Observe where the laser spot hits the wing former at the other end. Gently tap the intermediate wing former to bring it into position. You have alignment when the laser spot seems to disappear into the hole at the far end and appears on the wall of the workroom.

> *To find the laser alignment holes easily,* ***cover the holes with white computer sticky labels,*** *like the ones used used for envelope addressing. Poke a sharp pencil into each hole to make a smooth circle. Use a marker pen to make a cross mark over each hole.*

This operation is easy and quick with the large holes. Move to the intermediate 1/8-inch holes for finer adjustment, then finish with the 3/32-inch holes, which will take a bit more patience. (This method works particularly well when two people can work together - one to hold the laser, the other to tap the form into position.) Repeat with the second positioning hole. Re-check the first hole for movement. When all is in position, mount the wing former with self-drilling screws.

Align and mount the last intermediate wing form in the same manner. This completes the bending form.

Form the Molding Tool

Obtain a 4 x 8-foot sheet of solid plastic shower liner material, typically 0.090-inch thick with a smooth back and a fine pebble grain on the front, sold as ***white wall liner panel.*** (Note that plastic-faced Masonite is not satisfactory for this purpose; it is too rigid and is not reliably airtight.) Cut the sheet to 4 x 5 feet. See the figure on the next page.

(6) *PREPARE THE PLASTIC SHEET FOR BONDING:* Use a portable belt sander or an orbital sander with 120-grit sandpaper to ***"break the gloss"*** on the ***pebble grain side*** of the plastic (the normal "front").

(7) PREPARE THE PLASTIC SHEET SUPPORTS: Cut six pieces of steel studding to approximately 6 feet long. Set aside.

(8) PLACE THE PLASTIC MOLD SHEET: Place the plastic sheet over the bending form, with the smooth "back" side toward the forming ribs. Adjust the position so that the edges approximately correspond to the forming ribs. The semi-flexible material will bend a bit under its own weight, but will require external force to conform to the bending form, as shown below.

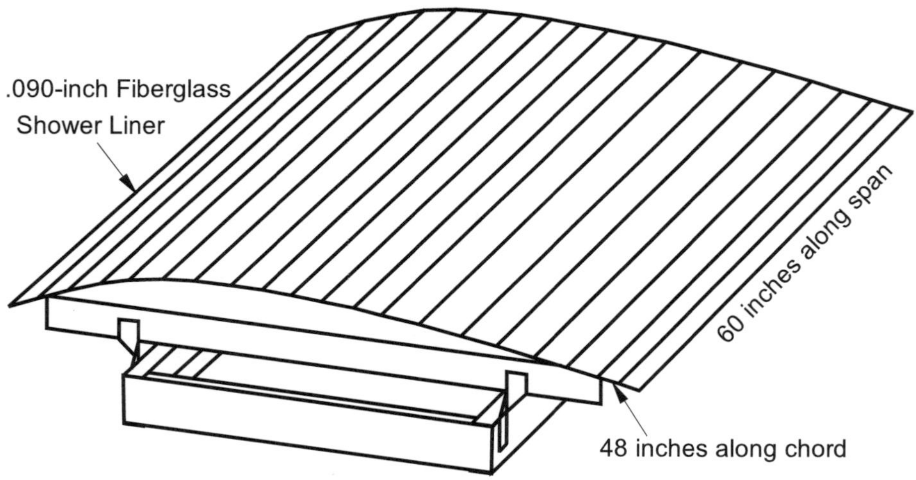

Bending Mold Sheet over Form

(9) ***PLACE THE STEEL SHEET SUPPORTS:*** This step will require a bit of ingenuity by the builder. The basic idea is to apply a force to the six steel supports ***(twelve parallel edges)*** of the steel studs so that the plastic mold sheet will accurately conform to the bending form.

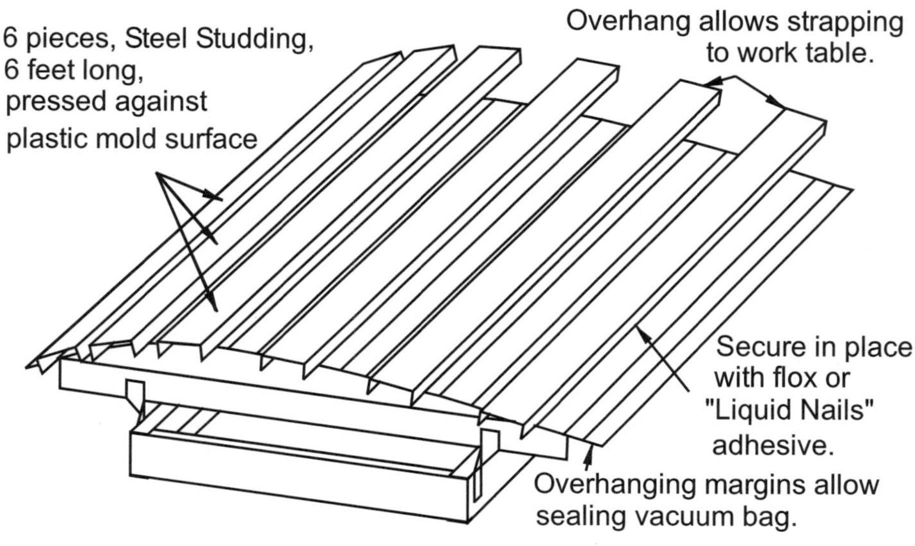

Attachment of Mold Supports

Put one of the steel studding supports over the plastic sheet, with the ***rolled edge toward the sheet***, approximately in the center of the wing form. Apply weights, such as sandbags, scrap metal, or heavy boxes, to force the plastic against the form. Add the other steel supports approximately as shown in the sketch above, temporarily securing each support as necessary. Use plenty of weight to assure that the plastic sheet is firmly in contract.

> ***Note:*** *As you approach the slope of the leading edge, it will be necessary to tie the steel supports back to prevent their sliding off. Packing twine tied to screw eye bolts or nails in the worktable may be convenient.*

Remove one steel support at a time, leaving the others in place. Use a calking gun to put cheap bathtub sealing/bonding compound, such as "Liquid Nails Project" compound all along the rolled edge. Replace the steel support and slide it slightly in position to distribute the glue. ***Weight with sandbags, boxes, or any other convenient heavy weights.*** Similarly, place and re-weight the remaining five steel supports. Allow the glue to dry or cure about 24 hours.

> ***Alternative glue:*** *cheap caulking gun bonding compound works well enough, but some builders may be more comfortable using flox, which is more expensive. Curing time is about the same, but flox is more secure.*

Remove the weights after the glue has cured thoroughly.

Mold Supports

(10) ***CUT OUT THE MOLD SUPPORTS:*** Use the wing form masters to mark two rough supports from 1/2 inch chipboard, as shown. Cut out the two supports. Approximate rough work is good enough.

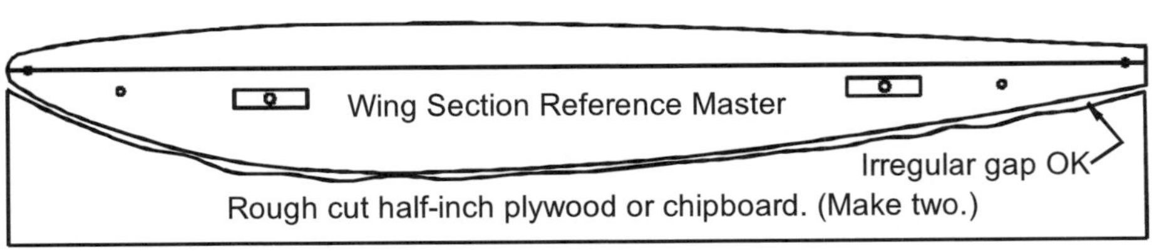

Mold Supports

(11) MOUNT THE MOLD SUPPORTS: Position one of the mold supports, as shown, about 1/4 of the way in from the end. Attach the mold support with ***steel right angle brackets***, such as the Simpson "Strong-Tie", Type A23. Attach the second mold support similarly. Adjust the bases of the mold supports so that they are approximately parallel.

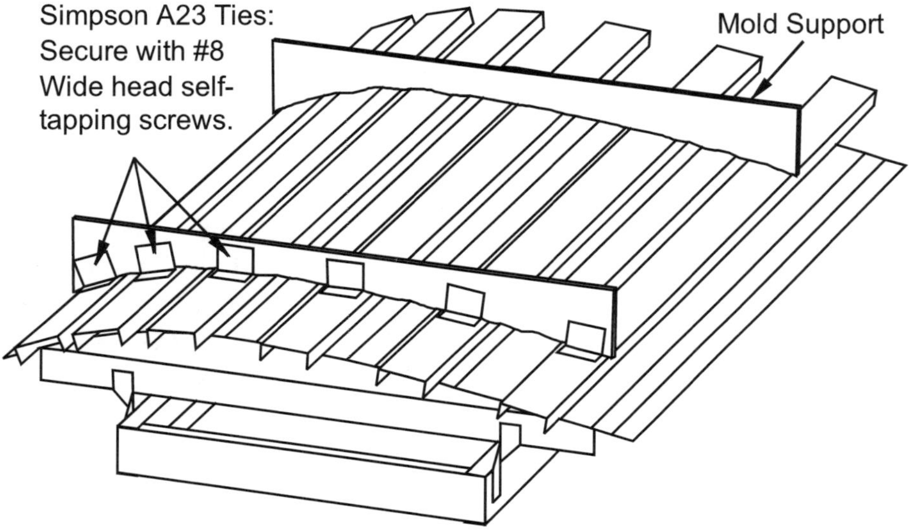

Securing Mold Supports

(12) FINAL LEVEL THE MOLD SUPPORTS: Find a scrap of wood about the length of a mold support and pin it with one screw. Clamp and adjust so that the mold supports are accurately parallel. Fix in position with a screw. (This operation avoids any twist in the mold when it is removed from the bending form.)

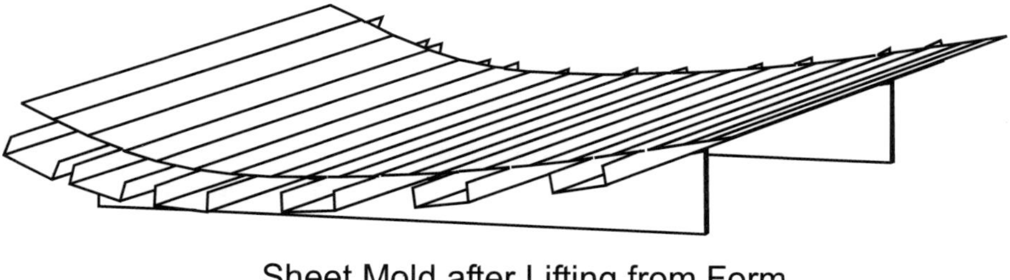

Sheet Mold after Lifting from Form

(13) ***REMOVE THE FINISHED MOLD:*** Lift the mold from the bending form, turn it over, and place it on the worktable. (One person can do this, but two is convenient. ***You will appreciate the light weight of the steel studs.***)

This operation completes the basic mold support.

14) *MARK THE WORK AREA:* Use a marker pen and a flexible straight-edge to mark ***directly on the mold face*** to define the work area. The ***wing-form master is the reference for the accuracy of the work.*** See the sketch on the following page. Place the wing form master in the mold and verify that a nearly perfect fit to the curve has been found. ***Transfer the key reference lines from the wing form master to the tool.*** See the figures on the following pages.

Mark the mold face for ***two joggle areas*** as shown in dashed lines and mark the boundaries of the four layers of glass, as shown in the following sketch.

15) PLACE THE JOGGLE STRIPS: Use any convenient plastic or cardboard material to make two ***2 inch by 25 inch*** joggle strips, ***1/8 inch thick.*** (Two layers of the plastic strips sold for insertions in chain link fences work very well. Art board or poster cardboard may be used.) Tape in position.

Forming the Tool Face

Ordinary 4-mil PE plastic sheet will be used as the working tool face. Essentially, the tool will be a stretched sheet of film, pulled tightly against the curved support sheet with vacuum. A ***toolside bleeder will NOT be necessary*** and the concave shape of the mold surface allows the ***tool surface to be formed without a "ring bleeder."***

(17) ***PLACE THE TOOL FILM:*** Cut a piece of 4-mil PE plastic to 4 x 5 feet and place over the tool sheet. "Tack" in place temporarily with short pieces of brown tape until the position is approximately correct. The intention is to have the tool film held in place smoothly and sealed to the tool support with a continuous line of brown tape. Work from the ends and sides, leaving one corner open for the vacuum pass-through.

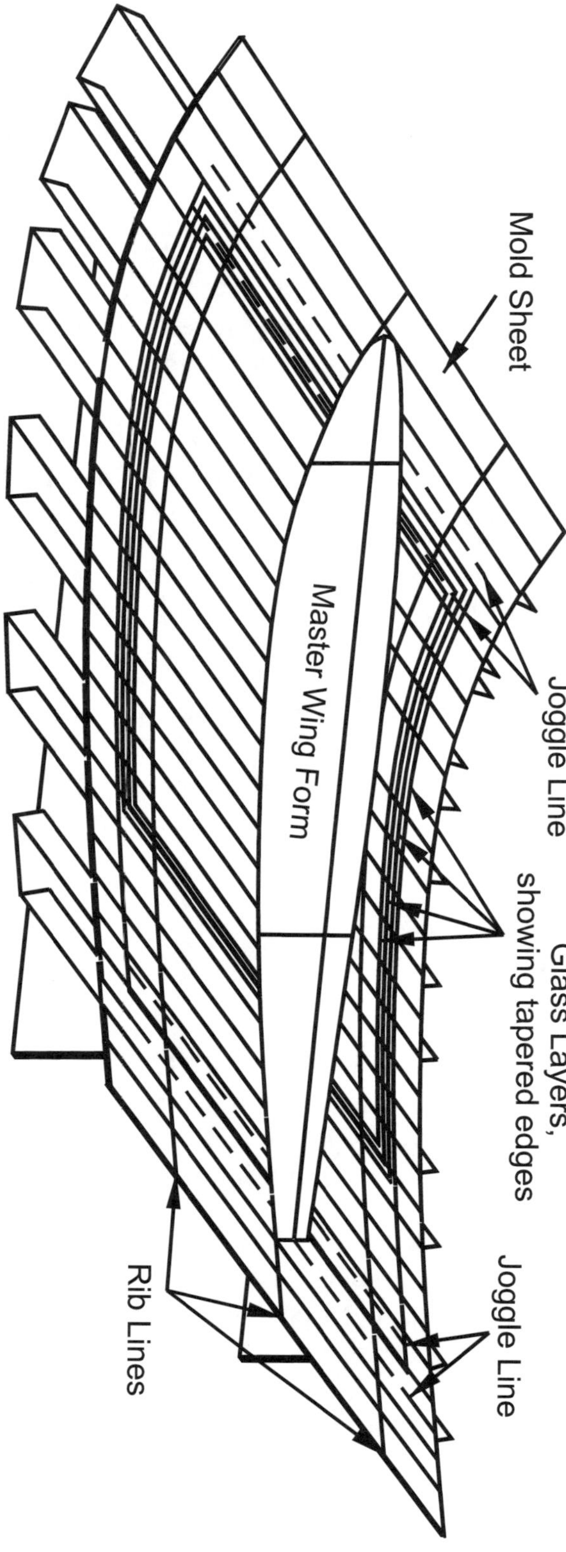

Marking the Wing Skin Sheet Mold

Exercise 13 - Single-curved Wing Skin Mold

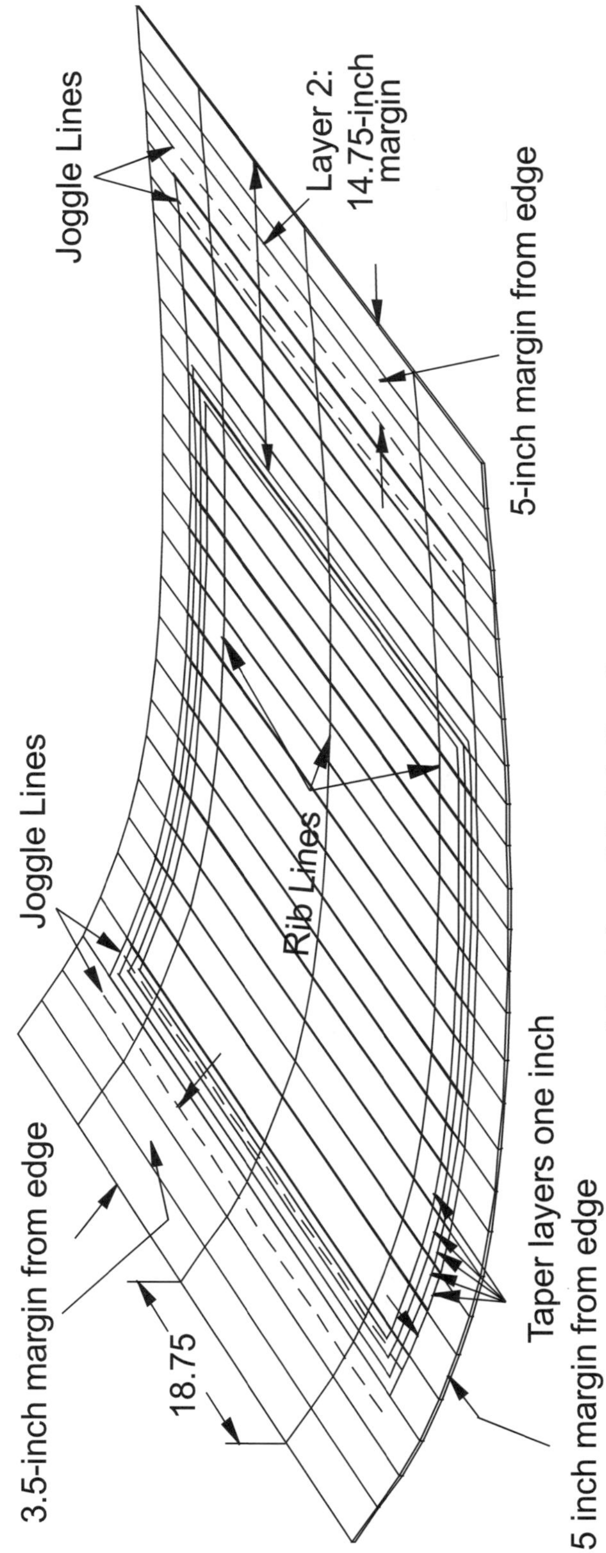

Marking Mold Surface

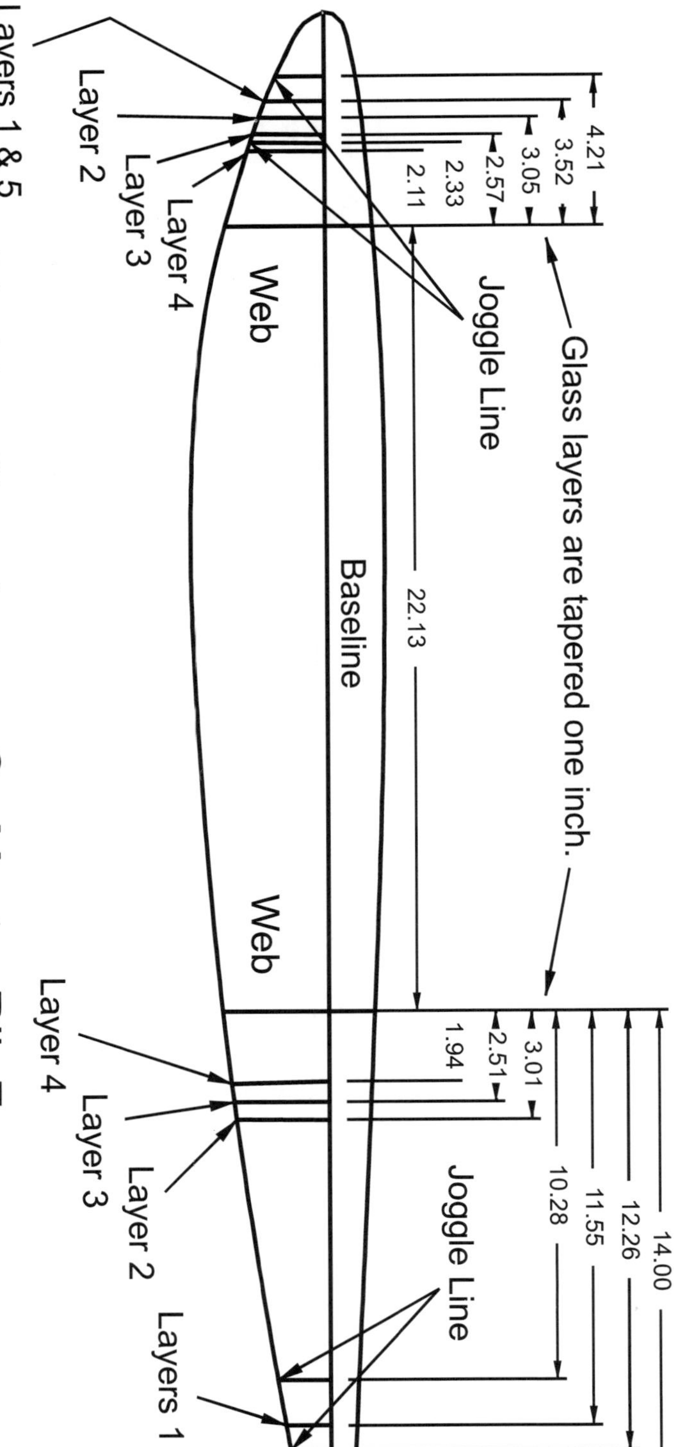

Marking Glass Layers On Master Rib Form

(18) ***PLACE THE VACUUM PASS-THROUGH:*** Lift one corner of the tool film and place a fairly thick pad of bleeder about three inches inside the edge of the film. Mark the bag and mount one pass-through, as before. Complete the seal with brown tape and inspect.

(19) ***TEST THE TOOL:*** Connect the vacuum rig, using the gauge manifold, and leave the ***bypass valve wide open to atmosphere.*** Start the pump. Close the bypass valve slowly and observe the pressure on the vacuum gauge closest to the pump. When the vacuum gauge starts to register, open the bypass valve so that the measured vacuum is very slight, 1 or 2 inches. Use your hand and/or rubber squeegee to ***sweep any wrinkles out of the work area*** so that the wrinkles are pinned near the ring bleeder by vacuum. Close the bypass valve a little more to increase the vacuum and continue to sweep the tool area until the ***tool surface is perfectly smooth.*** Close the bypass valve entirely.

Use a heat gun to warm the brown tape and smooth it down with your hand. ***At this point, the measured vacuum should be very close to the best that the pump can produce***, i.e., the leakage is very low. If PE film was used to form the bag, it may be that leaks will occur in the region of folds, probably not on the open surface of the film. It may be necessary to patch the film with brown tape or to choose another piece of PE film.

Actual testing may show that the bag should be repositioned to remove wrinkles. If so, lift the brown tape, make the necessary adjustments, and retest. ***DO NOT PROCEED WITH A LAYUP UNLESS THE TOOL VACUUM INTEGRITY IS EXCELLENT AND A PERFECTLY-SMOOTH TOOL FACE HAS BEEN PRODUCED.***

Turn off the pump. The tested tool surface will be restored later.

Exercise 14

Single-curved Wing Skin

Introduction

This Exercise will use the mold surface from Exercise 14 to make a nearly full-scale wing skin, using most of the techniques demonstrated previously, including handling large panels of wetted glass, placement of foam cores, and use of caul plates. Tapering layers will also be demonstrated.

*In the following Exercises, we will add the shear webs and ribs which will complete the vertical surfaces of the hollow "**wing box**" and we will close the "box" by adding the opposite wing skin. The wing box serves the function of the classical wing spar, except that the strong skin itself acts as the **equivalent to the "spar cap"** to carry compression and tension loads.*

These instructions are written under the assumption that the ***actual layup*** will be completed by ***one person in a single working session.*** The general dimensions and the number of glass layers used are within the capacity of a builder who has completed the previous Exercises, but the work will go much faster and with greater confidence if assistance is available. The builder will find that ***preparation time*** is very important and will be ***longer than the actual layup time***. Time taken to prepare well will be repaid doubly in the execution.

Even after having done the previous exercises, the beginning builder would be well advised to do the preparation in two work sessions and the execution in a third work session - with help. A "dress rehearsal" may prevent serious errors.

Completion of the steps necessary to go from initial wetting-out to final vacuum bagging will take more time than the normal one-hour pot life of the epoxy, especially if the builder is working alone. Still, the epoxy will have sufficient fluidity remaining to move under vacuum pressure to consolidate and remove excess. The less-experienced builder might be more confident if he used a ***two-hour hardener*** for this job, though the regular one-hour hardener will be satisfactory.

We will lay up the skin stack shown below:

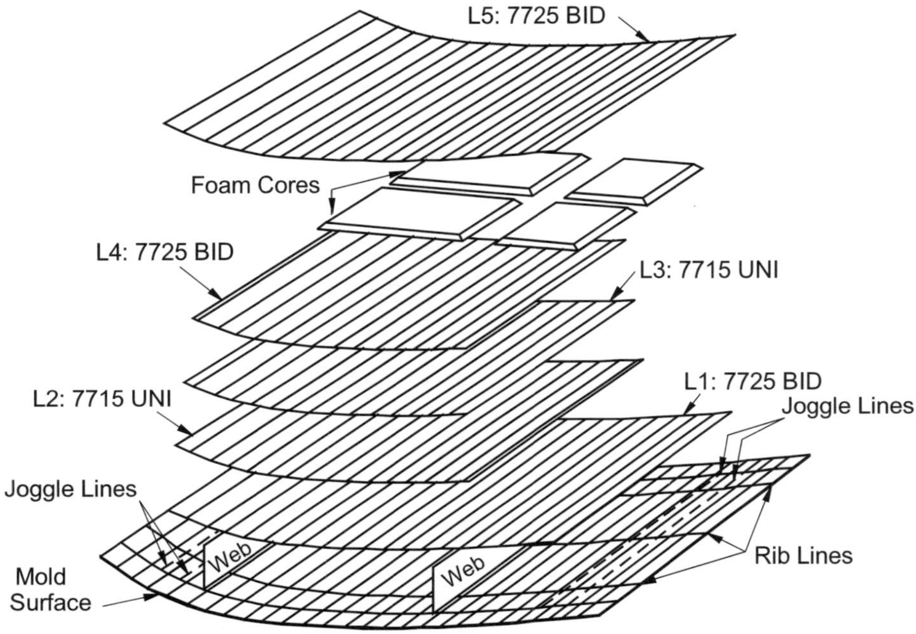

Skin Stack

Detailed Procedure

Prepare the Layup and Bagging Layers

Refer to the sketch above. See that three BID layers, two UNI layers, and four perforated core panels are called for. In addition, a bag, bleeder, and perf film will be required sufficient to cover the work area.

The ***first layer of 7725 BID may be placed in the dry mold***, using scraps of glass as convenient, making sure that an overlap of at least one inch is used.

It will be extremely difficult to handle and accurately position the following layers if the attempt is made with dry cloth. The ***"sandwich wetting" method must be used.*** Also, the total area to be covered will be difficult for a builder working alone, so these layers should be laid up in smaller sections, as described below.

1) MARK AND CUT THE GLASS LAYERS: The dimensions below allow about 4 inches oversize in length and 2 inches oversize in width, to allow for overlap and trimming. :

Layers 1 and 5, (overall) 7725 BID: 53 x 39 inches

Layer 2, 7715 UNI: 52 x 28 inches

Layer 3, 7715 UNI: 50 x 28 inches

Layer 4, 7725 BID: 48 x 28 inches

Place the cut glass in labeled grocery bags.

"Contingency" Glass: Even experienced builders will find that it will be very difficult to measure and cut the glass for an ideal situation and have it come out exactly as planned. Bias-cut glass seems to provide its own surprises when handled. It is a very good idea to cut a "protection" strip of 7725, bias-cut across the roll, about 24 inches wide. Use this glass, as needed, to fill out a layer. Place in a labeled bag.

2) PERFORATE, CUT, AND BEVEL THE FOAM CORE PANELS: Use half-inch 2-pound foam. Cut and perf two each 15.75 x 14 inches and 15.75 x 8 inches. Cut 40-degree bevels, as shown. (The core panels will be placed in the layup following glass Layer 4 and before the final glass Layer 5.) See the sketch. Set aside.

3) CUT THE PERF FILM, BLEEDER AND BAG: Cut the perf film and bleeder to 52 x 46 inches, which willl allow about 1 1/2 inches all around for trimming. Cut the bag to about 55 x 49 inches. Place in grocery bags.

4) CUT CAUL AND PEEL PLY MATERIAL: We will want smooth bonding areas for the shear webs and ribs. As molded, the bagside of the layup may have small wrinkles from the soft bleeder and small “pips” from the perforations in the release film. These can be avoided by using strips of smooth, semi-rigid material inside the bag (caul plates) which force a smooth surface on the bagside. As previously, 2-inch strips of ***peel ply will be used over the areas where the ribs and shear webs will be bonded.*** See the sketch.

A single layer of stiff poster card makes a surprisingly good caul plate. Perforate and cut poster card ***caul plate material to make about 22 running feet of 2-inch strips.***

5) CUT PEEL PLY STRIPS FOR THE JOGGLES: About 10 running feet of 2 1/2-inch strips of peel ply will be required.

6) FORM THE MOLD SURFACE: Start the vacuum pump. As in Exercise 14, start with very low vacuum and smooth the wrinkles in the bag away from the work area with your hand or a rubber squeegee. Close the bypass valve for full vacuum. ***Verify that the tool vacuum is very good.*** If necessary, warm the sealing tape with a heat gun and chase leaks with a mechanic's microphone.

7) APPLY DRY PEEL PLY TO THE JOGGLE AREA: Use brown tape to secure 2 1/2-inch strips of peel ply over the joggle areas. Be sure that the crayon-marked side of the peel ply is ***away from the layup***. The peel ply will be removed when the wing skin is bonded.

Make the Layup

8) PLACE AND WET OUT THE FIRST LAYER OF GLASS: Place the cut 7725 BID over the entire marked work area on the dry tool. Allow some glass to overlap the edge of the work area to allow for trimming. Overlap glass pieces as necessary. Mix about 7 ounces of resin and tint lightly with carpenter's chalk. Wet out as in previous exercises. ***Use plenty of resin and wet out very thoroughly.*** Work out bubbles with the edge of a squeegee and/or a grooved roller or bristle roller. It should be easy to do a very good job of wetting and getting intimate contact with the tool surface with this very wet thin first layer. Inspect the work by viewing at a low angle with a shop light. Use a ***heat gun*** with the grooved roller to assure that all bubbles are removed.

8) WET OUT AND PLACE THE SECOND AND THIRD LAYERS OF GLASS: Use the "sandwich" method to wet out these layers. Mark the ***outside surface*** of the wetting plastic so that the ***width of the wetted glass is 26 inches,*** as called out, but allow an extra three inches of length for overlapping and final trim. Wet evenly with tinted resin, driving excess out to the sides of the sandwich. ***Trim on the marked lines.***

Note that the placement of the intermediate glass layers is ***not critical.*** *It will be difficult to place the sheets of wet glass exactly and variations from the marked lines of as great as a half-inch are satisfactory, as long as no gaps are left within the layer. The "tapering" of the layers is approximate, intended to avoid too-perfect straight lines with a too-abrupt change in stiffness.*

This piece of wetted glass will be too large for one person to handle and place with the required accuracy. If working alone, divide the wetted layer as described below.

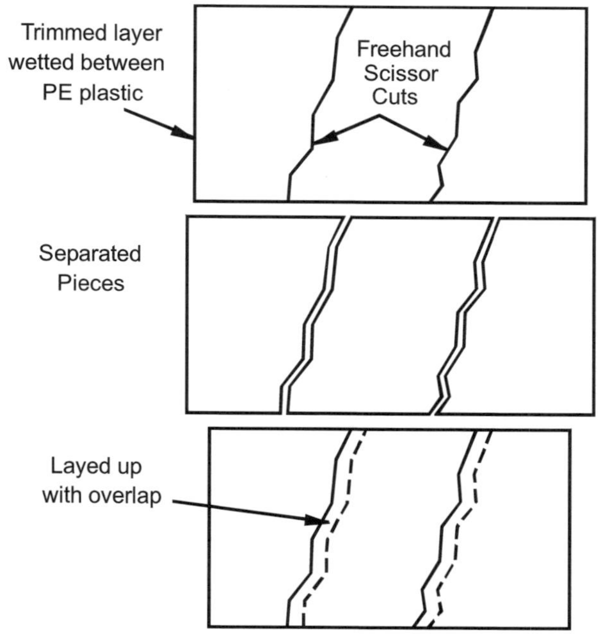

Handling Large Glass Areas

By eye, use scissors to make a diagonal cut in the wetted layer, approximately as shown in the sketch above. Peel away one layer of wetting plastic and discard it. Carefully lower the square edge of the glass over the ***marked work area*** and put in place. Remove the second layer of wetting plastic and discard it. Smooth and adjust with a squeegee.

Repeat the placement process with the third and fourth sections of glass, overlapping the cuts by about one inch. ***This method will place the wetted glass quickly and accurately and will make neat overlaps.***

9) PLACE THE PERFORATED FOAM CORE PANELS: Mix about 3 ounces of epoxy and make medium slurry. Paint the slurry on the bottom of all four core panels and squeegee it in. Add slurry as needed to get good filling and squeegee well. Remove excess with the squeegee. Place the panels over the marked areas and press into position. (The large panels will spring back slightly because of the curved form. This is not a problem because the panels will be forced into final position later, under vacuum.)

10) PLACE THE FIFTH (OVERALL) LAYER OF GLASS: Wet out this layer of 7725 BID in a plastic sandwich similarly to the previous layers. This layer will be too large to handle as a single piece, working alone. See that it will be necessary to ***conform the layer to cover the beveled edges of the core panels smoothly***, so some displacement of the BID weave will be required. This is not a problem if only one end of the core panel is covered at a time. If two people can work together, this layer can be cut into two sections and each section can be put in place very easily. (Again, allow a 1-inch overlap between sections.) As each section is placed, smooth out with a gloved hand and verify that the cores are correctly located.

If the builder must work alone, six sections can be used.

11) PLACE THE PEEL PLY STRIPS AND CAUL PLATES: See the figure. Press the 2-inch strips of peel ply over the areas where the shear webs and the ribs will be bonded. Squeegee out smoothly. Place the 2-inch poster board caul strips over the peel ply and roll out with the grooved roller. The ***poster board will soak up enough epoxy*** to stay in position.

12) PLACE THE PERF RELEASE FILM: Divide the perforated release film roughly by eye into four strips. Place over the work area and smooth out around the core plates with the gloved hand. (Ignore the small wrinkles that will occur around the corners.) Overlap the perf strips by about one inch. Trim around the work area.

13) PLACE THE BLEEDER: Cover the work area with bleeder. Use fresh bleeder next to the work, but previously-used bleeder scraps, if available, can be placed on top, to help vacuum distribution. Trim and/or adjust the position so that the margin of the work area is clear. Place a thick pad of bleeder under the vacuum pass-through position.

14) PLACE THE BAG: Position the bag. Small pieces of brown tape may help with preliminary positioning. Mark the position of the vacuum pass-through, cut an "x," and insert the vacuum pass-through. Trim as necessary. Seal the entire periphery of the bag with brown tape. Connect the vacuum pass-through to the gauge manifold. Use a heat gun to warm the tape and smooth it down carefully.

15) PUMP DOWN AND CURE THOROUGHLY: Slowly open the valve to the bag. Watch the bag gauge so that the tool vacuum does not drop below about 6 inches. Use a little patience until the bag has had a chance to evacuate. (This may take several minutes.) When the bag vacuum starts to register on the bag gauge, gradually open the valve a little more. Continue the process until the bag valve is open all the way. Verify that the vacuum is at least 10 inches. If not, look for leaks and repair them. When the vacuum is good, consider the job finished. Leave the pump on until the residual epoxy in the last mixing cup is quite hard, so that the partly cured layup will not be subject to creep. By this time, the builder may be glad of the end of continuous pump noise and eager to examine her completed work. ***Resist this impulse*** until the residual epoxy in the last mixing cup is quite firm and not subject to creep, ***at least 24 hours.***

16) REMOVE THE WORK: Start the pump again and ***apply vacuum to the tool side***, to hold the tool film firmly. Strip off the bag and bleeder. Work from one corner to start separation of the work from the tool film.

Before removing the new skin from the mold, mark with a marking pen so that the skin can be re-registered to the reference marks made on the mold in Exercise 14.

Lift the finished work from the mold sheet, invert it, and observe the excellent shiny surface.

Turn off the pump and strip off the sealing tape for the tool.

Strip off the peel ply and the poster board caul plates.

Return the new skin to the mold, which will now serve as an assembly jig for the wing section to be completed in the following Exercises.

Suggestion For The Next Exercise

This Exercise has required a great deal of laying up of fairly large areas of glass cloth. The following Exercise will require some flat glass stock to make the webs and ribs. If the builder is "in production" for flat layups and has all of the materials at hand, it would be efficient to lay up a sheet of three layers of 7725 BID glass, or its approximate equivalent, 52 inches long and 28 inches wide, with peel ply on both sides. Ideally, this sheet would be vacuum bagged, but if time and facilities are not favorable, an open layup on a side table will be quite satisfactory. This might be done while the main wing skin is curing (see Exercise 16).

Exercise 15

Wing Skin Ribs And Webs

Introduction

This Exercise will prepare and install the ribs and shear webs in the upper wing skin which were produced in the previous two exercises. These ribs and webs will complete the main load-carrying "box," The relatively thick skin that has been completed in Exercise 15 will carry very large tension and compression loads, but additional structure is required to carry shear. Ideally, the shear load could be carried by some uniform solid medium, such as a solid volume of foam, as seen in the Rutan family of designs, but practical foams will be heavier than necessary. In addition, the designer may wish to use the interior of the wing for other purposes, such as fuel or control mechanisms. Here, we will use two shear webs, one at each side of the "box," as shown below, to carry nearly all the shear load. (This approach is common in modern jet aircraft design.)

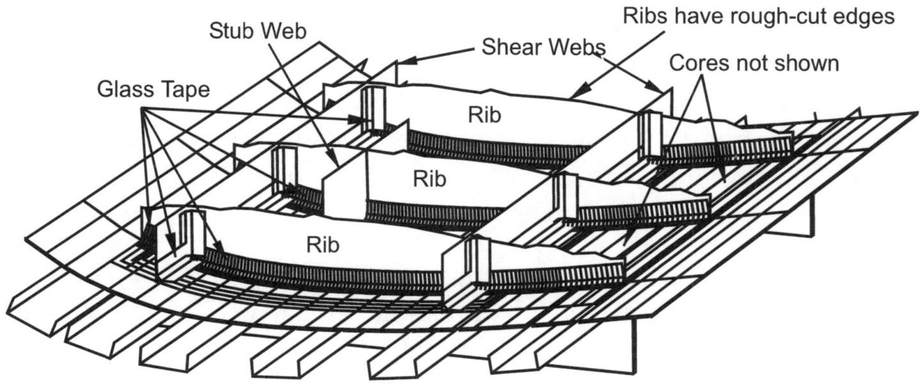

Wing Skins in Jig with Ribs and Webs

By laying up the several layers of the wing skin in a mold, we ***control the outside surface of the wing quite accurately.*** The consequence is that the ***variation in the inside surface is unpredictable.*** This Exercise will demonstrate techniques for handling and taking advantage of these significant uncertainties.

Strut Attachment

The *Crosswind* design requires a single strut to be attached at an intermediate point (approximately at the center of lift of the wing). Very large span-wise and lift forces exist at this location and need to be transferred between the strut attachment and the strong skins. One way to mount the attachment is to use a short span-wise shear web which could accept a simple steel fitting at the attachment point. The short web would then transfer loads to a large area of skin. ***We will install the short stub web, but, for simplicity, will not include the steel attachment fitting.*** (See the Comment at the end of this Exercise.)

To make things simpler and save some work in this training Exercise, we will cut the shear webs, the stub web, and the ribs from a single uniform flat sheet of cured glass. We will not bother to vacuum-bag this layup and peel ply will not be required.

The sketch on the previous page shows the general idea. As shown, the webs and ribs are rough-cut to approximate size.

Lay Up the Webs and Ribs

17) PREPARE A FLAT WORK AREA: Cover a 4 x 8-foot worktable with a sheet of 4-mil PE plastic film. Use a sheet of 1/8-inch Masonite, or other convenient smooth surface, of at least 48 x 53 inches, to use as a flat layup tool support. (This layup will not be vacuum-bagged and will not require a very smooth surface, so the "tool" is not critical and even the well-used work table itself may be a satisfactory surface.)

18) MAKE A PATTERN: Use wrapping paper or flip chart paper to make a layup guide pattern as shown on the following page. Tape to the tool support surface. (Note that the sketch shows three extra strips of glass which will be used in Exercise 17. It is easier to get that done now.)

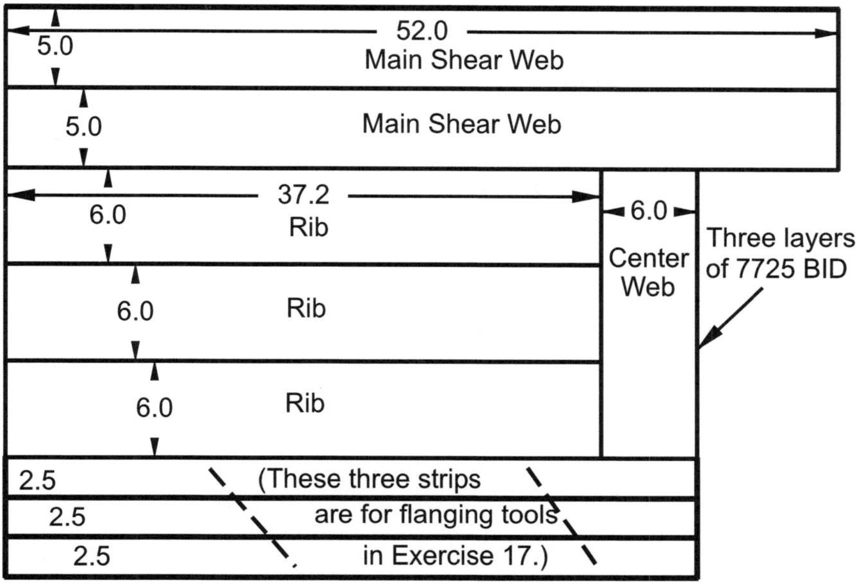

Rough Pattern For Sheet Layup for Ribs and Webs

Dimensions of the ribs and shear webs are not critical; final accuracy and most of the load-bearing material will be provided by the taped flanges.

19) PLACE RELEASE FILM: Cover the tool support with a sheet of 4-mil PE plastic. Pull from the edges to make it smooth and tape in place. If extra tension is needed to make the plastic smooth, use duct tape.

20) PLACE GLASS CLOTH: Cut and place three layers of dry bias-cut 7725 BID over the work area. Scrap glass, if overlapped and placed on the bias, will be quite satisfactory. To save some time and work, it will be convenient to cut the glass cloth straight, but place the cut pieces on the bias. Cloth that hangs over the edge of the work area can be cut in place and the trimmed pieces can be used to cover the layup area. Overlap the glass by at least a half-inch.

21) WET OUT: Mix and tint epoxy and wet out the layup. Use the squeegee, the grooved roller, and the heat gun to assure that the layup is completely wetted. Squeegee through a piece of PE plastic to chase the epoxy around as needed to quickly get good wetting.

Use ***"poor man's vacuum bagging"***: Cover the work area with perf film and squeegee and roll vigorously to drive air and excess epoxy through the perf holes. Leave the perf film in place. Allow to cure for 24 hours.

Remove the cured glass sheet and buff both cured surfaces gently with a hand-held disk sander, with 200 grit, to prepare the surface for bonding.

> *Note: **Best practice is to use peel ply** on both surfaces instead of the perf film as described. The method described saves a little time and a little peel ply.*

24) CUT OUT THE WEBS AND RIBS: We will cut the webs and ribs ***significantly undersize*** by approximately 1/8-inch all around to allow for variable skin thickness and uncertainties in cutting. Mark the cutting lines on the cured glass sheet by tracing over the pattern with a marking pen. Use a bandsaw or spiral saw to separate the sections. Cut the front main web to 3.9 by 50 inches and the rear main web to 4.0 by 50 inches. Similarly, cut the stub web to approximately 5.2 by 18 inches.

For maximum shear strength, the webs will be not be cut at the intersection with the ribs, but will be continuous in the span-wise direction. Instead, ***the ribs will be cut to fit the webs.***

Use the Master wing rib reference to mark the ideal outline of the rib sections, as shown. Mark a second cutting line roughly 1/8-inch inside the "ideal" line. Cut out the ribs on the cutting lines with a spiral saw or bandsaw. ***Wiggly edges are good enough.*** Smooth the edges with a belt sander.

"Tack" the Webs and Ribs

25) PLACE THE WEBS AND RIBS: We will use "5-minute flox" to "tack" the web and ribs into position. The original mold marks, made when the skin was formed, will be visible through the glass, even if tinting was used. Use small pieces of brown tape, scraps of wood, and moderate creativity to position the glass sheets over the marks, as shown.

The several glass sheets will support each other fairly well and a sufficiently accurate location for all of the parts can be found quickly. Mix up a small amount of 5-minute epoxy with about half a volume of flox to make "5-minute flox." Tack the ribs and webs into position with small globs of 5-minute flox as needed. Cure until firm enough to continue.

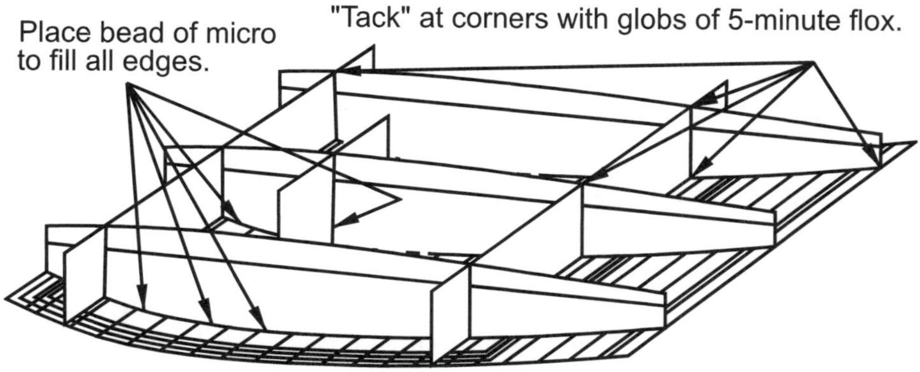

Tacking and Filleting

Tape the Webs and Ribs

Allowing for overlaps and cutting, about 52 running feet of 2 1/2-inch tape will be required. Use a single layer of 7725 BID, cut at 45 degrees, and wet out in a 4-mil PE plastic sandwich. ***The less-experienced builder should assume that this job will require two separate taping cycles*** (two cups of epoxy and two wet-outs).

27) PLACE THE MICRO FILLET: Obtain a set of expendable ***plastic pastry bags*** and several 1/8- to 3/16-inch (size 3 and size 7) nozzles and nozzle holders from a kitchen supply store. (This is the tool bakers use to write "Happy Birthday" on a cake.) Mix about 3 ounces of epoxy and add micro to make a medium-dry mixture about the consistency of cake frosting. Fill a pastry bag about one-third full. Squeeze the bag to place a bead along the intersection to be taped. Notice how the *use* of ***the pastry bag makes it very easy and quick to lay a clean bead*** without contaminating the bonding area with micro. (If any micro gets on the bonding area, wipe it off with a facial tissue.)

Smooth out the bead with the gloved finger to make a blended curve, ***filling the irregularities and gaps*** between the ribs and the cured skin. Continue with the following operations without waiting for the fillet to cure.

28) WET THE FIRST BATCH OF TAPE: Mark 3-mil PE plastic with a marking pen to define a working area of about 15 inches x 5 feet (6.25 square feet). Mark 6 strips 2.5 inches wide. Turn over and wet out 7725 BID tape with tinted epoxy. ***Tint with the color opposite to whatever***

color was used for the wing skin. Use about 1/3-volume of fused silica and about 1/6-volume of flox in the mixed epoxy.

29) PLACE THE TAPE: Use a paint brush to paint a thin coat of epoxy on the working area. With care, the wet bead will not be disturbed. Cut the marked strips of wet glass with scissors roughly by eye, peel and discard one plastic film, and place into position. The beginner will be more confident working with tape strips less than 8 inches long. Overlap adjacent strips by at least a half-inch and smooth down with the gloved hand. Use a narrow squeegee or a plastic pot scraper to be sure that the tape is driven ***all the way into the corner.*** Experience quickly teaches the technique.

Inspect each section of tape carefully to be quite sure that the entire area is bonded and that there is ***no air trapped in the interface.*** "Stippling" with a paint brush or use of the roller brush or grooved roller will help. This is a circumstance where the tinting is particularly valuable. *Consider that the tape is intended to carry a shear load across the bonded area and that unbonded tape is wasted.* Take time to get it right.

30) PEEL PLY THE TAPE EDGES: Use strips and scraps of peel ply to cover and smooth out the edges of the freshly-applied tape. The peel ply will absorb some excess epoxy and will ***prevent sharp spikes of glass.***

31) WET OUT AND PLACE THE SECOND BATCH OF TAPE: Similarly, complete the taping task for the entire job, using a second batch of epoxy. The less-experienced builder may find that it is necessary to work with three batches for a job this size, but this is the cheap way to gain the necessary experience.

32) CURE AT LEAST 24 HOURS: When cured, the webs and ribs will ***lock in the shape of the molded wing skin.*** If the work is left unsupported or is subjected to a significant sustained force, partially cured epoxy may creep and make it necessary to reject the part.

> *Note: In a "real life" curing situation, the cautious builder may wish to place **sandbags** on the skin in the gaps between the ribs to assure an accurate fit to the mold/jig. The structure will be very rigid after the tape cures and there will be no practical possibility of adjustment.*

33) REMOVE THE PEEL PLY: Strip the peel ply from the taped joints after cure is complete.

Comment on Strut Attachment

This Exercise has made a number of ***simplifications and omissions to reduce the amount of work*** necessary to demonstrate the essential principles. A practical wing would require some means of bringing out a strong attachment fitting for the wing strut. The stub spar described here is a ***span-wise web*** 18 inches long which is bonded to the main load-bearing skins along its full length.

The design requirement is to distribute the concentrated load developed at the strut attachment to a large area of skin. One way to do this would be to weld a steel tab to a flat steel strip and bond the steel strip to the web with several bolts. The steel tab would then carry the load outside the wing. The ***steel is stiffer than the glass,*** so the shear strain is passed to the skin by multiple bolts along the web, through the large area of bonding tape.

The reader may get an intuitive feel for a comparable structure by noting that the original *Tailwind* design makes the strut attachment in a spruce spar 1 1/4 inches thick, using two steel straps and two bolts (double shear).

Exercise 16

Bottom Wing Skin

Introduction

This Exercise essentially repeats the operations described in Exercise 14 to make the bottom wing skin (required to close the wing) and it is not necessary to repeat them in detail here.

A Compromise

We will require a bottom wing skin to close the wing section produced in this series of Exercises. This is a training Exerecise, so there is little point in repeating all the procedures taught in making the top skin, including making an accurate mold to produce the slight curve required. It will be sufficient to ***lay up the bottom wing skin on a flat tool, omitting the front cores, but including the rear cores, and making the rest of the skin the same.*** It will then be possible for the builder to bend the multi-layer skin to the small degree required.

Bottom Wing Skin Mold

It is not necessary to make a mold for the purposes of this series of Exercises, but the procedure is described here for completeness.

The procedure for making the bottom wing skin mold is exactly the same as for the top skin. The only difference is that the curvature required is much less and the builder, with the first experience behind him, will find the second mold even easier and quicker to make.

Use the reference wing section to define and shape the four forms, as previously.

It will be convenient to use the same steel bending frame that was used in Exercise 14. Remove the old wing forms, discard them, and replace with the corresponding forms for the bottom skin.

Bottom Wing Skin Layup

Repeat the procedure for Exercise 15, using exactly the same glass layup schedule and the same peel ply and poster board caul plate system. If the builder has chosen to skip the step of making a mold, but will be using a flat tool, the front skin cores should be omitted.

Exercise 17

Wing Closing Flanges

Introduction

This Exercise will produce the precision flanges which will support the bottom wing skin.

In Exercise 16, the shear webs and the ribs were attached directly to the upper wing skin by means of wet glass tape. The joint conformed exactly to whatever irregularities existed. Whatever gaps existed between the shear members and the tape were filled with micro and the tape provided a smooth transition radius between the surfaces.

The main problem in closing a large volume like a wing is the ***creation of accurately located bonding surfaces*** which match the irregularities of the surfaces to be bonded and allow for the thickness of the bond joint itself.

The process of locating a flange line and forming the flange was discussed in general terms in Chapter 10, *Flanges and Closing*. In this Exercise, detailed instructions will be given for the specific case of the *Crosswind* wing.

Finding the Flange Line

Make Flange Line Bridge Tools

1) Make a rough flange line reference tool. Use the wing section reference master to trace the line of the lower surface of the wing on a piece of 7/16 inch chipboard, as shown in the sketch. The bridge must be long enough to span the chord of the wing section. Cut the bridge support out with a table saw and/or a spiral saw. Rough freehand work is entirely satisfactory.

2) Set up the wing section reference master. Remove the upper wing skin from the mold/jig and set it aside. Place brown tape on the master rib, as shown, to serve as release film.

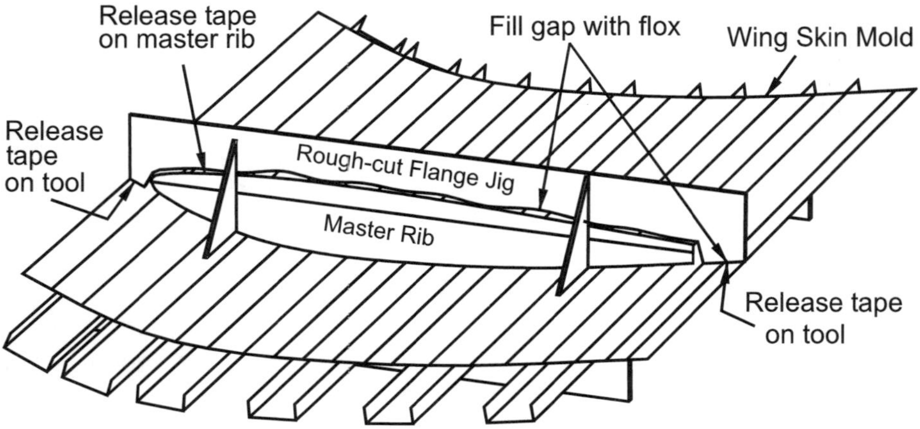

Making a Closure Reference Jig

Place the rough-cut bridge over the master rib, as shown, and tack as needed. Mix some fairly stiff filler with about half flox and half micro and fill the gap between the master and the bridges. Cure. The ***finished bottom surface of the wing is now accurately defined,*** relative to the tool/jig. This is ***not the flange line,*** because the thickness of the bottom wing skin and the required adhesive must be considered. Remove the cured reference jig and set aside.

3) Locate the true flange line: Return the wing skin to the tool/jig. Use the reference jig to test the clearance between the ribs and webs from Exercise 16 and the reference jigs. If there are high spots, grind them down with a belt sander so that there is about a 1/8-inch clearance. Assume that the bottom wing skin will have the same distribution of skin thickness as the top wing skin. We will need to modify the reference jigs to locate the true flange line by making spacers equal to the skin thickness. See the sketch on the following page.

Use a bandsaw to cut freehand small "offset chips" of scrap cured glass, to about 1 inch x 3/8 inch. About ten "offset chips" will be required foreach rib and about 20 more will be required for the webs. Cut 50 chips.

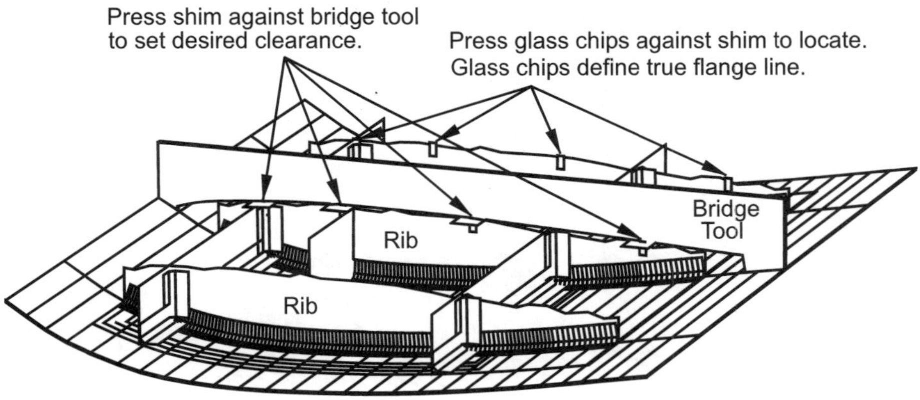

Finding the Flange Line

Position the reference jig over the wing skin at each rib position. (The quickest way to do this is to have two helpers hold each end of the reference jig while the offset chips are being applied, in the following operation. Mix a small batch of 5-minute flox.

Use a hand-held spacer (shim) equal to the desired offset and place a line of offset chips along each rib, approximately as shown. (A stack of ***common business cards make accurate spacers***; they can be added or subtacted as desired. For this non-flying Excercise, use about seven business cards as a typical offset.) Hold the stack of cards between the reference jig and the rib, wet a chip with 5-minute flox, and place the chip so that it touches the card stack. Repeat as shown until the ribs have a line of offset chips about every three inches.

Similarly, the chips for the webs can be set after the rib chips have cured by carrying the rib line across the webs with a straightedge. The edges of the ***chips now define the true flange line.***

3) Make rib flanging tools. Find the three extra 2.5-inch glass strips made in Exercise 16 at the same time as the ribs and webs. Trial fit the strips to the ribs as shown in the figure on the next page. Trim as needed. Cover one side of the strips with ***clear mailing tape to serve as a release surface.***

4) Place the rib flanging tools (see the figure). Place the glass strips over the ribs and center them by feel. Hold in place by weighting with any convenient straight object. As shown, steel studding is particularly convenient, but wooden 2 x 2's will work. The weighted strips will hold the glass strips flat to the ribs. Tack as needed with 5-minute flox.

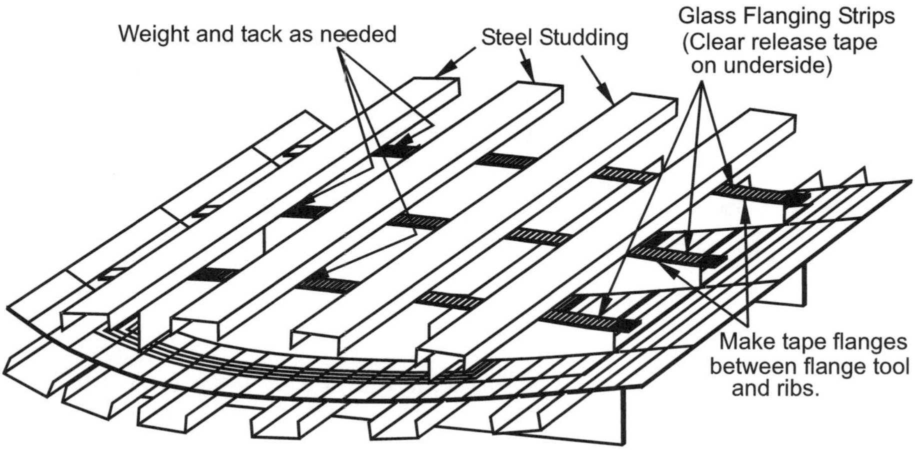

Making Closure Flanges

5) Wet and place the rib flanges. Mark wetting plastic (one side perforated) and lay up about 52 running feet of two-layer 7725 BID tape, 2 1/2 inches wide. Tint in "opposite color" and use about 1/2-volume of Cab-O-Sil and 1/4=volume of flox in the mix. Squeegee out thoroughly and inspect for thorough wetting, with excess epoxy driven out through the perf holes.

Ignore the small glass offset chips used to set the flange line. Place the flange material directly over the offset chips.

Use strips of tape 8 to 10 inches long and place the flanges. Be sure that the flange is well-bottomed and use a small squeegee edge or a kitchen pot scraper to assure contact. ***You will be able to see through the glass flanging tool, so you can correct any placement error immediately.*** Cure. Remove the rib flanging tools.

> *You will need to work around the weighting strips. If necessary, allow the work that can be easily reached to cure, and complete the job in a second work session, which will allow movement of the weighting strips.*

6) Make the shear web flanges. The existing rib flanges will establish the line for the web flanges. As before, place the offset chips held in place with 5-minute flox. Cut transparent glass flanging tool material and face with CLEAR mailing tape, as above. Position as needed with scrap wood and hot melt glue. Place the shear web flanges. Cure. Remove the flanging tools.

7) Fill the void under the flange. Inspect, trim and touch up as needed. Use a pastry bag with a small (#3) nozzle to place a bead of micro to fill the void caused by the curve of the flange. Be careful not to get micro on the bonding surface and clean up as needed. Cure.

Close the Wing

Trial Fitting

8) Place the lower wing skin over the main wing core structure and inspect the entire work carefully. Press down as needed to verify that the skin is very close to the bonding flanges. When the skin is in the best possible position, find a spot outside the thick skin area and drill a pilot hole for a #6 self-drilling screw. Place the screw. Inspect carefully and drill and place three more ***alignment screws.*** Mark the location of the alignment screws.

9) Drill for pull-up screws. With the skin accurately located, ***drill pilot holes*** for #6 self-drilling screws along the entire bond line at about 3-inch intervals. ***You will be able to see the tinted flanges.*** Place the holes so that each screw hole is approximately at the center of the flange. Remove the alignment screws, lift off the wing skin, and vacuum out the drilling debris. Open up the alignment holes in the skin by ***one larger drill size*** to allow the skin to slip over the alignment screws.

Final Closing

10) Prepare and place bonding tape. Buff the flange area with 200-grit sandpaper and wipe with clean alcohol. Wet out, under plastic, 26 running feet of single-layer 7725 glass, marked for 2 1/2 inch strips. Tint the epoxy "opposite" and use about 1/3-volume of flox in the mix. Assure that the tape is completely wetted. Place the wet tape in the bonding area and verify that the tape contacts the flanges over the entire area. Smooth out with the gloved hand.

11) Place and secure the wing skin. Lower the wing skin to the bonding area and secure the four alignment screws. (A sharp-pointed tool like an awl or steel scribe will be convenient to establish alignment.) Inspect the bond line, looking for areas where there is no contact or only partial contact. ***Insert "pull up" screws*** in these areas in the previously-drilled pilot holes and run the screws down. Inspect again, and add more pull-up screws as needed. If an area already has full contact, do not use a screw. Continue until there is full contact over the entire bonding area. Allow to ***cure until the epoxy is approximately half hard***, typically 12 to 24 hours. Remove the screws.

See that the pull-up screws are used only as necessary. The wet bonding tape will seal off any pilot holes which are not used. The pull-up screw holes will appear as small blemishes which must be filled at the time of final surface finishing.

Appendix 1

General Tools

Introduction

In previous chapters, when tools were required, they were briefly described in general terms. This Appendix will serve as a detailed reference for the special tools discussed.

Most composite construction can be accomplished with the ***common and familiar hand tools*** found in almost any home shop. Examples are hand drills, portable saws, sanders, and routers, like the ones widely used for household applications such as woodworking.

Composite work will also require some ***special commercially-available tools*** which may be less familiar to the conventional aircraft builder. Some of these special tools, such as plastic squeegees and sanding accessories, are sold for auto-body work, plumbing, or drywall construction. Others, such as vacuum pumps and associated plumbing, are available as standard commercial products but are not commonly used in aircraft shop work or found in most tool stores.

Other special tools are so handy to use in composite work that they may be considered practically essential for a project on the scale of building an airplane, but they ***must be constructed by the builder.*** This appendix will describe these special tools in sufficient detail for the builder to make his own. Examples are beam balances for measuring epoxy, various items of vacuum plumbing, and tools for cutting, beveling, and perforating foam.

Common Tools

Drills

Driver-Drills - Battery Powered

The 3/8-inch battery-powered driver-drill will be used so many times for drilling holes and setting and withdrawing screws that the builder should ***consider owning two of them*** to avoid the bother of changing setups during a work session. One of them, at least, should be of 12-volt rating, or more.

Other Hand Electric Drills

A smaller 1/4-inch electric drill, with or without battery power, will be useful in confined spaces and where higher rotation speed is desirable for smaller drill sizes.

A 1/2-inch cord-driven electric drill will be needed for occasional larger holes and for setting and countersinking screws in wood.

Drill Press

A drill press may be considered optional, but barely so. Most holes needed for the operations described in this book can be made with the basic driver-drill, but the builder will be grateful for the ability to make ***holes which are accurately square with a surface***, which is difficult to do with a hand-held drill. The drill press should have a 1/2-inch capacity with a 4-inch stroke.

Power Saws

Table Saw

A standard table saw of at least 8-inch diameter will be used frequently for cutting and trimming a wide variety of wood framing and jigging materials. The Exercises in this book require plywood or chipboard for many applications and the availability of a table saw with table extensions will save a great deal of time, especially when long, accurate, straight cuts are required. A ***carbide-toothed blade will be necessary*** when cured fiberglass is to be cut. ***Cutting laminated glass with a cir-***

cular saw tends to cause delamination and should be used only for rough trimming. ***Note that glass dust is hazardous.***

Portable Circular Saw

A standard 7 1/4-inch portable electric saw, with a ***carbide blade***, will be used frequently for a wide range of rough wood work. This saw will also be used with an ***abrasive wheel*** to cut steel framing material, which is recommended by the author for making stable, light-weight jigs.

Crosscut, Compound, or Chop Saw

A crosscut circular saw is not essential if the builder has a general purpose table saw, but is often useful when working with smaller parts and may be safer in some circumstances.

Band Saw

An 8-inch bandsaw will be very useful for trimming a clean edge on a cured glass layup ***without danger of delaminating the piece.*** This tool will have only occasional use except for trimming glass, but is ***nearly essential*** for that purpose.

Battery-powered 3 3/8-inch Circular Saw

This tool is not essential, but is a great timesaver when trimming sheets of plywood, chipboard, or Masonite. It is small enough to be stored in a drawer, yet can be pulled out quickly and used without the bother of rigging a power cord. With a ***carbide blade, it can be used freehand for rough trimming cured fiberglass.***

The depth of cut may be adjusted, so this saw may be used for ***slitting foam cores to facilitate bending to a curve*** (Chapter 5).

Rotary Tools

Spiral Saw

This high-speed tool is basically a small hand-held router which may be used free hand or used with a guide. The manufacturer calls this tool a "saw," but the cutting principle is that of a router. ***Special bits are available for cutting fiberglass.*** This tool is slow, compared to other saws and ***freehand bit-breaking is frustratingly common.*** The small size permits working in confined areas. This is the ***most general-purpose of the tools used for trimming fiberglass.*** It works quickly where nothing else will and is ***essential for composite work.***

The spiral saw has provision for setting the depth of cut, and may also be used for ***slotting foam for bending.***

Dremel Tools

This is a family of rotary tools which makes use of a very large variety of small saw blades, grinding wheels, rotary cutters, and sanding disks which are all driven by a small hand-held motor unit, much smaller than the motor used for the spiral saw. Several models are offered by the Dremel Corporation, but other manufacturers make similar devices. This type of tool is ***extremely general purpose for small work*** and sometimes is the only appropriate tool for a task. The main use will be as an alternative to the larger spiral saw for trimming cured composite, especially in confined spaces. One special task might be cutting a V-shaped groove in foam so that it may be bent into a concave curve. With a small abrasive disk, the tool may be used for ***cutting hard steel piano wire.***

Work Area, Benches, and Storage

Individual preferences and available space will be important here, so this section will describe some alternatives that the author has found workable in his situation - only as suggestions.

Workbench

Airplane subassemblies are large and need access from both sides, so the general purpose workbench should be narrower, longer, and somewhat lower than might be used by, say, a cabinet maker. If the builder intends to build a plans-built aircraft, the recommendation of the designer should be followed. The author uses a bench 34 x 90 inches, 30 inches high.

The workbench will be used to support subassemblies for precision alignment, so should be stable, but it is not practical to make it so accurate that it can be used as an alignment jig. (Accurate alignments will be done by other methods.)

The ***top of the bench should be an expendable, replacable surface*** and will be used for securing work, often using Bondo polyester body filler as a temporary attachment, or using nails or screws freely as needed. The ideal top material should be stable and without "grain" and allow for easily sanding off to a flat surface after filling holes and cuts. ***Usually, depressions in the surface are acceptable, but high spots are not. Plywood, therefore, is generally not preferable as a working surface.*** After hard use, the top should be easily and cheaply replaced. The author uses 5/8-inch particle board.

Situations will arise frequently when the builder will want to clamp work to the edge of the table top. If the top is made slightly oversize, so that a margin of about 1 1/4 inches is left on both sides, clamping will be quite easy.

Tool Cabinet/Table

The author uses a second, higher, table, the "tool bench," which is useful for smaller work, but that is basically a cabinet filled with drawer space for tool storage, which has had a chipboard top attached to it. This bench is a much larger (and heavier!) version of the roll-around tool cabinets used by professional aircraft mechanics.

Bench and Tool Movement

Most builders must compromise on workspace, so it is very desirable to be able to ***move heavy tools and tables*** around the shop. The tools must be positively positioned when in use, so any lifting or rolling mechanism must allow for the tools to stand firmly on the shop floor. Some tools make provision for easy movement, but most do not. Roll-around tool lifters are available commercially. The author uses a double-lever device made from welded steel which can be bolted to whatever load requires movement. The basic principle is to raise the heavy tool a very short distance so that the tool is carried by some kind of wheeled carriage, then lowered for positive location on the floor. (Note that the tool bench, when full of tools, is very heavy indeed and the lifting device must be strong.)

Temperature Control

Workroom

Workroom temperature will be dependent most of the time on the personal comfort of the builder, who might be perfectly happy working in an unheated garage while wearing three sweaters. The viscosity and curing time of aircraft epoxies are sharply dependent on temperature and require at least ***a temporary environment of 70 degrees F - higher if practical.*** The essential problem is how to provide adequate heating of the work area while still providing ***sufficient ventilation*** of chemical fumes and avoidance of combustion gasses. Direct electrical heating is expensive, but produces no combustion products. Gas-fired ceiling-mounted air heaters, as used in commercial buildings, are excellent, but are expensive and bulky for the typical small shop. The best compromise must be left to the ingenuity of the builder in her individual case.

A cheap outdoor thermometer should be hung in the workroom where it is easy to see and is outside the air stream of the heat source.

> *The author heats and cools a garage efficiently with a 1/3 HP-heat pump which served as a heater and air conditioner in a doctor's office. The doctor had "traded it in" for a larger unit and it became available through a local air conditioning dealer for a very attractive price. There are, of course, no combustion products, and the air conditioning function is nice for the builder, if not for the epoxy. A few phone calls might be rewarded.*

Resin Storage

It is very much easier to measure, mix, and spread epoxies when they are well warmed. Also, the builder will find that there will be occasional "small jobs," which take only a few minutes to execute and do not justify the bother of heating the entire work room. The builder can make the layup with warm epoxy, then move the work to a warm place to cure.

The author stores epoxy resin in well-rinsed 100-ounce ***plastic laundry soap bottles*** and hardener in similar 44-ounce bottles. Depending on the builder's requirements, it may be a good idea to keep two sets of filled bottles, one set "on deck," the other in ***warm storage for immediate use.***

The sketch shows a simple "warming box" as used in the author's shop.

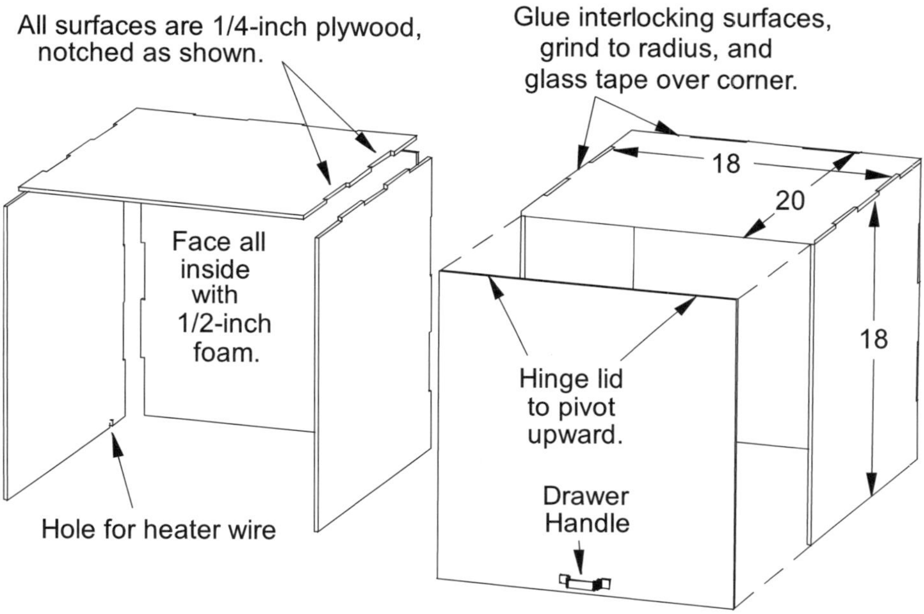

Construction of Warming Box

The reader will recognize that there are many ways to make an open-bottomed box, but this way is simple and quick. The notches can be very rough, using a table saw or a freehand spiral saw. Water-catalyzed wood glue can be used for initial assembly and for the matrix for the glass tape. Expensive epoxy need not be used. The resulting corner may not be pretty, but it is fast, cheap, strong, and takes up very little space.

The next sketch shows a simple self-contained scheme for heating the warming box with a thermostatically-controlled light bulb. Air circulation is provided by a computer fan. Exact dimensions will be determined by the specific components that the builder has available. The coffee can heat shield helps isolate the heat source from the thermostat. The author has found that 85 degrees F is a good setting for storage of epoxy intended for immediate use.

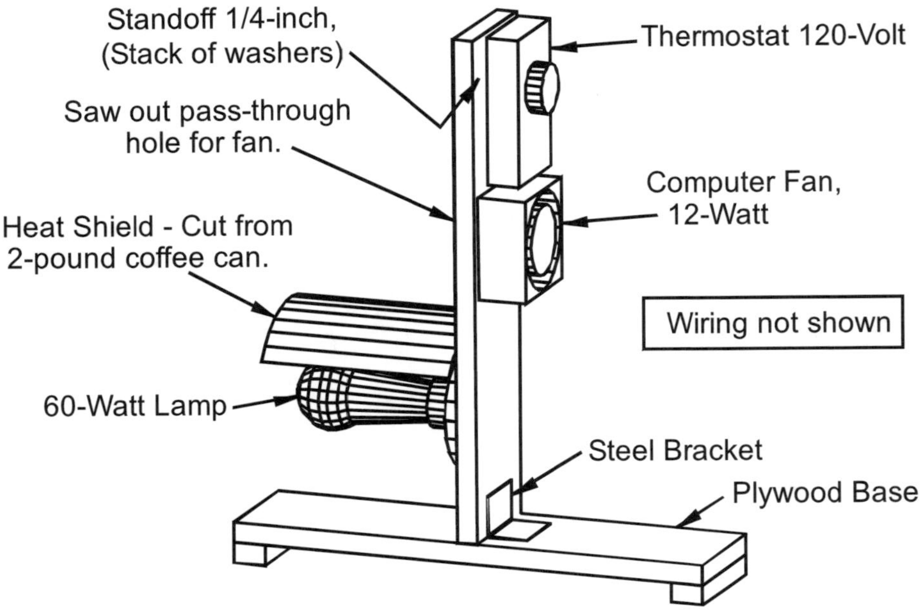

Thermostatically Controlled Epoxy Heater

> ***Warning:*** *Warm hardener has very nasty volitile components. It is necessary to provide excellent ventilation when opening and handling these materials. Unless it is quite cold outdoors, the author prefers to move outside to measure and initially mix warm epoxies. If it is not practical to move outside,* ***wear a mask.***

> ***Related Warning:*** *Working with pre-heated mixed epoxy increases the* ***risk of an exothermic reaction*** *by reducing the "time window." If the builder is making a large layup, such as a wing skin, and is mixing and applying larger quantities of epoxy, the batch of mixed epoxy must be put in place IMMEDIATELY so* ***that the epoxy will be cooled to the temperature of the work.***

The warming box is useful for curing small jobs. The epoxy bottles can be removed, the small layup inserted, and the thermostat turned up to maximum for a quick and efficient cure.

Direct Layup Heating

This book discusses flat tool layups under vacuum at some length. It is not necessary to heat the entire work room to cure a layup if there is some means of heating just the layup. One way to do this is to use a common household queen-size ***electric blanket spread over the work*** and covered in turn with another insulating blanket, such as a comforter.

The Temporary Curing Oven

Builders who must work in cold climates and who find that it is quite expensive to heat a very large work volume for the extended time required for cure may wish to consider the use of a temporary curing oven. The roofing trade uses 4 x 8-foot ***panels of heat insulating foam which have aluminum foil on both sides.*** These panels are cheap and relatively stiff and it may be convenient to make a temporary box over the work table.

The heat source can be a standard ***1500 watt space heater*** with a ***self-contained fan.*** The overtemperature device and/or ***temperature control must be disabled.*** Temperature control and safety then require a separate thermostat control. ***Electronic temperature controllers*** are commercially available which have separately-adjustable set points and control range. The temperature probe is on a long wire, which can be let into the box through a small hole. The temporary curing oven will be too small for the larger airplane parts, but there are many smaller parts for which the curing time saved will be worth the effort of setting up the oven.

As an alternative, where workroom conditions are cooler than ideal, the builder may accept a slow and partial cure in the shop until the parts can be handled, then move the work to the temporary oven to complete the cure.

Postcuring

Postcuring requires that the builder create an environment for relatively large cured composite parts which can be heated evenly to 140 to 150 degrees F for a period of about five hours. Depending of the specific epoxy system used, this operation will raise the "temperature of glass transition" to approximately 40 degrees above the curing temperature.

The ***elevated temperature used to postcure will cause a matrix cured at room temperature to soften***, which may allow the part to deform.

The builder must provide temporary jigging or constraint to prevent this.

Postcuring can be done at any time after the original shop cure, so most builders delay postcuring until several large parts have been completed. The temporary postcuring facility can be built when needed and dismantled after the large parts have been done. Few builders will want to have an object this large available permanently.

The reader who follows any of the composite builders groups on the Internet will find that there is a large range of ingenuity used by different builders. "Rube Goldberg" lives. Essentially ***the challenge is to build a chamber large enough to hold an airplane wing, jig the wing accurately, and heat the oven evenly to a high but reasonably controlled temperature.***

Electrically Heated Postcuring

Probably the easiest way to make a well-controlled postcuring facility is to make a well-insulated chamber and to heat it electrically. There are at least two practical problems: (1) It is hard to make a demountable chamber to operate at 150 degrees F which does not leak heat at a rapid rate. (2) Ordinary portable space heaters as sold in hardware stores require that their temperature control and safety devices be disabled.

The author has made a successful medium-sized postcuring oven using aluminum-skinned 1-inch foam roofing insulation. Gaps were covered with 4-inch strips of the insulation, held in place with #10 machine screws and fender washers. A single 1500 watt heater, with the controls disabled, was sufficient to bring the oven to 160 degrees F. The fan built into the heater was used to circulate the air. Temperature control was provided by a commercial electronic thermostat with the "dead band" set to about 5 degrees and the probe hanging down into the center of the chamber. A larger oven would probably require two such heaters and require that two separate power circuits be used.

> ***Note:*** *This system has an element of risk which the builder should consider. The built-in fan will be operating well out of its rated environment. The cautious reader might consider driving an automobile fan with a belt, using a small electric motor outside the chamber.*

The aircraft supply houses offer a "***postcuring kit***" which consists of ordinary thermometers with the appropriate range and a simple fixed-setting snap-switch thermostat. The thermostat is of the type used as a safety switch for home heating systems and typically has quite a large "dead band." The basic problem of creating the temporary oven that is to be managed remains up to the ingenuity of the builder.

Solar-heated Postcuring

Some builders have reported using a system in which galvanized sheet steel has been used to built a tightly-closed shed. Such a shed will reach 180 degrees F in the sun in still air. To be safe and practical, it will be necessary to provide for good air circulation and a means of limiting the temperature rise. Successful systems have been built with a belt-driven fan and limited temperature rise by using a thermostatically-controlled cool-air bypass valve. Again, a reliable electronic thermostat controller is required.

One alternative is for the builder to monitor the process directly, reading the thermometers from time to time, and adjusting fan air flow as necessary to contol the temperature.

Propane Heating

Another alternative is to use a general-purpose single burner propane stove as the heat source. Such stoves are self-contained and include the regulator, ready for connection to a standard propane fuel container. The propane burner is typically about six inches across, supported in a cast iron frame. These burners are available quite cheaply at discount import tool stores as they are widely used for households in less-developed countries.

Standard Composite Tools

The catalogs of aircraft supply houses which cater to homebuilders show widely used tools gathered together under the heading of "Composite Tools," and looking through such a catalog is a very good idea. Some of these tools are general purpose cutters, files, knives, and even razor blades, which will be found in the tool department of any home-supply store and will be mentioned only briefly here. Others, like plastic and rubber squeegees, are standard items in the auto-body field. A few are unique to the needs of composite layups. "Starter" kits are offered in the aircraft supplier catalogs.

Squeegees

The yellow ***plastic squeegee*** is the most commonly-used tool in composite construction, used to spread wet epoxy over a layup or to spread filler. Plastic squeegees are sold in a wide range of widths as auto body and drywall tools in paint departments. They are cheap enough to be considered expendable, but they can be cleaned in alcohol and reused. (The author often places a really wet squeegee on a scrap of PE plastic and lets it cure. Two days later, the mess can be flexed and the cured epoxy will crack off, leaving the squeegee clean.) ***Rubber squeegees*** work a bit better in some situations which are best taught by experience and the builder should perhaps own two.

Brushes

Cheap expendable paint brushes in 1-inch to 2-inch widths will be used for a wide range of applications, including applying mixed epoxy and slurry and driving bubbles out of a layup. They can be cleaned in vinegar, alcohol, and/or acetone, but the busy builder can consider this an option.

Plastic Notched Spreader

This item is essentially a squeegee with coarse notches cut in two sides. It is very useful in leveling epoxy, especially for larger layups, where more than one worker is handling and spreading epoxy. A notched spreader will be useful in distributing structural adhesive.

Rotary Glass Fabric Cutter

This item is sometimes called a "pizza cutter" or a dressmakers cutter. It is essentially a very sharp-edged wheel which is rolled against a plastic backing to cut glass fabric. It is fast and accurate for cutting single layers of dry glass, but sometimes fails to cut all the fibers at one pass. (Uncut fibers should be cut with a razor blade.)

Grooved Rollers

This very useful tool is special to the composite technology area and must be ordered from a specialty supplier or from an aircraft supply house. Grooved rollers are basically aluminum rollers, straight or slightly crowned, with ***grooves about 1/8-inch deep,*** leaving a web of about 1/32-inch. The rollers are available in a wide range of sizes, including smaller "detail rollers," which are used for getting very close to corners.

The builder will probably find that it is ***sufficient to own a 1 by 3-inch laminating roller and a 1/2-inch corner roller.*** It is possible to get along without rollers, but there will come a time when the builder needs to get a layup done with all deliberate speed, possibly inspecting the work of less experienced assistants, and speed will be important.

> *Grooved rollers are hard to clean. The author* ***gave up on cleaning the "stock" models,*** *threw away the original handle, and used the closest standard paint roller handle drilled to accept a piece of aircraft .040 safety wire at each end of the shaft. The roller is confined between two washers, held in place with the safety-wire pin. After each use, the safety wire is clipped and the roller is disassembled for cleaning.* ***The grooves must be cleaned in vinegar, then alcohol, with a toothbrush and the bore cleaned out with a scrap of rag.***

Bristle Rollers

This item is similar to the grooved roller, consisting of a roller brush instead of a metal roller, but is even harder to clean. It is useful for driving bubbles out of finely woven fabrics, like peel ply. The builder will find that he will avoid the bristle roller unless nothing else works.

Measuring Epoxy

This subject was discussed in general terms in Chapter 3 and supplemental information will be given here.

Ratio Pumps

A ratio pump delivers ***two separate fluids into a common output in a precisely defined volume ratio***, usually a fixed ratio set by the particular epoxy system chosen by the builder, and is the most widely used system. The major aircraft supply houses sell several kinds of ratio pump, including expensive pumps in which the ratio can be adjusted for different epoxy systems.

Ratio pumps have one great advantage: ***They are fast.*** They deliver almost any quantity of material, even very small quantities, with essentially no preparation time. A lesser advantage is that they are available for immediate purchase.

They have several disadvantages: (1) They are ***hard to clean***, especially the parts associated with hardener. (2) They ***do not isolate the stored materials from the air,*** allowing contamination. (3) They are tall and bulky and take up precious shop space, especially if the builder has stored the pump in a "warm house." (4) If unprotected, they are ***at risk of tipping over*** and spilling unreacted epoxy components on the shop floor, a prospect to be avoided. (5) They are relatively expensive.

Ratio Balances

A ratio balance enables the builder to pour out a desired quantity of resin in a cup, then add exactly the required quantity of hardener to complete the desired mix ratio. The ratio balance is not quite as quick to use as is the ratio pump, but cleaning is not a factor and the device is small enough and light enough to be kept on a window sill, out of harm's way.

The user must prepare ***several pairs of carefully-calibrated weights*** representing the range of epoxy quantities he is likely to use. Each pair consists of one weight representing a quantity of resin and a second weight representing the exactly corresponding weight of resin. One widely-used epoxy system requires 27 parts by weight of hardener for every 100 parts of resin, so if the builder made a 100 gram weight to represent the resin, he would need to make a 27 gram weight to represent the hardener. (Note that the required measurements by weight are not the same as the corresponding measurements by volume.)

Note that it is ***very convenient to use more than one epoxy system.*** The builder must only provide a different set of measuring weights if a different ratio is to be used.

To use, the builder places an empty cup on the pan, then balances the weight of the cup. She next places a "resin weight" on the balance and pours a quantity of resin into the cup to balance. The busy builder will usually overpour the resin a bit, so the balance must be adjusted slightly to rebalance. Next, the "hardener weight" is added and hardener is added (more carefully) to balance. The measured epoxy is then removed from the pan and stirred. With experience, the process goes quite quickly; a typical batch can be measured and be ready for stirring in less than a minute.

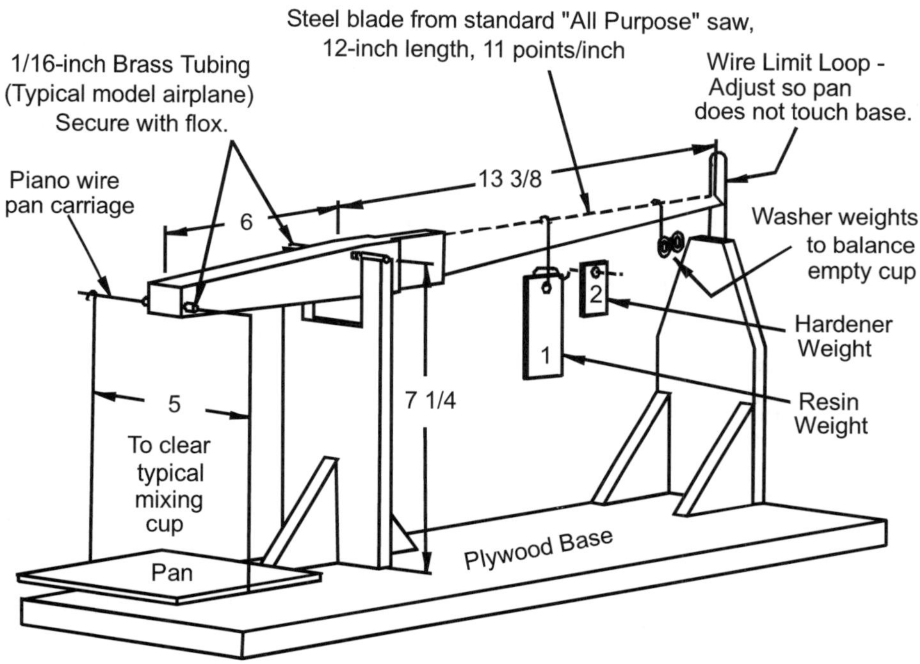

Saw-Blade Ratio Balance

Ratio Balance Construction

The ratio balance is ***not a commercially available product***, so the builder must make his own. This section gives the necessary information to reproduce the simple balance used for years by the author.

See the figure for basic dimensions and references to the following construction notes:

(A) The ***balance beam*** is a common ***steel "utility" saw blade***, 12 inches long, 10 teeth to the inch. Remove and discard the handle. Glue the blade between two pieces of 1/2-inch plywood and trim as shown after the glue cures. A belt sander will be useful. Note that the teeth must face upward. Try to make the "pan end" light by drilling out lightening holes.

(B) The ***bearings are model airplane brass tubing*** which will just clear 1/16-inch steel piano wire. A Dremel tool with an abrasive wheel will work well to get a clean cut on the tubing. Brass tubing bearings will be required in the beam and on the center support, as shown. ***Place the center support bearing on the line determined by the teeth of the saw. Place the pan-end bearing SLIGHTLY BELOW THIS LINE, about 1/4-inch, so that the balance will have a very slight positive stability.*** Align with a piece of steel piano wire and secure with flox.

(C) The 5-inch square ***pan*** should be as light as is convenient for the builder. The author used a sandwich of .004 Style 120 glass over 1/8-inch 2-pound foam, sliced with a hot-wire cutter.

(D) The ***pan bail*** is a single piece of 1/16-inch piano wire, easily bent after heating with a propane torch. Bend the wire while red hot. The pan support is a "zig-zag" bend hidden under the pan. The pan is secured to the zig-zag bend with flox. Note that the closure is a simple hook.

(E) Make up a ***set of pairs of measuring weights*** with integrated piano wire mounting hooks, as shown. Flat steel stock, drilled as shown, works well. The "resin weight" includes the mounting hook with an additional hook for the hardener weight; the "hardener weight" is drilled for mounting on the "resin weight." ***The author has found resin weights of 30, 45, 85, and 165 grams to cover a wide range of requirements.***

(F) The "cup balance" weights are an assortment of ***ordinary washers*** to be hung on an "S" hook, as shown. Usually, the builder will choose to place the "cup" weights near the far end of the beam, though these weights can go anywhere on the beam that does not interfere with the basic weight measurements.

Calibration of the Ratio Balance

It is ***not necessary to have access to precision weight standards to calibrate the measurement weights.*** Common finishing nails, all made by the same machine at the same time, are extremely uniform. The exact weight of each nail can be unknown. As long as the weights are uniform, large numbers of nails can be used with the beam balance to calibrate the reference weights.

Using the example above, where pairs of weights in the ratio of 100 to 27 are required, 100 nails can be counted out and balanced against a "resin" weight. Twenty-seven additional nails can then be counted out and added to the pan. A second weight, the "hardener" weight, can then be added to the "resin" weight and adjusted until the system balances. Trimming the "hardener" weight can be done by drilling holes and/or grinding the edge until an accurate balance is found. ***The calibrated weight pairs can then be marked.***

In practice, to cover the wide range of measuring weights required, it will be best to use two sizes of finishing nails to calibrate the weights.

The author keeps measuring weight pairs in a cup and keeps washers for balancing empty cups on a plastic cup lid.

Foam Tools

Hot Wire Cutters

Foams are excellent heat insulators. Polystyrene foams are very easily and accurately cut with a wire heated to just below red heat and passed through the foam at about 1/2-inch per second. The process works by melting a very small volume of foam around the hot wire, collapsing the foam cells around the wire, and leaving the rest of the foam undisturbed. This cutting technique is so convenient that it is one of the main reasons for using polystyrene foams.

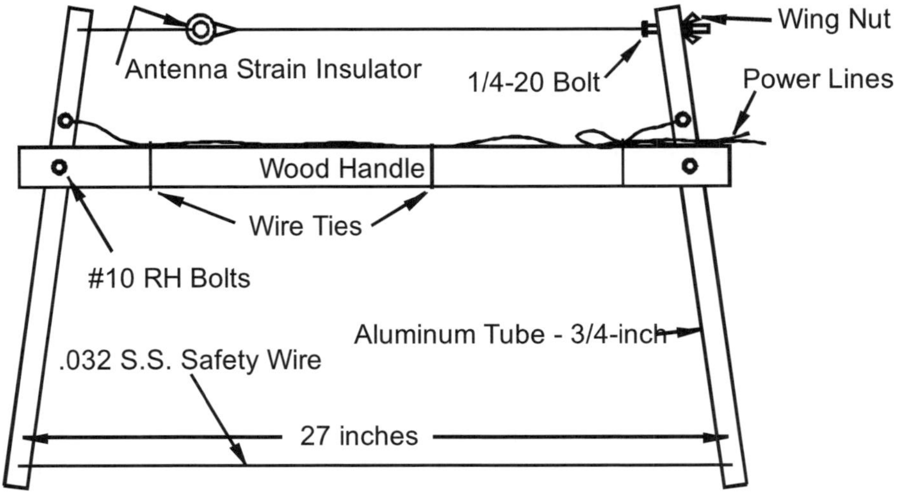

Typical Hot Wire Foam Cutter

> ***Warning:*** *Do not use hot wire cutting for urethane foams. Melting urethane foam releases a very poisonous gas.*

A frame, as shown above, will be necessary for supporting the hot wire for cutting. The author keeps two frames, one cutting about 21 inches and a second for 33 inches, to cover most requirements.

Ordinary ***stainless steel aircraft safety wire, 0.032-inch***, can be used as the resistance wire in the cutting tool for most applications. The builder will find that the wire, when heated to the cutting temperature, will tend to soften and ***stretch ("sag") when pulled through the foam.*** This effect is not usually a problem for straight cuts, but will cause inaccuracies when curved cuts are required. Worse, safety wire tends to break at awkward moments and may cause foam to be spoiled.

The aircraft supply houses offer ***special inconel wire in 0.032-inch and 0.041-inch diameter which does not stretch significantly at cutting temperature.*** This wire is expensive, but the builder will not need much and the reliability and saving in hassle is well worth the price.

Hot Wire Elecrical Controllers

An adjustable source of low voltage is necessary to drive a hot wire cutter. A ***general purpose variable ratio autotransformer*** can be used without modification and such transformers are offered by electrical supply or aircraft supply houses. Autotransformers have the disadvantages that the "ground" terminal is common with the line, raising a possible safety issue, and fine control at the low-voltage range is difficult. They are also relatively expensive.

The aircraft supply houses also offer a ***"hotwire kit"*** which consists of a standard hardware store ***electronic light dimmer, a low-voltage transformer, and instructions*** for application. The transformer may be hard to obtain, so it is worthwhile to buy the kit. The builder may make a small ventilated box to contain these components, including a standard light socket and a switch. The author uses a box 4 by 4 by 6 inches, not including a handle on the top.

Flat Foam Slicer

It may be convenient and/or cheaper to obtain blue polystyrene foam in large blocks, so the builder will need a ***quick method for accurately***

slicing the foam to a desired thickness. The sketch shows a very simple slicing board made by attaching two 1/2 inch aluminum rails to a piece of 5/8 inch particle board. The aluminum rails are standard drywall edging available at any home supply store.

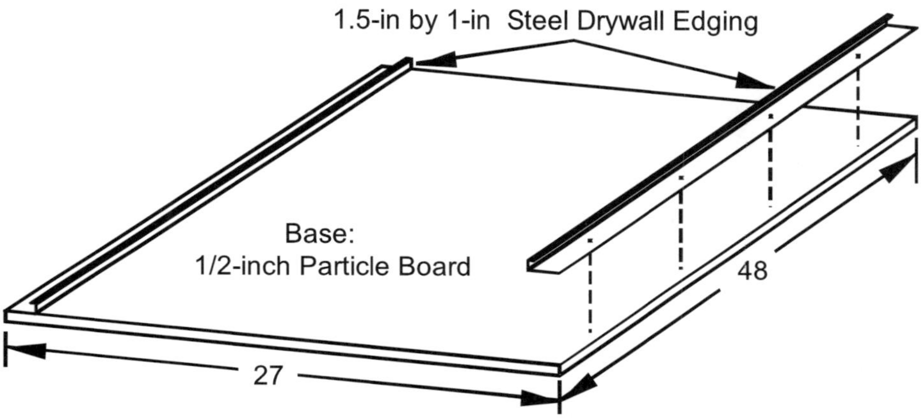

Foam-splitting Board

Assuming that a slice of foam exactly 1/2-inch thick is desired, the builder places the block of foam on the slicing table and places weights, such as sandbags or scrap metal, on the foam to hold it down firmly. Using the rails as guides, the builder passes a hot wire through the block of foam at about 1/2 inch per second. If a thickness smaller than 1/2-inch is required, the builder places a spacer, such as a sheet of 1/8-inch Masonite or 1/4-inch plywood, on the table, then makes the cut as before.

One advantage of keeping a slicing table handy is that foam can be split on an as-needed basis, reducing the risk of shop damage.

Right-angle Foam Slicer

The sketch on the following page shows a simple tool which will be useful for cutting accurate 90-degree angles. It is also useful as a steady guide for making freehand cuts on smaller pieces of foam.

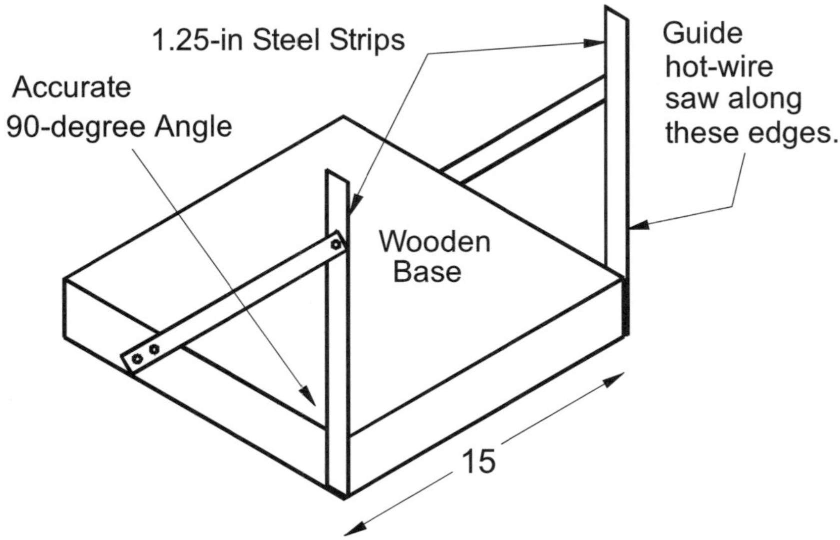

"Baloney Slicer" Foam Trimming Table

Perforating Foam

Throughout this book, most sandwich panels are assumed to be of quite large area compared to their thickness and are also assumed to be cured under vacuum to assure very good bonding and removal of excess epoxy. This means that a ***path must be provided through a foam core*** so that excess epoxy or any trapped air will not need to flow all the way around a block of solid foam. Experience shows that a grid of holes on approximately 7/8-inch centers is a good general-purpose spacing for foams and perforated release films.

The drawing on the following page shows a very successful perforating tool used by the author for several years. This tool uses a nailboard to make an array of 36 holes (6 x 6) about 1/16-inch diameter in foam. A stripper plate quickly removes the work from the nails. The choice of 36 holes at a pass is partly determined by the ability of the stripper plate to remove the tool from the foam with reasonable force and reasonable deflection of the stripper.

> *This tool may also be used to perforate plastic films, such as polyethylene (PE) film or the several films used for bagging. Note that the nail holes are pierced, not punched, so the holes are actually torn. This is favorable in films, because the torn edges tend to return to the flat. When used as a release film, the perforated film strips off of a*

cured surface easily because most of the column of cured epoxy has been cut off.

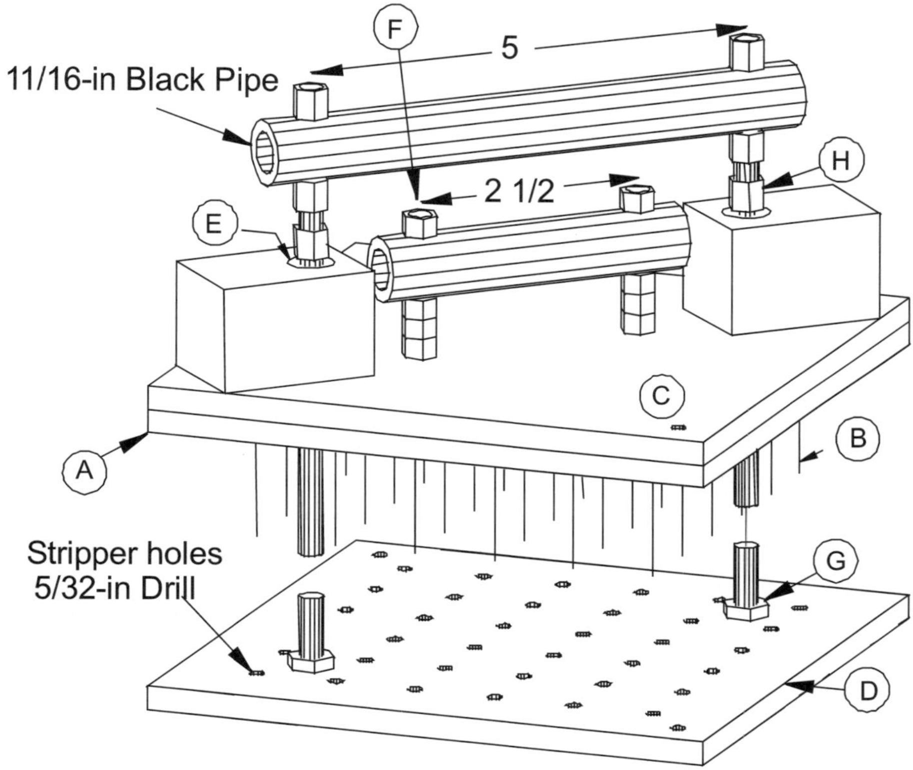

Perforation Tool

To use, the builder places the raw foam sheet on a ***soft backing, such as a piece of carpeting***, then presses firmly on the top handle to drive the nails through the foam. (As shown, the edge of the tool is on the same pitch as the array of nails, so the edge of the tool can be used as a sight line to align successive passes.) After the holes are made, the builder squeezes the two handles together, driving the stripper plate down and separating the tool from the foam. The builder then advances the tool to the next position and repeats the operation. Each pass and separation requires about five seconds.

Perforator Construction

This section gives the necessary information to reproduce the general purpose perforation tool used by the author. As described, the tool will perforate ***clean holes up to 3/4-inch deep.***

See the figure for basic dimensions and references to the following construction notes:

(A) The ***nailboard base*** is a sandwich of two pieces of 1/4-inch plywood. Make the two base pieces and the stripper plate 6 5/8-inches square. Trim all together to 6 1/8-inch square after assembly. Drill the alignment holes in the stack as one operation.

(B) The ***nail array*** is 1 1/2-inch bright box nails, 1/16-inch diameter, 6 x 6 array on 7/8-inch centers. Drill nail holes in bottom plate only, for snug fit. ***Use the alignment holes (C) to temporarily attach the stripper plate and drill the pilot holes for the stripper plate in the same operation as the nail holes in the nail plate so that they align perfectly.***

(C) The ***stripper plate*** is 1/4-inch plywood, ***skinned both sides*** with a single layer of 7725 BID glass. The ***glass layer is required for adequate stiffness.*** Drill out the holes to 5/32-inch, using the pilot holes from note (B).

(D) The sliding ***guide bearings*** are 1/4-inch I.D. ***black steel pipe.*** Place the wood support blocks as shown and glue with flox. Drill out the blocks ***and the nail plate*** to clear the O.D. of the pipe by approximately 1/8-inch. Pot the steel pipes in the drilled holes, ignoring any flox that gets into the pipes. Cure. Assemble the stripper plate, using the alignment holes. Drill out 9/32-inch, using a drill press to ***assure an accurate vertical.*** Disassemble.

(E) Place the 36 nails in the bottom half of the nailboard and glue the two parts together with flox.

(F) Drill and countersink 1/4-inch holes, as shown, and ***assemble the small handle*** with 2 1/2-inch 1/4-20 FH bolts.

(G) Countersink the stripper plate for two 5-inch 1/4-20 steel flathead bolts and ***assemble the stripper plate (D)*** with "thin" 1/4-20 nuts, as shown. Lock the nuts with epoxy or a thread locking compound.

(H) Complete the entire assembly. Use 1/4-20 nylon stop nuts, as shown, to adjust the stroke length. Secure the top (large) handle between stop nuts. ***Adjust the stroke length*** so that the maximum extension of the stripper plate just barely covers the ends of the nails.

> ***Note:*** The stripper plate requires a full glass skin as called out in (D), above. While not strictly a requirement, a glass skin on the bottom of the nail plate will help assure long-term alignment of the nails. Assuming that the builder will start with a single piece of bare plywood, it is easier to ***glass that uncut sheet in a single operation and use the glass-skinned plywood for all three plates.***

Foam Beveling Tool

The builder will find that the simple clamping tool shown in the following figure will ***greatly speed up working with foam cores.*** As shown, the dimensions chosen are optimized for use with 1/2-inch core material. The tool allows rapid production of very accurate beveled foam panels.

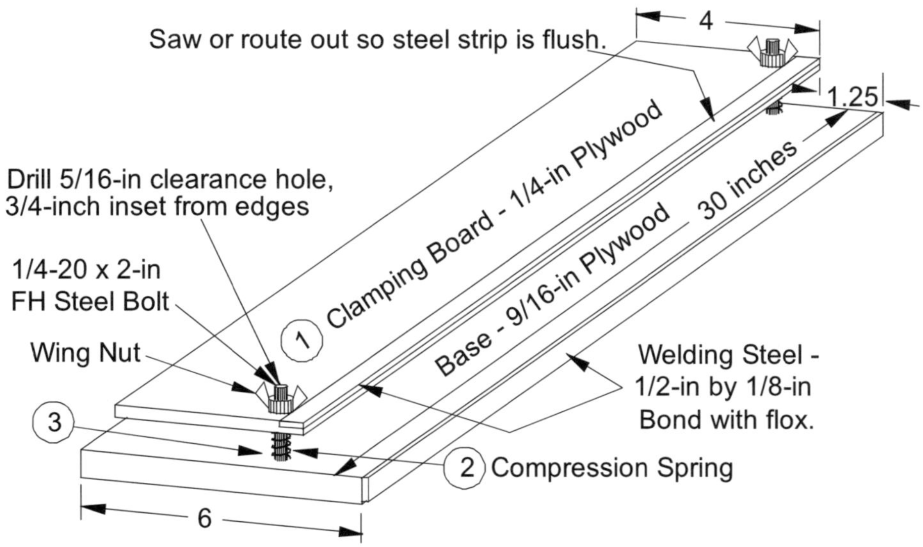

Foam Beveling Tool

To use, the builder loosens the wing nuts, slides the uncut foam between the clamping boards to the desired cut line, and tightens the wing nuts. If polystrene foam is used, the cut is made with a hot wire cutter, guided by the two steel edges. Typically, marked foam can be placed in the loosened clamp, adjusted to the line, and cut in well under a minute.

The builder will find that it is often not necessary to adjust the wing nuts to "reload" the tool. With smaller pieces of foam, the builder can squeeze the bowed spring upward with thumb and finger enough to slip the new work into cutting position. The coil springs keep the tool open, making quick adjustment quite easy.

Exactly the same setup procedure is used for urethane foam, but the foam to be beveled must first be cut off square, then beveled with a "cheese grater" tool, using the steel edges as guides. (If the builder presses only moderately, the wear of the steel on the "cheese cutter" will be acceptable.) Urethane foam cuts so easily that the builder will find the time is almost the same as if it had been hotwired. The powdered urethane that results from the cut should be considered hazardous.

Beveling Tool Construction

Note 1): The top clamping board should be ***bent into a slight downward bow*** so that the board serves as a gentle ***leaf spring***, greatly facilitating holding small work. An easy way to do this is to skin both sides of a slightly oversized, untrimmed piece of 1/4-inch plywood with 7725 BID. Before the skin cures, use any convenient weights to bend the board so that it has about a 3/4-inch bend. Allow to cure completely. The board will spring back about half-way. Trim, then rout the edge to accept the steel edge as shown in the sketch.

Ordinary hardware store welding steel can be bent slightly so that it approximately matches the curve of the top board. When the curve is slightly less than the curve of the board, attach the bent steel strip with flox. After cure, the completed top board will have approximately 1/4-inch to 1/2-inch of bow.

Note 2): Use a 1-inch compression spring, 3/8-inch in diameter, available at any hardware store.

Note 3): Drill a 5/16-inch clearance hole, inset 3/4 inch from the edges.

Appendix 2

Very Elementary Beam Theory

Reason for This Appendix

The techniques discussed in this book emphasize strongly the application of what the mainline industry would consider to be ***very thick cores and very thin skins***. The author has found that builders who have some previous familiarity with current composite construction are often not "comfortable" with what they see as a departure from accepted methods.

This appendix will briefly discuss simple beam structures so that the reader will have an intuitive grasp of the advantages of using "thick" sandwiches in the construction of a light-weight structure.

A rigorous treatment of the actual stress forces in a sandwich of composite skins over a light-weight core is not possible, but an understanding of how a simple truss works will be helpful in understanding why "thick" sandwiches are efficient. The reader who wishes to pursue the subject further will find beams and trusses discussed in any standard engineering handbook.

Trusses and Beams

A ***beam*** is a bar or structure ***subjected to transverse loads*** that tend to bend it. A log bridge over a ditch is a beam. A lifting airplane wing is a beam.

A ***truss*** is a beam made up of a system of links or ***members***, where the members are attached by pivoting pins, so they can ***only transmit pure tension or compression forces*** along their length. Further, ***a truss can be analyzed to compute the force in any member***.

An ***ideal truss*** is made up of members whose weight is very small compared to the forces that they carry and the "pivots" are ideal bearings.

Highway bridges are frequently built as trusses, and, in fact, we can often see where actual pivoting joints are used. Aircraft fuselage frames are often built with welded steel tubing trusses where the ***members are very long compared to the welds at the ends***; effectively, the designer has ignored the limited ability of the small-area welds to carry useful bending forces and the ***joints are treated as pivots.***

A simple truss is ***determinate,*** which means that the designer can calculate the forces in all the members and assign structural material to carry the calculated forces. The force in each member in the truss can be worked out by using the fact that all the ***forces at a pin must add to zero*** and whatever force is found to be in a member must be transmitted to the other end of that member.

A Very Simple Example

The sketch on the following page shows a simple truss, essentially a bridge, which carries a 100 pound downward load between two fixed supports. The truss is "ideal," so the two fixed supports must each react upward against half the load (50 pounds), as shown, and all forces are balanced. Engineers call this a "free body" diagram, in which all the forces acting on a stationary object are shown in magnitude and direction and where all the forces must balance to zero. Our task is to find the forces inside the free body (the truss) which meet that condition.

The 100 pound load can only be reacted by an ***upward force provided by members C-D and C-E***. (Members A-C and B-C are horizontal and cannot provide any vertical component of force.)

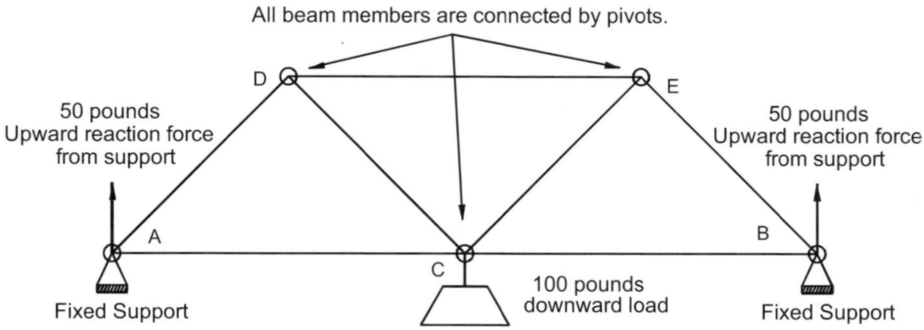

A Very Simple Truss Bridge

The next sketch shows the the resolution of the forces in members C-D and C-E which must add to equal the 100 pound reaction force. We find that the force along the member C-D is 70.7 pounds.

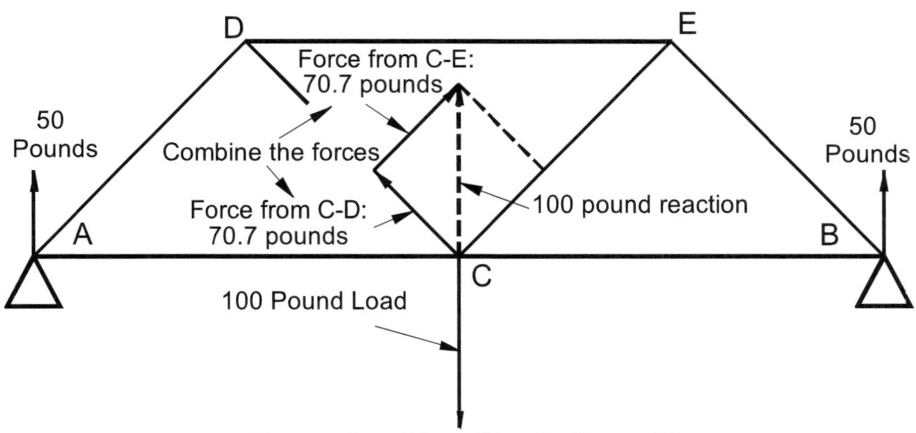

Forces in a Very Simple Truss Beam

The next sketch shows the forces in one side of the symmetrical truss. The forces at the support, pin A, are resolved as shown and we find that Member A-D must carry 70.7 pounds and A-C carries 50 pounds. With the forces in these members known, the forces at pin D can be calculated. The truss is symmetrical, so we ***know all the forces at all the pins*** and the truss is analyzed.

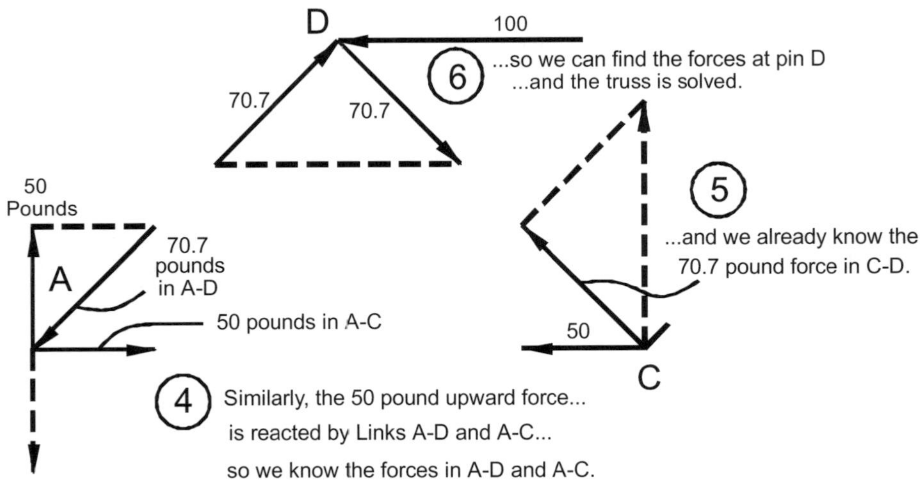

Solution of a Simple Truss

The following figure summarizes all the forces at all the pins of the "free body," with the ***force vectors drawn to scale.*** Note that the compression force in member D-E is equal to the 100-pound impressed load, for this particular case.

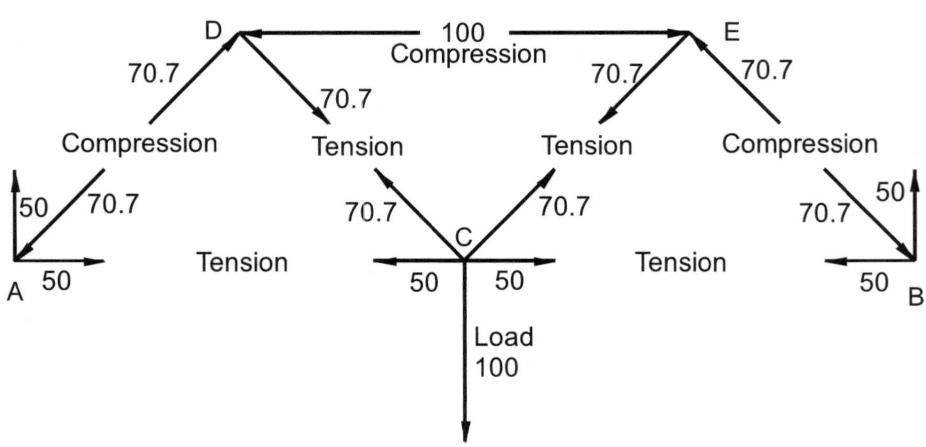

Summary of Forces in the Truss

A Second Example

Consider a modification of the truss of the previous example, as shown in the free-body diagram on top of the following page. The angles and proportions of the members of the truss are the same, but we have ***added one section*** to the truss, ***reducing the height*** of the truss and off-setting the 100-pound load, as shown.

Using the methods of the previous example, the modified truss has been solved and the magnitude and directions of the forces in the members are shown. Pins A, B, C, D, and E are labeled to correspond to the pins in the previous example. See that the ***load in member D-E has risen to 133 pounds.***

A Third Example

Let us modify the example one more time to keep all the angles, proportions, and load the same, but ***reduce the height to half*** the previous height, which requires that we double the number of sections. See the figure on the botttom of the following page.

As in the previous example, the forces in the members have been worked out, with interesting similarities and differences.

Comparing the two examples for the right side and left side of the load, see that:

> ***(1)*** The ***forces in the diagonal members are the same*** on the corresponding side of the load (47 and 94 pounds)..
>
> ***(2)*** The ***horizontal forces have increased sharply.*** The maximum horizontal forces are adjacent to the load. Note that the compression force in member D-E has doubled (267 pounds) as the height of the truss is reduced by half.
>
> ***(3) The forces in the sections at each end of the two trusses are the same.***
>
> ***(4) All members of these determinate structures are essential.*** If any member is removed, the structure must collapse.

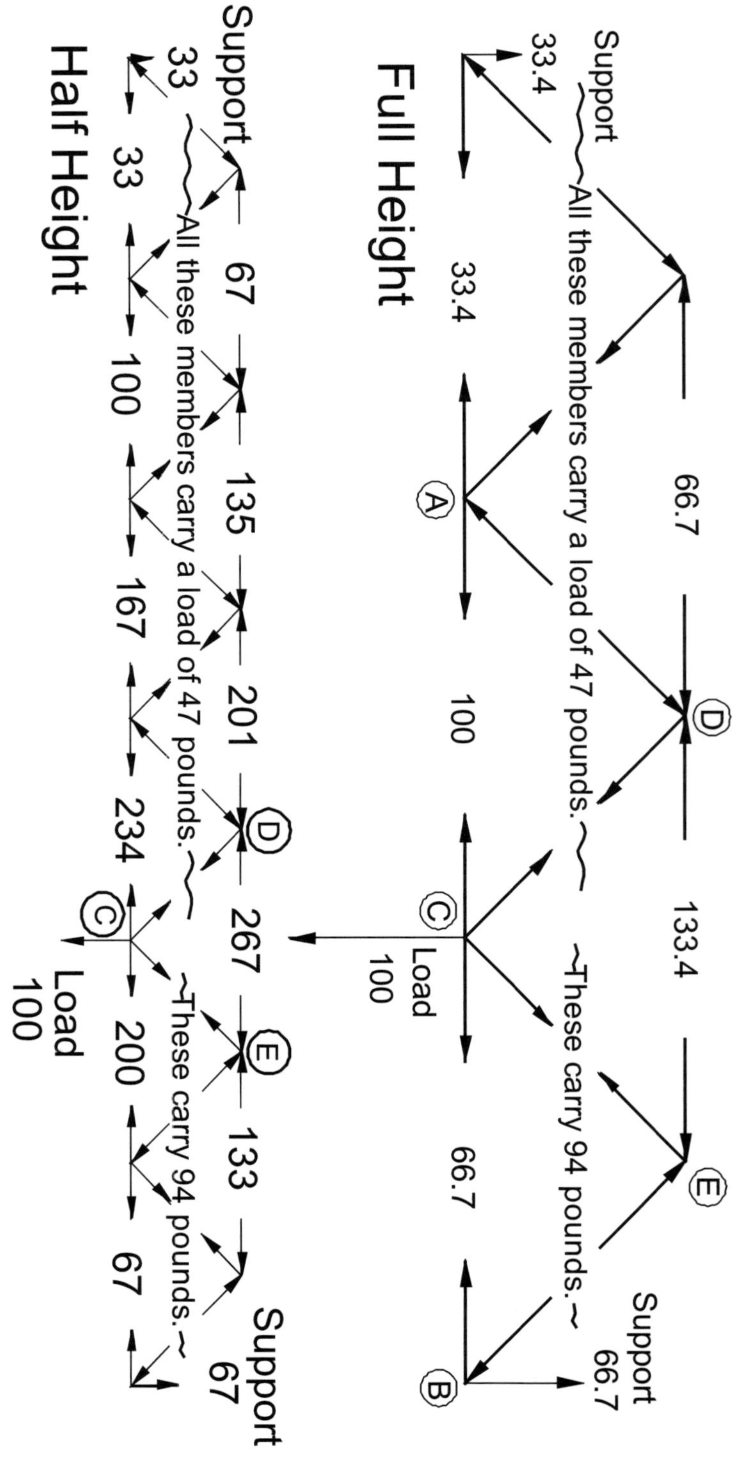

Comparison of Two Trusses With the Same Load

Arrows show direction of force at a pin, not magnitude.

Numbers show tension or compression load in a member.

Relationship to a Practical Cored Panel

The conclusion is that the truss of the first example is a more efficient beam than the truss of the second example because each member carries a much smaller load. Restated, ***thickness is a better investment than material properties.*** We intuitively understand that an unsupported sheet of thin aircraft aluminum bends quite freely and must be stabilized against bending and compression by an external means. We also understand that a sheet of thick aircraft aluminum that is stiff enough to resist bending will be entirely too heavy to use as a covering for a small aircraft.

From the analysis of the above simple trusses, it is clear that ***significant forces are carried in the diagonal members.*** In general, the diagonal forces are ***considerably less than the horizontal forces***, but they are essential for the integrity of the beam. The designer can choose materials for the "diagonal members," or core, which meet requirements but do not exceed them. (Engineers refer to these diagonal forces as ***shear forces.***)

We see from the analysis above that ***a solid rectangular slab of any material, whether metal, wood, or fiberglass, cannot be an efficient beam***; the limiting stress will be reached at the surface of the slab and the solid material in the middle of the beam is wasted. (This is the reason that steel "I-beams" are used in building construction, instead of solid bars of the same external dimensions.)

As a "thought experiment," we might consider that the beam/trusses of the above examples are replaced by a stack of 100 parallel similar structures, each carrying 1/100 of the load. The outside members then appear as an approximation of a sheet and the dense network of inside members approximate a foam. In "real life," such very thin members would buckle under a significant compression, but also in "real life" the solid core medium, whether wood or foam, will not buckle.

Practical three-dimensional sandwich structures use continuous solid cores, like foam or plywood, and do not use discrete members, like our ideal trusses. Still, the analogy with the truss is a good one. If we build our example beam (or panel) with glass skins (horizontal members), the ***same diagonal shear forces must be transmitted between the skins*** in some way and the chosen material must be capable of carrying the required shear stress.

The aircraft supply houses offer bulk rigid foam material in a range of densities. Essentially, the different foams are different trade-offs between weight and shear strength. Obviously, the designer would prefer to specify the lightest core material that would have the required shear strength.

> ***Most emphatically, the foam core is not "just a spacer."*** *The author has seen parts intended for aircraft being constructed which used the same low-strength expanded polystyrene pellets used for coffee cups and packaging. The intent was to use a "light-weight spacer." Also, some builders do not understand the difference between the soft polystyrene foam sold for heat insulation and the true rigid foams which can carry stress. Corrugated cardboard gets its stiffness not from the "spacing," but because the folded paper core makes a large number of distributed shear webs. Hollow tubes and hollow-cored skis can carry bending loads because the shear loads have a path at the* ***sides*** *of the structure.* ***USE PROPER CORES. SHEAR IS REAL.***

Appendix 3

Fibers, Weaves & Matrices

Why We Use Structural Fibers

Theoretical and Actual Strength

When materials scientists examine structural materials at the atomic level, they find that the interatomic forces that bind the atoms together are very large indeed. When these materials are made into practical test specimens and broken on a testing machine, they find that the actual breaking stress is orders of magnitude less than it "ought" to be, based on the interatomic forces. When we use materials in finite sizes, ugly realities arise, like surface flaws and internal flaws and the need to pass stresses from one part of a structure to another. We find that real materials reach nowhere near the theoretical strength and our problem becomes understanding why real, non-perfect, materials are so weak, not why perfect materials are so strong.

Griffith's Experiments

The British engineer A. A. Griffith (1893-1963), published a paper in 1920 on the problem of crack propagation in real materials. He wished to understand the huge difference between theoretical strength and test strength and chose to study common glass as a "model" material. He had calculated the theoretical stress limit (strength) of ideal glass to be about 2 million psi, which would place glass considerably above the best commercial steels. He needed to measure the true interatomic tensile strength of glass under ideal conditions.

In a series of experiments, he drew molten glass into thin rods and tested the thin rods to breaking. He found that the breaking stress increased as the rods were made thinner, and that this measured stress increased very quickly for very fine fibers. He was able to make glass fibers about .0001-inch in diameter, and found that the fibers reached nearly one million psi, if tested immediately after preparation. (The test strength dropped to about half and stabilized at a very high strength after a few hours.)

Griffith's explanation for his results is that real bulk materials cannot be produced in practical dimensions without flaws -- such as surface cracks, inclusions, or small discontinuities, which cause stress concentrations. By drawing very small fibers, he was spacing out the inevitable flaws, leaving sections of nearly ideal material in between.

Since Griffith did his work, the principle has been shown to be quite general and many materials other than amorphous (glass), or crystalline (steel), have been found to have great strength when used as fibers.

Strain and Stiffness

Strength is the measure of the force needed to break an object, without asking how far the object deflected before it broke or deformed past some limit. As discussed in Chapter 1: ***Stress*** is the ***force*** applied to the test object ***per unit area.*** If a 2,000-pound force breaks a test object of 2 square inches in cross section, we find that the breaking stress is is 2,000 pounds/2 sq. inches = 1,000 psi.

All objects must deform to some extent, however small, when a load is applied. For a uniform tensile specimen like a steel rod, the ***elongation*** will be uniformly distributed all along the rod. ***Strain*** is the fraction of the elongation of an object per unit of length. ***Strain is a dimensionless ratio***, in the same units, such as inches/inch or meters per meter. If the elongation of a test specimen is small when a large stress is applied, we say that the material is ***stiff.*** Steel and glass are typically stiff materials and nylon is not. (A nylon rope is strong because it will carry a large load with a small cross-sectional area, but it stretches a great deal under that load, therefore ***nylon is strong but not stiff.***)

The ***ratio between an applied stress and the resulting strain*** is called the ***modulus of elasticity***, or just the "modulus," or the ***stiffness***, and has its own symbol: ***E***. Stiffness of a material is a very important design parameter and is generally given in published specifications. The ***specific modulus*** is the modulus divided by the density of the material and is often included in materials specifications.

Fiber Materials - Summary

A wide variety of structural fiber materials are used by the mainline composites industry and by the amateur builder. Those used by the amateur, and available through the aircraft supply houses, include ***glass, carbon (or graphite), and aramids (known generally by the trade name Kevlar).*** Each of the fiber materials is available in a range of diameters, groupings, and surface treatments. ***Cellulose***, in the form of individual loose cotton fibers (or "***flox***"), is used frequently in amateur construction, so should be included among the fibers used. Other fiber materials are coming into use by the large-volume automotive and aircraft industries and may become available to the amateur builder.

The cost-no-object aerospace applications use fibers made of quartz, certain ceramics, even boron, and achieve very high performance at corresponding high cost. We will stop here with only their mention.

In the near future, aircraft homebuilders may see fiber materials offered made from ***basalt***, the same material from which the Hawaiian islands are made. At present, the material appears in some sporting equipment, such as snowboards. It has very attractive mechanical properties and the raw material is certainly plentiful. It has the disadvantage that it is not transparent, which would complicate amateur layups.

Strands, Yarns, and Roving

Manufactured fiber filament materials are supplied in groups of continuous individual fibers called ***strands***. A group of strands, if ***twisted*** together, is called a ***yarn***. If a group of strands is ***not twisted***, it is called ***roving***.

High production composite product manufacturers may obtain the raw fiber product in bulk in the form of yarn or roving. ***Yarn is usually woven into cloth*** to facilitate handling, as we have been doing throughout this book. ***Roving is often used in continuous machine-winding processes*** like composite pipe, tanks, pressure vessels, or rocket casings. Most industrial high volume production tonnage is in the form of roving.

The commercial market offers a wide range of filament sizes in glass, ranging from 1.5/10,000-inch (not much larger than Griffith's best effort!) to 5/10,000, and from 50 to 2,000 filaments per strand.. Aerospace ***carbon is considerably finer***, with filaments around .03/10,000-inch and well over 12,000 filaments per strand. The amateur builder will not be getting into this level of detail, however, and will use material configurations established by successful designers and available as-is from the supply houses.

Sizing

For a composite system to work, it is absolutely necessary that there be a secure bond between the fibers used and the less-stiff matrix that imbeds the fibers. Sizing is a thin coating which is applied by the manufacturer immediately after drawing the strands to (1) facilitate bonding, and (2) protect the fine fibers from damage during the weaving process. Sizing is required for both glass and carbon filaments and different sizings are preferred for different applications and matrices. If the builder obtains fiber material from an aircraft supply house, the proper sizing will have been used. Kevlar does not require sizing.

Sizing dissolves in the binding resin during the layup.

Sizings have a finite shelf life in air. Typically, ***materials intended for high-stress applications should be used within a year*** of manufacture, or as recommended by the manufacturer.

Design Strength of Yarns

A designer of an aircraft structure would like to have a simple set of rules by which she can easily choose the ideal material and ideal material form and apply it in an ideal manner to achieve an ideal design. Unfortunately, there is a ***great difference between a published number for the ideal tensile breaking strength of a fiber configuration and the final test performance*** of a real aircraft component under complex loads.

The supply catalogs do the best they can to satisfy demand and they usually give stiffness, tensile breaking strength, and weight figures for the several fiber forms they offer. Breaking strength of forms such as cloth or tape is expressed in terms of pounds per inch of tape width. Test methods are rarely given, however, and it is nearly impossible to compare accurately materials quoted by different authorities as used in a practical layup. The reader will find that existing printed authorities agree only approximately on quoted performance figures, often differing by a factor of two for the same material.

(The figures below for fiber materials are from a single engineering textbook with the hope that they are at least mutually consistent.)

The reader will recognize that these materials will be always be used with a resin matrix which will be assumed to have zero strength by itself, so the ideal catalog figures will be reduced by approximately half for typical amateur construction. The only thing that finally counts is the test of the completed composite structure. It is easy to overdo this search for the ideal material.

Some very general statements can be made, however, which will allow a useful comparison:

E-Glass and S-Glass

Ordinary window glass or bottle glass is by far the most common source of structural fiber material used by both mainline industry and by the amateur builder who obtains his material from the aircraft supply houses. The most widely used formulation of ***common glass fiber is known as E-glass because of its excellent electrical properties.*** It is, of course, transparent, which is extremely convenient for the home aircraft builder, greatly increasing the confidence that a layup is free of air bubbles.

Pure ***E-glass***, as ideal fibers without a matrix, has a test tensile strength of roughly 500,000 psi and a stiffness modulus of about 10 million psi. ***S-glass is a special high-performance glass*** which has a tensile strength of about 650,000 psi to as high as 700,000 psi and a modulus of about 14 million psi with toughness superior to E-glass. ***S-2 glass*** is another variation which offers approximately the same performance at lower weight. Other things being equal, S-glass would be the preferred choice, except that it is considerably more costly than E-glass and is available in only a few weave configurations.

Carbon (Graphite)

Carbon fiber formulations have only recently become cost-effective and generally available for amateur aircraft construction and is now widely used in high-end aircraft construction and sporting equipment. The material is still considerably more expensive than glass and has significant advantages and disadvantages which the builder should consider.

Graphite is very strong in tension, at roughly 350,000 psi, and has a stiffness range from 30 to as high as 140 million psi, depending on the formulation.

If the reader runs some numbers with the engineering data, he will find that the ***specific strength*** (strength per unit of weight) of carbon fiber is actually slightly ***less*** than the value for E-glass. The parameter that gets the attention of a spacecraft designer is the ***specific modulus***, or stiffness per unit of weight, which is approximately ***five times that of glass.***

The material has some significant disadvantages, other than price, which should be considered:

> 1) Carbon fiber is quite brittle, with a published elongation figure which is the lowest of the aerospace fibers, about one fifth that of glass. Presumably, the reason for choosing to use carbon in a structure is to save weight, so we can assume that the structure will see loads near its design limit. If that limit is locally exceeded, the structure will shatter in that area, releasing tiny carbon filaments which are exceedingly sharp and nasty for the skin and for risk of inhaling. Care must be taken in the shop to guard against airborne particles.
>
> 2) Simple design methods may not be adequate to realize the true potential when the builder can only do a one-off design and much of the weight benefit must be spent on large safety factors.
>
> 3) Carbon is not transparent and it is difficult to be sure that the yarns are wetted out completely. Vacuum bagging, of course, greatly helps.

The use of carbon fiber in highly-engineered products like commercial aircraft and sporting equipment is now widely known to the non-technical public, which produces a general view that carbon must be "best." We see it, therefore, in trivial applications such as automotive door handles and we even see false carbon fabric where the pattern is printed on sheet steel to give the appearance of "cool" carbon fabric.

Aramid (Kevlar)

Kevlar is extremely tough, unlike brittle carbon, and is widely used as tire cord, body armor, and for commercial and military aircraft. As a thin outer layer, it has been used successfully to protect very light-weight sea kayaks from chafing on sand and gravel. It is so tough that it is hard to cut.

Kevlar must be used with epoxy matrices, never polyester.

Mechanical properties are excellent: Tensile strength is around 500,000 psi and the stiffness is about 20 million psi. The density is even less than that of carbon, so the specific modulus is high.

Weaves

Nonwoven Fibers

Ideally, we would place fibers exactly parallel with the load paths, so that the loads are carried efficiently. In the case of simple beams, like wing spars, these load paths are known, the stresses are known, and material can be precisely placed to carry the stresses. Similarly, for pressure tanks and rocket casings, the bursting forces are known and winding machines can be used to place wetted roving where required.

Plain roving is not practical for the amateur to handle when placing fibers on a large surface, so suppliers offer ***tapes and sheets of roving*** which have been held together with a bare minimum of cross stitching. The product is entirely unidirectional and is referred to as UNI.

Glass ***mat*** is essentially a thick, non-woven felt-like configuration of randomly oriented glass fibers. It is widely used in the boat industry. Basically, the material provides a ***quick and easy way to make a thick core*** of arbitrary shape, which is useful in making repairs to damaged structures or in making tooling where weight is not a concern. When soaked with epoxy and cured, it makes a strong core. The primary disadvantage to mat as a core is weight (though the cost of all the absorbed epoxy might be a concern). Plain mat, without a skin, does not provide good fiber orientation at the critical surface, so mat is inefficient as a beam.

Woven Fibers

Woven cloth is easy to place and wet out without disturbing the orientation of the fibers and is the configuration most used by amateurs and in the mainline industry. Most cloth is woven with approximately equal quantities of fiber in the horizontal and vertical directions and is referred to as ***bidirectional, or BID***.

The basic cloth weave, in which horizontal and vertical yarns go over and under each other at every crossing, is called a ***plain weave*** or ***square*** weave. This weave tends to stay flat if the weave is tight, with as much yarn per area as possible, because the yarns cannot slide easily past each other. Plain weaves, therefore, ***do not conform well*** to a curved surface that requires the plane of the cloth to be deformed.

Manufacturers supply other weaves, called ***satin weaves***, in which the yarns go over from three to as many as seven yarns before going under a crosswise yarn. These weaves have the advantage that the ***yarns will easily slide to conform to a curve***, or "***drape***," but still provide a maximum of yarn per area. The supply catalogs describe the weaves for each product, often using the traditional names from the weaving industry. One very widely used general-purpose configuration is "crowfoot," in which the yarn goes over three yarns, then under one.

Obviously, any cloth weave is a compromise with theoretical perfection, because the yarns must have a curve, or "crimp" to get past each other, so the fibers cannot align exactly with the load. The weaves with the maximum amount of yarns skipped and the least crimp have a moderate performance advantage over the tighter weaves.

Knitted fabrics have become available to the amateur. Essentially, a knitted fabric appropriate for aircraft use consists of layers of unidirectional fibers which cross each other at 90 degrees, like woven BID, except that the fibers are not woven, but are held in place by a small amount of knitted cross stitching. This product provides about a ***20% increase in test strength***, compared to a cloth woven from the same base fiber. It improves the fiber to resin ratio by reducing the size of the voids between adjacent yarns which would otherwise fill with wasted resin. The product can be a ***significant time saver*** for the amateur because placement and wet out is made easier. Special-purpose knitted fabrics and fabric stacks have been used for years in high-volume industry for such products as snowboards.

Matrix Materials

Polyesters

By far the greatest tonnage of commercial fiberglass products is made with the styrene-based polyester family of matrix materials, primarily because it is ***cheap***, and secondarily because it is ***easy to work with***. The characteristic "fiberglass smell" one finds when one enters a store containing new boats is unreacted styrene. We will briefly discuss it here only because the aircraft builder will need to explain to her friends just what she is doing with her "fiberglass" project and why it doesn't "smell like a boat."

The polyester family of resins all are hardened the same way: The liquid base resin is mixed with a catalyst which greatly speeds up the hardening process. To allow normal storage before use, the manufacturer adds an ***inhibitor which permits storage*** times of many months. When the builder wishes to use the material, he ***inhibits the inhibitor*** by adding a very small amount of material called the "***catalyst***" or ***hardener***. This allows the builder to adjust the cure rate, which can be convenient.

Polyester resins are available for the commercial trade in a wide range of formulations which trade off hardness against flexibility. The formulations offered for the automotive, boat, and aircraft applications are a compromise between these extremes.

The base resins in commercial polyester matrix resins are very viscous and are unsuitable for wetting out a layup. To make the material fluid enough to wet cloth, it is diluted about 40% with pure styrene monomer. Pure styrene is a clear, somewhat volitile, syrup-like liquid, which is the source of the powerful "boat store" smell. The added styrene combines chemically with the rest of the mix to become the final solid matrix. With the styrene dilution, the mix is quite fluid and wets out easily. It also flows down a vertical surface easily, and the builder needs to be aware of this fact.

Disadvantages of Polyesters

Pure, unreinforced, polyester matrix shrinks about 7% in volume when it cures. This shrinkage corresponds to a linear shrinkage of about 2%.. When bonded to glass fibers, the stiff fibers resist the shrinkage, so ***sig-***

nificant strain and, therefore, a powerful shear force exists between the glass and the matrix. This pre-strain reduces the amount of strain-to-failure available in the matrix which would otherwise be available to carry a useful load.

If cured in open air, plain polyester resin is inhibited by the air, so that the surface remains semi-cured and gummy while the interior cures properly. This gummy surface bonds well to layers which may be added shortly afterwards and completes its cure with the added layers. Plain resin is sold as ***laminating resin***, for this reason.

If plain resin is cured for the last layer in the presence of air, it will be necessary to remove the film of gummy material. This ***can be avoided*** by using resin which has had a ***small amount of wax added.*** The wax does not dissolve, so it moves to the surface of the layup to form a protective film, allowing the cure to proceed without contact with the air. After cure, it then becomes necessary to remove the wax completely. Plain resin with added wax is called ***surfacing resin.***

After a polyester layup has fully cured, it ***does not make a good secondary bond*** to additional layers which may be required. Polyester resin is ***not a good adhesive.***

Epoxy Matrices

For aircraft work, we will prefer to use the epoxy family of resins, which outperform the polyester family in almost every parameter. Shrinkage is less than half that of the polyesters. The only significant disadvantage in application is that the epoxies are more viscous and it is harder to get a good wet-out. The addition of local heat and patience will overcome this minor nuisance.

The epoxies are shipped as two components, "A" and "B," generally called "resin" and "hardener," though the distinction is not important. The two components are required in exact ratio to combine to form a solid matrix, therefore ***measurement must be accurate and mixing must be thorough.*** If unreacted molecules cannot reach each other in the final stages of curing, they will remain unattached forever, even though the mixture has been properly measured.

Shrinkage of the epoxy family in curing, depending on which engineering handbook the reader refers to, is very approximately one fifth of the shrinkage found in the polyester resins.

Index

A

B

C

G

M

O

P

T

V

W

Y